··························
Cardiovascular Disorders in Hemodialysis

Contributions to Nephrology

Vol. 149

Series Editor

Claudio Ronco   *Vicenza*

KARGER

Cardiovascular Disorders in Hemodialysis

Volume Editors

Claudio Ronco *Vicenza*
Alessandra Brendolan *Vicenza*
Nathan W. Levin *New York, N.Y.*

58 figures, 6 in color, and 28 tables, 2005

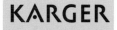

Basel · Freiburg · Paris · London · New York ·
Bangalore · Bangkok · Singapore · Tokyo · Sydney

Contributions to Nephrology

(Founded 1975 by Geoffrey M. Berlyne)

· ·

Claudio Ronco
Department of Nephrology
St. Bortolo Hospital
I–36100 Vicenza (Italy)

Alessandra Brendolan
Department of Nephrology
St. Bortolo Hospital
I–36100 Vicenza (Italy)

Nathan W. Levin
Renal Research Institute
207 East 94th Street, Suite 303
New York, NY 10128 (USA)

Library of Congress Cataloging-in-Publication Data

Cardiovascular disorders in hemodialysis / volume editors, Claudio
   Ronco, Alessandra Brendolan, Nathan W. Levin.
     p. ; cm. – (Contributions to nephrology ; v. 149)
   Includes bibliographical references and index.
   ISBN 3-8055-7938-1 (hard cover : alk. paper)
   1. Hemodialysis–Complications. 2. Cardiovascular system
   –Pathophysiology. 3. Kidney failure–Complications.   I. Ronco,
   C. (Claudio), 1951–   . II. Brendolan, Alessandra. III. Levin,
   Nathan W. IV. Series.
     [DNLM: 1. Renal Dialysis–methods. 2. Cardiovascular Diseases
   –complications. 3. Renal Dialysis–adverse effects.   WJ 378
   C267 2005]
   RC901.7.H45C36 2005
   617.4'61059–dc22

2005010445

Bibliographic Indices. This publication is listed in bibliographic services, including Current Contents® and Index Medicus.

© Copyright 2005 by S. Karger AG, P.O. Box, CH–4009 Basel (Switzerland)
www.karger.com
Printed in Switzerland on acid-free paper by Reinhardt Druck, Basel
ISSN 0302–5144
ISBN 3–8055–7938–1

Contents

Lessons from International Trials

Dialysis Schedules and Techniques

Contents

Contents

Preface

Hemodialysis therapy represents today a well-established form of renal replacement. Technology has continuously improved over the years, resulting in a series of new biomaterials and techniques for the clinical routine. With regard to new areas of investigation in hemodialysis, special attention is paid to epidemiological and outcome studies, the effect of practice patterns on clinical results and the impact of comorbidities on the final mortality data. These issues represent very important aspects that require further elucidation in the years to come.

Nevertheless, there is no doubt that cardiovascular disorders in uremia and hemodialysis represent one of the most important aspects of modern nephrological care. For this reason, this year's International Vicenza Course has dedicated several sessions to the cardiovascular risk factors, the role of the endothelium, the problems related to mineral disorders and finally to the possible impact of new technologies on the final outcome in hemodialysis.

This book represents a complete synthesis between an educational effort for the newer generations of physicians and nurses in the field of hemodialysis and a state-of-the-art description of the most advanced technologies in the field of extracorporeal therapies. At the same time, important contributions deal with the lessons learned from large international trials, either already completed or to be completed in the years to come.

As usual, an outstanding group of experts has honored the Vicenza course with their participation. Thanks to their effort to complete their written contributions beforehand, the tradition of having the proceedings ready before the course itself could be continued. This book will therefore be an invaluable tool for the participants of the course, allowing them to follow the presentations with special attention. At the same time, this publication will outlast the course as a state-of-the-art appraisal of today's technology and current issues in hemodialysis with special regard to cardiovascular disorders and risk factors.

We are grateful to all contributors who made possible such a complete and informative collection of papers and to Karger for the usual quality and efficiency in the timely publication of the book.

Once more, the International Vicenza Course with its long tradition (since 1982, www.vicenzanephrocourses.com and www.nefrologiavicenza.it) represents an invaluable tool for solid education and advanced information in the field of renal replacement therapy. We wish to thank all individuals and institutions who have made this course possible. Among them, special thanks go to the International Society of Nephrology and COMGAN, which have strongly supported and sustained this initiative. The course has been endorsed with a significant number of CME credits both for physicians and nurses. The tradition of quality and scientific rigor continues at its best.

Claudio Ronco
Alessandra Brendolan
Nathan W. Levin

Ronco C, Brendolan A, Levin NW (eds): Cardiovascular Disorders in Hemodialysis.
Contrib Nephrol. Basel, Karger, 2005, vol 149, pp 1–9

···················

Hemodialysis – From Early Days to Tomorrow

Eberhard Ritz, Ralf Dikow, Vedat Schwenger,
Marie-Luise Gross

Department Internal Medicine, Ruperto-Carola-University,
Heidelberg, Germany

Abstract

The Nestor of chronic dialysis, Belding Scribner, was well aware of most problems
which plague hemodialysis to this day, but the major problem we are confronted with today
has emerged much later, i.e. excessive cardiovascular mortality. Apart from modulating duration and frequency of dialysis, there is currently no logistic or technological solution immediately apparent to solve this problem. Elucidating and correcting the cardiovascular risk is
the most important challenge to clinical nephrology today.

'The longer you can look backward, the further you can look forward'
Winston Churchill
London, March 1944
Address to the Royal College of Physicians

The Road to Maintenance Hemodialysis – What Can We Learn?

It is useful to take note of what our predecessors thought because it provides a clearer perspective on the problems that plague us today.

In March 1966 Belding Scribner put the first person ever, Clyde Shields, on
maintenance hemodialysis. No more than 8 weeks later he was asked to report
the findings to the American Society of Artificial Internal Organs and to submit
a paper. This is a jewel of renal literature. Ten weeks after he had started the first
patient ever on dialysis, he stated modestly (note the absence of overboarding

optimism) 'preliminary experiments suggest that the useful life of patients dying of chronic uremia can be prolonged' [1]. It is impressive to see how he clearly identified 4 of the problems which still plague hemodialysis today: nutrition, hypertension, anemia and deranged action of drugs and hormones.

I think we can learn from Scribner the wisdom of deriving logical scientific conclusions based on simple, but well-structured, clinical observations. He was convinced that solutions will not come from pure speculation. Rather, when he received the Lasker award in 2002 he stated in his acceptance address: 'Although we have accomplished much, we still have much to do to improve the well being of our patients … we owe them continued research into dialysis methods and improvement' [2].

Some of Scribner's statements in his first publication on maintenance hemodialysis have prophetic foresight. On hypertension: 'As in the case of nephrectomised dogs hypertension appears to be influenced by the size of the extracellular space. A combination of dietary sodium restriction and ultrafiltration during dialysis permits regulation of extracellular volume'. This simple message had meanwhile been almost completely forgotten [3]. Today only a miniscule minority of patients are truly normotensive on conventional hemodialysis. Scribner's original schedule of long slow dialysis had been abandoned by shortening of dialysis sessions and by neglecting restriction of dietary intake of salt. To solve the problem of hypertension wrong tracks were tried, e.g., by sodium modeling. Winston Churchill is on record to the statement: 'However beautiful the strategy you should occasionally look at its results'.

On anemia Scribner made the following statements: 'The problem of the nature of the anemia with progressive decrease in renal function needs elucidation. The relative roles of erythropoietin deficiency and the possible use of this hormone in therapy should be investigated' – astonishing anticipation of today's discussions which reminds of the aphorism of Johann Wolfgang von Goethe: 'All that is wise has been thought before, it is only necessary to think it once again'. The task of the scientist is to amalgamate the knowledge of the ancients and the insights obtained with today's superior methodology.

The simple question of how frequently and how long patients should be dialyzed has recently taken a surprising turn. Nephrologists have been seduced by the idea that the simple quantitation of urea removal by Kt/V was the ultimate yardstick to judge adequacy of treatment. Today we are still confronted with a shocking mortality of dialyzed patients, in Europe currently 10–15% per year [4], similar to the life expectancy of patients with some types of malignancy. In our opinion much of the risk to which the patient on dialysis is exposed has been acquired prior to dialysis. If this is so, even the best modality of dialysis will not reverse the damage that has accumulated in the predialytic stage. This point is well illustrated by the data of the USRDS. In nondiabetic patients with chronic kidney disease the risk to die from cardiovascular causes is 10 times higher than to wind up on maintenance hemodialysis. Such death is caused by cardiovascular events as documented by several recent studies. Go et al. [5] documented that in early stages of chronic kidney disease the risk of cardiovascular events and of hospitalization for cardiovascular disease rises dramatically. How early the increment in risk is demonstrable was shown by the study of Wright et al. [6]: in the patient with an acute cardiac ischemic event the risk is double if the estimated creatinine clearance is 80 instead of 110 ml/min and it rises dramatically when renal function decreases further. Certainly one cause is undertreatment: patients with impaired renal function and myocardial infarction receive less frequently aspirin, β-blockers, thrombolytic agents, angiotensin converting enzyme (ACE) inhibitors or percutaneous transluminal coronary angioplasty, but this is only part of the explanation.

What are the risk factors underlying the excess cardiovascular mortality in renal patients? Are some of these cardiovascular risk factors unique to renal disease? How can the cardiovascular risk be counteracted?

Undoubtedly classical cardiovascular risk factors are much more frequent in renal patients than in the background population. Renal dysfunction is also associated with risk factors which are not specific for, but the intensity of which is amplified by renal dysfunction. Such risk factors comprise excessive sympathetic activity [7], increased concentrations of asymmetric dimethylarginine [8], high lipoprotein(a) [9], abnormal apolipoprotein patterns [10] – all of which are demonstrable even when GFR is still normal. Presumably the most important factor, however, is accelerated atherogenesis which is clearly demonstrable in experimental models and related to increased endothelial cell oxidative stress [11–13].

But certainly nonconventional risk factors have been neglected in the past. In the DOPPS study, no less than 20% of patients on hemodialysis were 'depressed' [14]. Their cardiovascular risk was increased dramatically as is well known in depressed cardiological patients. Depressive syndromes predict the risk of later appearance of cardiovascular disease by a factor of 1.7–4.5 [15–17] and depression is an independent factor predicting higher cardiovascular mortality. We are not necessarily talking about 'endogenous depression'. More frequent is minor depression which our ancestors called melancholy, i.e., depressive mood. Obviously nephrologists are not very good at handling this because patients considered themselves much more frequently 'depressed' than did the physicians [14]. It is not known, however, whether reversing such depressed mood actually improves cardiovascular prognosis.

Can we correct classical cardiovascular risk factors and when should such treatment start? There have been some unpleasant surprises in this field.

Angiotensin II undoubtedly plays an important role. Observational studies reported that the survival was better when patients received ACE inhibitors [18]. It therefore came as a complete surprise that in a prospective randomized double-blind placebo-controlled trial in elderly hemodialysis patients with left ventricular hypertrophy no effect whatsoever on primary cardiovascular events or all-cause mortality was seen when patients received the ACE inhibitor fosinopril [19]. Thus, matters are not as simple as we thought they are. Nevertheless, since residual diuresis is a powerful predictor of survival [20], and since both ACE inhibitors [21] and angiotensin receptor blockers [22] cause longer persistence of residual diuresis, it still makes sense to block the renin-angiotensin system – at least if the patient is not yet anuric.

Another example is provided by the statins. In individuals without renal disease there is an inverse relationship between cholesterol and survival: the higher the serum cholesterol, the better survival. In dialyzed patients the relationship is inverse as first reported by Degoulet [22] – an example of confounding factors causing so-called reverse epidemiology. The underlying mechanism has been documented by a recent study: overall the relationship between cholesterol and all-causes as well as cardiovascular mortality was inverse with a slight tendency of a U-shaped curve, but if only patients without markers of inflammation were considered, the relation was perfectly as that seen in the general population [23]. The question arises then whether lowering lipids with statins is effective. Indeed, posthoc subgroup analysis of major studies showed that patients with mild and moderate impairment of renal function derived similar benefit from statins as did nonrenal patients [24]. Furthermore, observational studies in hemodialyzed patients showed that patients using statins had better actuarial survival than nonusers [25].

A first note of caution was provided by the study of Fahti et al. [26]. He compared the effect of massive lipid lowering on carotid intima-media thickness in nonrenal patients with coronary artery disease and in patients with chronic kidney disease. Although LDL cholesterol was lowered to a similar extent in the 2 groups, the maximum intima-media thickness decreased only in the patients with coronary artery disease. That matters are more complex and that pathomechanisms other than those in nonrenal patients are operative in terminal renal failure was shown by the 4-dimensional study. Hemodialyzed type 2 diabetics received atorvastatin and LDL cholesterol was effectively lowered. The primary composite endpoint, i.e., cardiovascular events, was not significantly modified [27]. It appears that in terminal renal failure a point of no return is reached at which classical interventions fail to exhibit their ordinary beneficial effects. One can speculate that this is the result of advanced glycosylation end products, of vascular calcification or more intense inflammatory insult to vessels (see fig. 1).

a   b

Fig. 1. Immunohistologic staining for CRP (magnification ×300). *a* Aortic wall of a 60-year-old man with no renal disease. There is no immunohistologic staining for CRP visible. *b* Aortic wall of a 61-year-old uremic male patient with marked staining for CRP.

After such pessimistic observations a note of hope indicating that effective interventions are not impossible. With Koch et al. [28] we had observed that in hemodialyzed type 2 diabetic patients only 4% of those who died from cardiovascular causes had β-blocker treatment, whilst 12% of those who survived did so. In the DOPPS study mortality was less by 13% in dialyzed patients on β-blockers [29]. The 4-dimensional study shows, however, how misleading observational studies may be. An interventional study, however, compared carvedilol versus placebo in a placebo-controlled randomized prospective design in hemodialyzed patients with dilated cardiomyopathy [30]. Cardiovascular death as well as hospitalization were significantly reduced by carvedilol. Based on medical care data, it has recently also been shown that heart failure developed less frequently in hemodialyzed patients who received β-blockers [31].

One potential cardiovascular risk factor may be specific for advanced renal failure. In a small underpowered but nevertheless interesting study, Tepel reported that in hemodialyzed patients the risk to die was less by 67% if they had been treated with calcium channel blockers [31]. Ishani et al. [32] noted in a random sample of 3,141 hemodialyzed patients studied at baseline in 1996 that mortality was not influenced by ACE inhibitors and surprisingly also not with β-blockers or statin treatment. The use of calcium channel blocker was associated only with reduced risk of events, although one cannot completely exclude confounding by indication. In the pioneer experiments of Fleckenstein, vascular calcification could be prevented by calcium channel blockers and this may explain the efficacy in hyperphosphatemia dialysis patients prone to vascular calcification.

Another specific risk factor may be parathyroid hormone which apparently impacts on cardiovascular risk. This is well known in primary hyperparathyroidism [33] and recently less long-term mortality has been reported in dialyzed patients after parathyroidectomy [34]. Furthermore in experimental studies parathyroid hormone was permissive for the development of cardiac pathology [35]. Therefore, we look forward to the long-term results with the calcimimetic cinacalcet which lowers parathyroid hormone and – in contradistinction particularly to vitamin D – lowers the calcium and phosphate product, a major risk factor for vascular calcification. At least in experimental studies a calcimimetic was as beneficial as parathyroidectomy to lower blood pressure, to reverse dyslipidemia, and to improve cardiac pathology [36].

But much remains to be done. An incomplete list of some areas where we urgently must conduct research as postulated by Scribner includes vitamin E. The results of large intervention trials in nonrenal patients were negative throughout, particularly in the large HOPE study. In contrast in dialyzed patients Boaz et al. [37] found in the small SPACE trial less cardiovascular events, although not less death as a hard endpoint. To our great surprise in uremic rats we found that vitamin E almost completely abrogated the changes in cardiac pathology in uremic rats [38]. There may be a reason why uremic patients may differ from nonrenal patients: oxidative stress is higher, water soluble bioactive metabolites of vitamin E cumulate. Currently we are still confronted with the dilemma: 'To E or not to E, that is the question' [39]. This and other questions require intervention trials and as recently pointed out by Strippoli et al. [40], we do not have sufficient high-class intervention trials in nephrology in contrast to cardiology. The kidney has remained the Cinderella of intervention trials.

In my view even the best technical improvements of dialysis will not be sufficient to deal with the issue of end stage renal failure. The recent analysis of Lysaght predicted a dramatic increase of the global and US dialysis population between 1970 and 2000 and if one extrapolates to 2010 the number of dialysis patients will have been doubled. In parallel with the number of patients the health care cost, caused by maintenance hemodialysis, have risen and will rise as well [41]. Thus the main challenge will be to prevent terminal renal failure. There are a number of interesting approaches which cannot be discussed here because of space restriction.

We emphasized here the importance of the medical management and did not emphasize technological advances. We do not wish to leave the impression that technology is unimportant. Unfortunately, in academic medicine, including nephrology, a tendency has prevailed (and in places does still prevail) that dialysis is more a craft than a science. Such arrogant attitude is reminiscent of the famous statement of Hutchinson in 1938 in the Lancet: 'It is unnecessary – perhaps dangerous in medicine to be too clever'. I am convinced that heroes such as Kolff and Scribner who failed to study ionic transports or gene expression, have done more to promote the health of our renal patients than the champions in the respective areas.

References

1 Scribner BH, Buri R, Caner JE, Hegstrom R, Burnell JM: The treatment of chronic uremia by means of intermittent hemodialysis: A preliminary report. 1960. J Am Soc Nephrol 1998;9:719–726; discussion 719–726.
2 Scribner BH: Lasker Clinical Medicine Research Award. Medical dilemmas: The old is new. Nat Med 2002;8:1066–1067.
3 Dorhout Mees EJ: Volaemia and blood pressure in renal failure: Have old truths been forgotten. Nephrol Dial Transplant 1995;10:1297–1298.
4 Pozzoni P, Del Vecchio L, Pontoriero G, Di Filippo S, Locatelli F: Long-term outcome in hemodialysis: Morbidity and mortality. J Nephrol 2004;17(suppl 8):S87–S95.
5 Go AS, Chertow GM, Fan D, McCulloch CE, Hsu CY: Chronic kidney disease and the risks of death, cardiovascular events, and hospitalization. N Engl J Med 2004;351:1296–1305.
6 Wright RS, Reeder GS, Herzog CA, Albright RC, Williams BA, Dvorak DL, Miller WL, Murphy JG, Kopecky SL, Jaffe AS: Acute myocardial infarction and renal dysfunction: A high-risk combination. Ann Intern Med 2002;137:563–570.
7 Koomans HA, Blankestijn PJ, Joles JA: Sympathetic hyperactivity in chronic renal failure: A wake-up call. J Am Soc Nephrol 2004;15:524–537.
8 Kielstein JT, Boger RH, Bode-Boger SM, Frolich JC, Haller H, Ritz E, Fliser D: Marked increase of asymmetric dimethylarginine in patients with incipient primary chronic renal disease. J Am Soc Nephrol 2002;13:170–176.
9 Kronenberg F, Kuen E, Ritz E, Junker R, Konig P, Kraatz G, Lhotta K, Mann JF, Muller GA, Neyer U, Riegel W, Reigler P, Schwenger V, Von Eckardstein A: Lipoprotein(a) serum concentrations and apolipoprotein(a) phenotypes in mild and moderate renal failure. J Am Soc Nephrol 2000;11:105–115.
10 Kronenberg F, Kuen E, Ritz E, Konig P, Kraatz G, Lhotta K, Mann JF, Muller GA, Neyer U, Riegel W, Riegler P, Schwenger V, von Eckardstein A: Apolipoprotein A-IV serum concentrations are elevated in patients with mild and moderate renal failure. J Am Soc Nephrol 2002;13: 461–469.
11 Buzello M, Tornig J, Faulhaber J, Ehmke H, Ritz E, Amann K: The apolipoprotein e knockout mouse: A model documenting accelerated atherogenesis in uremia. J Am Soc Nephrol 2003;14: 311–316.
12 Bro S, Bentzon JF, Falk E, Andersen CB, Olgaard K, Nielsen LB: Chronic renal failure accelerates atherogenesis in apolipoprotein e-deficient mice. J Am Soc Nephrol 2003;14:2466–2474.
13 Bro S, Moeller F, Andersen CB, Olgaard K, Nielsen LB: Increased expression of adhesion molecules in uremic atherosclerosis in apolipoprotein-e-deficient mice. J Am Soc Nephrol 2004;15: 1495–1503.
14 Lopes AA, Bragg J, Young E, Goodkin D, Mapes D, Combe C, Piera L, Held P, Gillespie B, Port FK: Depression as a predictor of mortality and hospitalization among hemodialysis patients in the United States and Europe. Kidney Int 2002;62:199–207.
15 Ferketich AK, Schwartzbaum JA, Frid DJ, Moeschberger ML: Depression as an antecedent to heart disease among women and men in the NHANES I study. National Health and Nutrition Examination Survey. Arch Intern Med 2000;160:1261–1268.
16 Pratt LA, Ford DE, Crum RM, Armenian HK, Gallo JJ, Eaton WW: Depression, psychotropic medication, and risk of myocardial infarction. Prospective data from the Baltimore ECA follow-up. Circulation 1996;94:3123–3129.

17 Frasure-Smith N, Lesperance F, Talajic M: Depression and 18-month prognosis after myocardial infarction. Circulation 1995;91:999–1005.

18 Efrati S, Zaidenstein R, Dishy V, Beberashvili I, Sharist M, Averbukh Z, Golik A, Weissgarten J: ACE inhibitors and survival of hemodialysis patients. Am J Kidney Dis 2002;40:1023–1029.

19 al ZFe: The effects of Fosinopril on cardiovascular morbidity and mortality in haemodialysis patients. The Fosdial study. Abstract 2A1, 2004.

20 Maiorca R, Brunori G, Zubani R, Cancarini GC, Manili L, Camerini C, Movilli E, Pola A, d'Avolio G, Gelatti U: Predictive value of dialysis adequacy and nutritional indices for mortality and morbidity in CAPD and HD patients. A longitudinal study. Nephrol Dial Transplant 1995;10: 2295–2305.

21 Li PK, Chow KM, Wong TY, Leung CB, Szeto CC: Effects of an angiotensin-converting enzyme inhibitor on residual renal function in patients receiving peritoneal dialysis. A randomized, controlled study. Ann Intern Med 2003;139:105–112.

22 Suzuki H, Kanno Y, Sugahara S, Okada H, Nakamoto H: Effects of an angiotensin II receptor blocker, valsartan, on residual renal function in patients on CAPD. Am J Kidney Dis 2004;43: 1056–1064.

23 Liu Y, Coresh J, Eustace JA, Longenecker JC, Jaar B, Fink NE, Tracy RP, Powe NR, Klag MJ: Association between cholesterol level and mortality in dialysis patients: Role of inflammation and malnutrition. JAMA 2004;291:451–459.

24 Tonelli M, Isles C, Curhan GC, Tonkin A, Pfeffer MA, Shepherd J, Sacks FM, Furberg C, Cobbe SM, Simes J, Craven T, West M: Effect of pravastatin on cardiovascular events in people with chronic kidney disease. Circulation 2004;110:1557–1563.

25 Seliger SL, Weiss NS, Gillen DL, Kestenbaum B, Ball A, Sherrard DJ, Stehman-Breen CO: HMG-CoA reductase inhibitors are associated with reduced mortality in ESRD patients. Kidney Int 2002;61:297–304.

26 Fathi R, Isbel N, Short L, Haluska B, Johnson D, Marwick TH: The effect of long-term aggressive lipid lowering on ischemic and atherosclerotic burden in patients with chronic kidney disease. Am J Kidney Dis 2004;43:45–52.

27 Wanner Ch KV, März W, Olschewski M, Mann JFE, Ruf G, Ritz E: Statin treatment in patients with type 2 diabetes mellitus on dialysis may come too late. submitted, 2005.

28 Koch M, Thomas B, Tschope W, Ritz E: Survival and predictors of death in dialysed diabetic patients. Diabetologia 1993;36:1113–1117.

29 Bragg JL, Mason NA, Maroni BJ, Held PJ, Young EW: Beta-adrenergic antagonist utilization among hemodialysis patients. J Am Soc Nephrol 2001;12:321a.

30 Cice G, Ferrara L, D'Andrea A, D'Isa S, Di Benedetto A, Cittadini A, Russo PE, Golino P, Calabro R: Carvedilol increases two-year survival in dialysis patients with dilated cardiomyopathy: A prospective, placebo-controlled trial. J Am Coll Cardiol 2003;41:1438–1444.

31 Abbott KC, Trespalacios FC, Agodoa LY, Taylor AJ, Bakris GL: beta-Blocker use in long-term dialysis patients: Association with hospitalized heart failure and mortality. Arch Intern Med 2004; 164:2465–2471.

32 Ishani A, Herzog CA, Collins AJ, Foley RN: Cardiac medications and their association with cardiovascular events in incident dialysis patients: Cause or effect? Kidney Int 2004;65:1017–1025.

33 Stefenelli T, Abela C, Frank H, Koller-Strametz J, Globits S, Bergler-Klein J, Niederle B: Cardiac abnormalities in patients with primary hyperparathyroidism: Implications for follow-up. J Clin Endocrinol Metab 1997;82:106–112.

34 Foley RN, Li S, Liu J, Gilbertson DT, Chen SC, Collins AJ: The fall and rise of parathyroidectomy in U.S. Hemodialysis patients, 1992 to 2002. J Am Soc Nephrol 2005;16:210–218.

35 Amann K, Ritz E, Wiest G, Klaus G, Mall G: A role of parathyroid hormone for the activation of cardiac fibroblasts in uremia. J Am Soc Nephrol 1994;4:1814–1819.

36 Ogata H, Ritz E, Odoni G, Amann K, Orth SR: Beneficial effects of calcimimetics on progression of renal failure and cardiovascular risk factors. J Am Soc Nephrol 2003;14:959–967.

37 Boaz M, Smetana S, Weinstein T, Matas Z, Gafter U, Iaina A, Knecht A, Weissgarten Y, Brunner D, Fainaru M, Green MS: Secondary prevention with antioxidants of cardiovascular disease in end-stage renal disease (SPACE): Randomised placebo-controlled trial. Lancet 2000;356:1213–1218.

38 Amann K, Tornig J, Buzello M, Kuhlmann A, Gross ML, Adamczak M, Ritz E: Effect of antioxidant therapy with dl-alpha-tocopherol on cardiovascular structure in experimental renal failure. Kidney Int 2002,62:877 884.

39 Friedrich MJ: To 'E' or not to 'E,' vitamin E's role in health and disease is the question. JAMA 2004;292:671–673.

40 Strippoli GF, Craig JC, Schena FP: The number, quality, and coverage of randomized controlled trials in nephrology. J Am Soc Nephrol 2004;15:411–419.

41 Lysaght MJ: Maintenance dialysis population dynamics: Current trends and long-term implications. J Am Soc Nephrol 2002;13(suppl 1):S37–S40.

Prof. E. Ritz
Department of Internal Medicine, Division of Nephrology
Bergheimer Strasse 56a
DE–69115 Heidelberg (Germany)
Tel. +49 (0) 6221 601 705/189976, Fax +49 (0) 6221 603 302,
E-Mail Prof.E.Ritz@t-online.de

Ronco C, Brendolan A, Levin NW (eds): Cardiovascular Disorders in Hemodialysis.
Contrib Nephrol. Basel, Karger, 2005, vol 149, pp 10–17

....

Mechanisms of Solute Transport in Extracorporeal Therapies

Claudio Ronco[a], *Nathan W. Levin*[b]

[a]Department of Nephrology, St. Bortolo Hospital, Vicenza, Italy;
[b]Renal Research Institute, Beth Israel Medical Center, New York, N.Y., USA

Abstract

Diffusion and convection are the main mechanisms involved in the membrane separation processes occurring in extracorporeal hemodialysis.

Operational parameters should be optimized in hollow fiber hemodialyzers to achieve the maximal efficiency.

The nature of blood which is a non Newtonian fluid, requires specific attention in the design of dialyzers to ensure that the blood compartment operates properly. Similar attention must be placed in the design of the dialysate compartment to ensure a homogeneous distribution of the fluid and to prevent blood to dialysate flow mismatch.

Finally, the membrane represents the third component of the hemodialyzer. Membrane performance depends on the used biomaterial, its biocompatibility, the thickness, the hydrophilic-hydrophobic mixture, the hydraulic permeability and the number and diameter of the pores.

In this setting, diffusion and convection tend to reciprocally interfere, producing a final result that depends on the prevalence of one or the other mechanism for every specific solute.

Introduction

Various techniques of renal replacement are designed to obtain the removal of uremic retention solutes sufficient to maintain serum level of toxins compatible with life. At the same time, homeostatic correction is achieved by managing electrolyte, acid-base and fluid balance with the extracorporeal techniques that only partially mimic the fine regulation and control provided by native kidneys.

All solute and water exchanges are accomplished in hemodialysis by membrane separation processes.

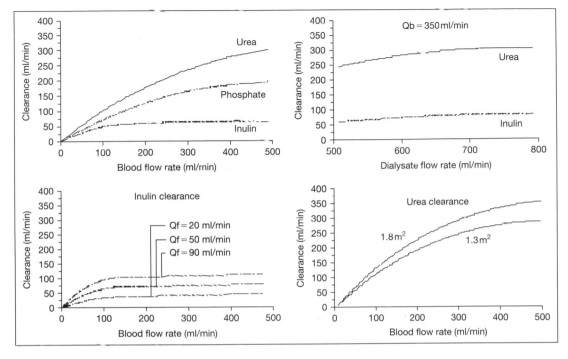

Fig. 1. Factors affecting solute clearances in extracorporeal treatments.

Since the beginning of dialytic therapy, diffusion and convection have been combined in an attempt to replace renal function [1]. The knowledge of diffusion came from the industrial chemistry and dialyzers were designed to be ideal countercurrent exchangers [2]. Only later, convection was used in clinical practice showing potential advantages [3, 4]. Although ultrafiltration (UF) was employed first to treat overhydrated patients [5], convective solute removal was subsequently employed to enhance solute removal [6–8]. Figure 1 summarizes the factors influencing the solute transport across semipermeable membranes. Blood flow greatly affects the clearance of small solutes like urea while (UF) rate primarily affects the removal of larger solutes like inulin. Increases in dialysate flow rate become important only with large surface area dialyzers and mostly affect the clearance of small solutes. Finally, dialyzer surface area determines the maximal solute clearance at a given blood flow. All these aspects must also consider the type of membrane utilized and the hydraulic conditions within the hemodialyzer. Theoretically speaking, each hemodialyzer should have a maximal diffusive permeability described by the KoA parameter. This is a condition in which blood and dialysate flows are unlimited and thus not affecting the system. Based on this concept, KoA has been assumed to be a fixed parameter for each dialyzer. Nevertheless this concept has been recently challenged and KoA has been demonstrated to vary according to flow distribution within the filter and modifications of other operational conditions.

Diffusion and Convection

Diffusion is a process in which molecules randomly move in all directions. Statistically this movement results in a passage of solutes from a more concentrated area to a less concentrated one. Besides the concentration gradient (dc), the solute diffusive flux (Jd) through a semipermeable membrane depends on the temperature (T), the surface area (A) and the diffusivity (D) of the solute, while it is inversely proportional to the membrane thickness (dx).

$$Jd = D \cdot A \cdot T \, (dc/dx)$$

The convective process requires a fluid movement caused by a transmembrane pressure gradient. Therefore, the convective flux of a solute (Jc) will depend on the (UF) rate (Qf), the solute concentration in plasma water (Cb) and the solute sieving coefficient (S),

$$Jc = Qf \cdot Cb \cdot S$$

being in ideal conditions: $S = 1 - \sigma$, where σ is the reflection coefficient of the membrane. These definitions present convection and diffusion as two separate phenomena. However it is impossible to precisely define the contribution of each single process in the removal of solutes because of their continuous interactions.

Membranes, Diffusivity and Sieving

Different membranes are utilized in extracorporeal therapy. Cellulosic membranes are considerably hydrophilic, with wall thickness values in the 5 to 15- μm range. Such membranes offer remarkable diffusive performances with limited solute sieving properties. Original synthetic high-flux membranes had an internal skin layer surrounded by a microporous structure with a total thickness up to 100 μm. The polymer was hydrophobic and its efficiency in diffusion was poor. Only in recent years, synthetic membranes with a reduced wall thickness, nano-controlled porosity and hydrophilic microdomains have been developed permitting the combination of diffusion and convection, as in the case of high-flux dialysis or hemodiafiltration.

As solute molecular weight increases the diffusivity coefficient tends to decrease. Thus, the characteristics of the solute are extremely important and diffusion of solutes in the range of 5,000–20,000 Da may be poor even in the presence of a very permeable membrane. In this case, transport is mainly limited by the low diffusivity of the molecule rather than by the sieving characteristics of the membrane. In addition to the hydrophobic nature of the membrane, the membrane wall thickness and considerable amount of unstirred fluid inside the support structure remarkably slow down the solute transport. The structure of the recent synthetic high-flux membranes partially avoids the above-mentioned problems combining a relatively less hydrophobic nature with a reduced wall thickness and a more homogeneous structure.

Solute diffusivity plays an important role in blood and dialysate also. The resistances generated by blood, dialysis fluid and membrane can be reported as a percent of the overall resistance to solute transport (total resistance = R blood + R membrane + R dialysate). At the cutoff value, the resistance of the membrane represents 100% of the total resistance. This resistance progressively decreases for smaller solutes while the resistances in the blood and dialysate compartments become increasingly important.

The resistance to the transport of larger solutes, due to their poor diffusion coefficients can be overcome by the use of convection. The convective flux is influenced by the permeability of the membrane, which is characterized by the observed sieving coefficient. The sieving coefficient is the ratio between the solute concentration in the filtrate and the solute concentration in plasma water, in the absence of a gradient for diffusion. However, solute distribution in the blood compartment is not homogeneous depending on polarization and other phenomena. As UF increases, part of the solute tends to accumulate at the blood/membrane interface, thus creating gradients for diffusion both towards the bulk region inside the hollow fiber and towards the dialysate compartment across the membrane. As a consequence diffusion is continuously interfering with convection and the sieving coefficient can be overestimated. In fact, the concentration in the bulk region (which is the value measured empirically) is generally lower than that at the blood/membrane interface. Therefore, the difference between the observed sieving coefficient (So) and the true sieving coefficient (St) can be significantly affected by the amount of convection used. With low UF values, So and St tend to be equal while large differences can be observed at high UF rates.

Membrane and the Blood Compartment

For all membranes to some extent but particularly in the case of synthetic membranes, a protein layer is deposited on the internal surface of the fiber. This slightly reduces the membrane sieving coefficient with a rather constant trend. However, in case of high UF rates or high filtration fractions, a thick

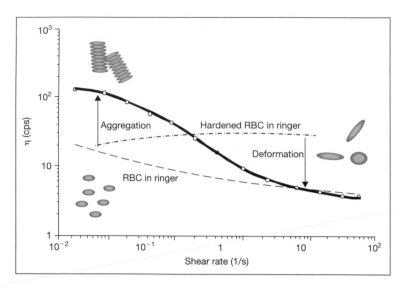

Fig. 2. Viscosity (η) of blood changes in relation to shear rate.

protein deposit on the membrane is induced by the additional phenomenon of polarization. This progressively reduces the membrane permeability and So becomes proportional to a new reflection coefficient (σ_1) of the membrane. This layer is a function of several variables, and above all the value of 'shear rate' at the wall. As the blood enters the hollow fiber, the shear stress generates different layers of blood from the bulk phase to the membrane interface flowing at different velocities. The ratio between the differential velocity of the fluid threads in the fiber and the differential distance from the center of the fiber (shear rate, expressed in sec^{-1}) is a function of blood viscosity and shear stress. The shear rate is also proportional to the blood flow per single fiber. The thickness of the protein layer at the blood-membrane interface depends on the wall shear rate value, and is extremely important for the membrane performance. The shear rate value linearly correlates with the shear stress in case of Newtonian fluids and the velocity profile is regularly parabolic. Blood approaches Newtonian behavior only at shear rates higher than 20 sec^{-1} (fig. 2).

UF and solute sieving coefficients are considerably influenced by the wall shear rate because it contributes to keep the polarization layer very thin. This is particularly important for solutes in the middle-high range. Diffusion is also affected by the value of shear rate since high shear rates contribute to maintain the diffusion distance from blood to dialysate within minimal values. This is because concentration polarization and the secondary layer of proteins lead to the formation of a pseudomembrane whose thickness is added to that of the original membrane. In clinical practice, high wall shear rates are obtained with high blood flows and appropriate device geometry, with relative preservation of UF rates and solute clearances. In a recent study carried out with dye injection in the blood compartment of different hollow fiber dialyzers, we could demonstrate that in peripheral fibers, blood flows and shear rates much lower than those observed in the central fibers of the bundle are observed [9, 10]. This is even more evident when hematocrit rises above 35%. Based on the above-mentioned issues we could speculate that blood flows higher than 350–400 ml/min must always be utilized in the presence of a 1.8–2.0 m² dialyzer, if an optimal performance of all the fibers is to be achieved.

Dialysate Compartment

While several attempts have been made to optimize the blood compartment, by creating adequate blood ports and flow distributors at the inlet of the dialyzer, very little attention has been paid to the dialysate compartment. The dialysate distribution may in fact be asymmetrical inside the dialyzer, presenting a nonhomogeneous distribution within the fiber's bundle and consequent phenomena of channeling. This may prevent the optimal performance of the dialyzer and may affect the final performance of the treatment.

Some attempts to avoid dialysate channeling have been made by new filter design. For example, an increased length of the fibers reduces contact because of a waived configuration of the fiber. An external irregular surface makes it possible to avoid the perfect contact of adjacent fibers. Different systems of nonparallel orientation of the fibers or the use of tissue structures within the bundle may further help to maintain adequate distances between the external surface of adjacent fibers. The most recent approach is the use of spacing filaments ('spacer yarns') between the fibers or the creation of waved fibers (Moire' structure). We have carried out a complex evaluation using helical CT scan to achieve a detailed imaging of the dialysate distribution pattern after dye injection. The modified dialysate compartment with the spacing filaments between the fibers and the Moire' structure displayed a more homogeneous distribution of the dye, as compared to the standard dialyzers in which a typical channeling effect was displayed in the peripheral regions [11, 12].

Interference between Diffusion and Convection

Although convection and diffusion are described as two separate phenomena, in practice we cannot distinguish the single contributions given by the two mechanisms separately. Moreover, especially in treatments that present a combined utilization of diffusion and convection, there is a continuous interference between the two transport mechanisms [13]. In such circumstances, enhancement of one type of transport can produce effects on the other mechanism of transport that may be beneficial or detrimental.

In hemodiafiltration, solutes are carried across the membrane at the same concentration as in plasma water because of high UF rate. This phenomenon mostly takes place in the proximal side of the filter and reduces the driving force for diffusion. In this case convection negatively affects diffusion that becomes more important in the distal side of the filter where UF approaches zero. This emphasizes the importance of the surface area for the diffusive performance in hemodiafiltration. However, in hemodiafiltration the backdiffusion of substances such as buffers from dialysate into the blood may also be negatively affected at least in the proximal side of the filter where UF is higher. In high-flux dialysis a typical filtration-backfiltration profile occurs. The minimal interference between convection and diffusion is achieved in the central part of the dialyzer at which the water flux in both directions is near zero. In the region near the blood ports, convection may interfere with diffusion both in the filtration and backfiltration modes.

All these factors should be considered in prescribing and practically executing the dialysis procedure. Their knowledge makes it possible to prevent system malfunction and to obtain the desired clinical results.

References

1 Alwall N: On the artificial kidney. I. Apparatus for dialysis of blood 'in vivo'. Acta Med Scand 1947;128:317–321.
2 Kolff WJ: First clinical experience with the artificial kidney. Ann Intern Med 1965;62:608–612.
3 Henderson LW, Besarab A, Michaels A, Bluemle LW Jr: Blood purification by ultrafiltration and fluid replacement (diafiltration). Trans Am Soc Artif Intern Organs 1967;17:216–221.
4 Henderson LW, Colton CK, Ford C: Kinetics of hemodiafiltration. II. Clinical characterization of a new blood cleansing modality. J Lab Clin Med 1975;85:372–375.
5 Maher JF, Schreiner GE, Waters TJ: Successful intermittent hemodialysis – Longest reported maintenance of life in true oliguria (181 days). Trans Am Soc Artif Intern Organs 1960;6:123–126.
6 Bergstrom J: Ultrafiltration without dialysis for removal of fluid and solutes in uremia. Clin Nephrol 1978;9:156–161.
7 Babb AL, Farrel PC, Uvelli DA, Scribner BH: Hemodialyzer evaluation by examination of solute molecular spectra. Trans Am Soc Artif Intern Organs 1972;18:98–105.
8 Babb AL, Strand MJ, Uvelli DA, Milutinovich J, Scribner BH: Quantitative description of dialysis treatment: A dialysis index. Kidney Int 1975;7(suppl 2):23–28.
9 Ronco C, Ghezzi PM, Metry G, Spittle M, Brendolan A, Rodighiero MP, Milan M, Zanella M, La Greca G, Levin NW: Effects of hematocrit and blood flow distribution on solute clearance in hollow-fiber hemodialyzers. Nephron 2001;89:243–250.
10 Ronco C, Brendolan A, Cappelli G, Balestri M, Ingaggiato P, Fortunato A, Milan A, Pietribiasi G, La Greca G: In vitro and in vivo evaluation of a new polysulfone membrane for hemodialysis.

Reference methodology and clinical results (Part 1: in vitro study). Int. J Artif Organs 1999;22: 604–615.

11 Ronco C, Brendolan A, Cappelli G, Balestri M, Ingaggiato P, Fortunato A, Milan M, Pietribiasi G, La Greca G: In vitro and in vivo evaluation of a new polysulfone membrane for hemodialysis. Reference methodology and clinical results (Part. 2: in vivo study). Int J Artif Organs 1999;22: 616–624.

12 Ronco C, Brendolan A, Crepaldi C, Rodighiero MP, Everard P, Balestri M, Cappelli M, Spittle M, La Greca G: Dialysate flow distribbution in hollow fiber hemodialyzers with different dialysate pathway configurations. Int J Artif Organs 2000;23:601–609.

13 Henderson LW: Biophysics of ultrafiltration and hemofiltration; in Maher JF (ed): Replacement of renal function by dialysis. A text-book of dialysis, ed 3. Kluwer Academic Publishers, 1989, pp 300–326.

Claudio Ronco, MD
Department of Nephrology
St. Bortolo Hospital, IT–36100 Vicenza (Italy)
Tel. +39 0444 993650, Fax +39 0444 993949, E-Mail cronco@goldnet.it

Ronco C, Brendolan A, Levin NW (eds): Cardiovascular Disorders in Hemodialysis.
Contrib Nephrol. Basel, Karger, 2005, vol 149, pp 18–26

....

Machines for Hemodialysis

Hans-Dietrich Polaschegg

Köstenberg, Austria

Abstract

Basic functions of hemodialysis machines are described. The paper focuses on essential treatment parameters and safety aspects. The cause of safety hazards and protective systems for amelioration of these hazards are described. With the exception of hemolysis caused by obstructions in the extracorporeal circuit and blood losses caused by user errors machine related accidents are rare. Foreseeable improvements of next generation hemodialysis machines will reduce the likelihood of accidents further. The accuracy of adjusted or monitored treatment parameters that may influence outcome (dialysate concentration, ultrafiltration, blood flow and on-line measured clearance) is discussed.

Introduction

The basic design of hemodialysis machines has not been changed since the end of the 1960s when the first single-patient machines for home hemodialysis became available. This basic design was influenced by the then available technology and by the pressure to provide devices quickly in order to save human lives. Missing knowledge about physiologically relevant parameters influenced the design as well. The technology available today may allow the design of more cost-effective and less complicated hemodialysis systems. In this paper I will address the basic design parameters common to all hemodialysis machines and the basic risks related to the hemodialysis process.

Hemodialysis Machine Overview

Hemodialysis machines are part of a treatment system consisting of water treatment, concentrate or dialysate supply, dialyzer and dialysis machine including the extracorporeal circuit.

For hemodialysis, blood is taken from the 'arterial' blood access, circulated through the dialyzer and returned to the 'venous' blood access.

In the dialyzer, blood is exposed to dialyzing fluid (also called dialysate) and fluid is withdrawn from blood by ultrafiltration (UF). Exposure to dialyzing fluid results in an exchange of substances and thermal energy following concentration and temperature gradients, respectively. In addition an electrical connection between the hemodialysis machine and the patient is established through the dialyzing fluid and the blood. The substances that are exchanged are water-soluble organic substances, electrolytes and dissolved gases.

Under fault condition (machine or system malfunction or user error) the patient and/or the user is exposed to hazards that in a worst-case condition may result in death or serious injury. In order to reduce the likelihood of such events, hemodialysis machines contain additional monitors, so-called 'protective systems'. The minimum requirements for safety of hemodialysis machines are described by standards (AAMI RD5 [1] for the US and IEC601–2-16 [2] for the rest of the world).

Extracorporeal System

The extracorporeal system is used to transport blood to the dialyzer and back to the body. All commercial machines today use blood pumps, which, with a few exceptions, are rotary peristaltic pumps.

Figure 1 shows the principle of the extracorporeal system: Basically, only two pieces of tubing are required ('arterial' tubing from the blood access to the dialyzer and 'venous' tubing from the dialyzer back to the blood access), but there is only one filtration machine employing such a simple system [3]. Most hemodialysis machines use more complicated blood tubing systems. Over the years components have been added for various reasons; some of them have no relevant function anymore, others may become redundant with foreseeable technological changes. The ends of these tubing systems are connected to the blood access devices (cannulae or catheters) and to auxiliary devices (heparin pump, pressure monitor, etc.) with luer connectors.

In the peristaltic pump, tubing is compressed between a cylindrical roller and the pump bed. Two or three rollers are arranged in the pump rotor to guarantee occlusion at any time during rotation.

The blood flow adjusted and/or displayed is usually calculated from the rotary speed of the blood pump and the displaced volume per revolution [4].

When the blood pump tubing is exposed to negative pressure, it will collapse resulting in reduction of the cylinder volume and in reduced blood pump flow. The typical flow reduction at -200 mm Hg is -5 to -15%. Some

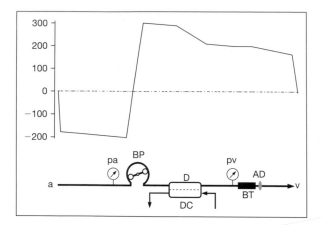

Fig. 1. Extracorporeal circuit. Major components and pressure. a = Arterial access; pa = arterial pressure monitor; BP = blood pump; D = dialyzer; DC = dialysate circuit; pv = venous pressure monitor; BT = bubble trap; AD = air detector; v = venous access. Air can only enter in the negative pressure part between a and BP. Blood can only be lost in the positive pressure part between BP and v.

dialysis machines display flow corrected for prepump pressure but this applies only for blood pump tubing specified by the machine manufacturer.

Pumping fluid through the extracorporeal system causes pressure drops which are functions of the flow resistances of the system, the blood flow and the viscosity of blood. Most of the pressure drops are caused by the blood access devices (cannulae, catheters). Flow in cannulae is turbulent with the result that pressure drops are almost independent of cannula length (for nominal cannula lengths between 15–25 mm) and dependency on cannula diameter is ~d^3 rather than d^4 expected from Hagen-Poiseuille's law. The pressure drop causes prepump pressure to become negative relative to atmosphere and, in case of a leak will allow the ingress of air into the system. The pressure drop postpump is positive which will result in blood loss to the environment in case of a leak. In order to mitigate the hazards related to leaks and obstructions in the tubing system, pressure monitors and air detectors are added to the system.

Dialysate Delivery System

The dialysate delivery system conditions water and mixes it with one to three concentrates. Conditioning of water means heating and degassing which is not done by the water purification system; although, this is possible and would be more cost effective. Water degassing is required for safety and performance reasons. Tap water is usually air saturated at a temperature of 5–10°C and contains 25 ml of air. Heating to 37°C and adding electrolytes reduces the solubility and dialysate becomes oversaturated. Air would diffuse to the blood side and would form gas bubbles and/or would coalesce in the dialyzer reducing the efficacy. Water is usually degassed by exposing it to low (negative) pressure.

When acetate was used as buffer substance, dialysis concentrates used to contain all ingredients of dialysate in concentrated form (usually 35-fold concentration). When, for medical reasons, acetate was replaced by bicarbonate, it was no longer possible to provide all ingredients of dialysate in a single form because carbonate ions would react with calcium and magnesium ions and precipitate out of solution as carbonates. In solution, bicarbonate (HCO_3^-) is in equilibrium with CO_2 and with carbonate (CO_3^{--}). This equilibrium is controlled by the pH value of the fluid. pH cannot be reduced sufficiently to reduce CO_3^{--} concentration because the CO_2 pressure in the concentrate would exceed atmospheric pressure. Most of dialysis treatments today are done with dialysate mixed from two concentrates. The 'acidic concentrate' contains sodium, potassium, calcium and magnesium as chlorides, some acid to control the pH of the final solution (usually acetic acid) and, optionally, glucose. The 'bicarbonate concentrate' is usually a 1-molar solution of sodium bicarbonate (the saturation temperature of the 1-molar solution is 15°C). A single substance concentrate can be produced on-line by continuous dissolution with water. This form of on-line production of bicarbonate concentrate is very common today. Because the concentration of the bicarbonate fluid produced by continuous dissolution is temperature dependent, the subsequent dilution with water must be appropriately controlled. Supplying concentrate in powder form for continuous dissolution has several advantages. For this reason some companies also provide sodium chloride in powder form which requires a dialysate mixing system for three concentrates plus water. The concentrates are sodium chloride powder, sodium bicarbonate powder and a liquid concentrate containing the rest.

Redundant conductivity and temperature monitors are universally used as protective systems to prevent dialysis with hyper- or hypo-osmolar fluid. In case the concentration is out of range ($\pm5\%$ of the set value) dialysate is bypassed. A blood leak detector is positioned downstream of the dialyzer to detect any blood that may leak from the blood side to the dialysate side in case of a defect in the dialyzer (fig. 2, table 1).

UF Control

Fluid removal in hemodialysis is performed by UF. Modern dialysis machines control the balance of fresh and spent dialysate. The flow or volume

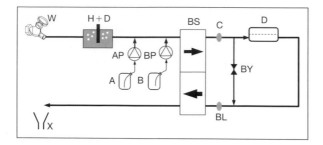

Fig. 2. Dialysate circuit. W = Water; H + D = heating and degassing; A = acidic concentrate; AP = acidic concentrate pump; B = bicarbonate concentrate; BP = bicarbonate pump; BS = balancing system; C = conductivity + temperature monitor; By = bypass; D = dialyzer; BL = blood leak monitor; X = drain.

difference between spent and fresh dialysate is replaced by ultrafiltrate from the patient. The flow difference causes the pressure on the dialysate side to decrease resulting in a pressure difference between the blood side and the dialysate side of the dialyzer membrane which is called the transmembrane pressure. This transmembrane pressure is the driving force for UF. Various technical solutions for precise controlling of UF have been developed. Technical differences between UF control systems are, however, of little relevance for the user.

Disinfection

The dialyzing fluid path must be disinfected between treatments. This is done by exposing it to hot water (>85°C) and/or chemicals. Using hot citric acid has become popular for this purpose because residues of citric acid remaining in the fluid path are harmless as opposed to more aggressive chemicals. Citric acid also dissolves carbonates that often precipitate in the dialysate circuit. Most other chemicals used are aggresive oxidants.

Risks, Safety and Essential Control Parameters

Basic definitions and requirement for risk analysis are described by ISO14971 [5].

Hazard = potential source of harm (harm = physical injury and/or damage to health or property)

Risk = combination of the probability of occurrence of harm and the severity of that harm

Table 1. List of major hazards

Hazard	Harm	Protective system(s)	Remark, max. sensitivity or deviation under fault condition
Dialyzing fluid composition	Osmotic hemolysis, hyper- or hyponatremia, hyper- or hypokalemia, acidosis, alkalosis	Conductivity monitor(s), volume monitor(s), bypass	Typical sensitivity ~3% for total conductivity
Dialyzing fluid temperature	Thermal hemolysis	Temperature monitor max. 42°C	Hemolysis threshold = 45.6°C
Ultrafiltration	Hyper- or hypovolemia	Redundancy, pressure monitoring, cyclic tests	~250 ml
Extracorporeal blood loss to the environment – caused by blood line rupture	Exsanguination	Venous pressure monitor – pressure-limited pump occlusion	~10% of blood flow. Cannula slipped from access usually not detected.
Blood leak	Exsanguination	Blood leak detector	0.5 ml/min
Air infusion	Air embolism	Air detector	Bubbles >50 µl Safe limit 100 µl/kg body weight
Blood line obstructions	Mechanical hemolysis	Venous pressure monitor	Does not detect obstructions between blood pump and venous pressure monitor

The operation of a dialysis system is related to unacceptable risks for the patient unless

- the system is properly designed and
- protective systems reduce the risks caused by malfunctions of water treatment, the dialysis machine, the blood tubing set and the dialyser but also the risks caused by foreseeable user errors.

Not all risks can be sufficiently ameliorated by automatic monitors. In these cases appropriate warnings and guidance must be supplied by the manufacturer of the device. Table 1 contains a list of hazards and the most common protective systems used for amelioration of the related risks.

Fatal accidents with hemodialysis machines caused by exsanguination or mechanical hemolysis have been reported in recent years. Blood loss to the environment remains undetected when the blood access cannula slips from the fistula. Typical fistula pressure is 20 mm Hg. When the cannula slips from the fistula the change of pressure at the venous pressure monitor is not sufficient to trigger an alarm. More sensitive monitors have been described in the patent literature but are not yet available. It must be pointed out that these accidents are caused by user errors rather than device malfunction. Nevertheless, future machine designs must take these events as 'foreseeable user errors' into account. Mechanical hemolysis is typically caused by an obstruction in the blood tubing between the outlet of the blood pump and the pressure sensor downstream of the pump. This is usually the venous pressure sensor although some machines have additional postpump arterial pressure sensors. The obstruction may be caused by a kink in the blood tubing which is a user error and can be detected by careful inspection of the extracorporeal circuit at the beginning and during the treatment. Manufacturing errors of the blood tubing set have been identified as cause of mechanical hemolysis as well. It is generally assumed that erythrocytes are damaged at the point of obstruction. It may be hypothesized, however, that the damage is caused by the loss of occlusion of the blood pump.

Air embolism was a frequent cause of fatal accidents in the beginning of dialysis. A few small bubbles becoming visible in the blood tubing system do not cause harm but indicate a malfunction of the system. Blood does not degas normally in the extracorporeal system. Air that becomes visible on the prepump side of the extracorporeal system comes in through leaks. Air cannot enter on the positive pressure side. Bubbles becoming visible downstream of the bubble trap on the venous side may indicate residues of disinfectants that have not been completely removed from the dialysate circuit after disinfection or they may indicate malfunction of the degassing system.

Accuracy of Essential Control and Monitoring Parameters

Dialyzing Fluid Composition. Modern dialysis machines usually allow adjustment of dialysate sodium (Na^+) and dialysate bicarbonate concentrations. The other electrolytes (K^+, Ca^{++}, Mg^{++}) are changed proportionally to Na^+. Three-concentrate mixing machines allow choice of K^+, Ca^{++} and Mg^{++} independent of Na^+ by using 'individual' bags. The accuracy of concentrations in dialysate can be expected to be within ±3.5% for sodium and ±6% for potassium, calcium and magnesium. This estimate results from the addition of the allowed tolerances of concentrates [6] and calibration accuracy of mixing systems. Under worst-case conditions the deviation can be higher. The accuracy of clinical laboratories is usually not better. In addition, it must be taken into account, that analyzers used in the clinical laboratory are calibrated for plasma and that for some clinical analyzers the results must be corrected for dilutional errors.

On-Line Clearance, Kt/V and Plasma Sodium. Some machines today allow the measurement of effective dialysance by varying the dialysate concentration and measuring dialyzer inlet and outlet conductivity [7, 8]. The statistical error of this method compared with clearance calculation from blood side measurements is ~5%. This is equivalent to the error that can be expected as sum of the laboratory error and blood flow measurement error. Classical urea kinetics uses the urea reduction ratio as input for the calculation of Kt/V. Neither clearance K nor dialysis time t or urea distribution volume needs to be known. When on-line measured clearance is used for calculating Kt/V, time t is measured precisely by the machine but volume V must be measured with an independent method or must be estimated. This results in an additional error.

The mathematical derivation used for calculating on-line clearance also provides a solution for the plasma electrolyte concentration. This is used in some machines to display plasma sodium. The accuracy of this value is determined by the sum of the error for dialysate sodium plus the error related to the on-line clearance measurement (3–5% or 4–6 mmol/l).

It is important to understand that the accuracy of all three parameters is usually not guaranteed by the manufacturer and the deviation can exceed the estimates made above. This may change in the future when technical standards for dialysis machines are adapted to recent developments.

Blood Flow. The rotational speed of the blood pump is precisely controlled. Blood flow inaccuracies are due to blood tubing diameter tolerances, imperfect compensation of prepump negative pressure influences and effects of tubing fatigue. For compensated blood flow displays the tolerance is estimated to be ±5%. If no compensation for negative pressure effects is used, blood flow may be 10–20% lower than displayed, which may result in under-dialysis because blood flow is a major determinant of clearance.

Some machines calculate total processed blood volume, which can be used for quality control.

Dialysate Flow. Dialysate flow is usually controlled within 2% or better. Because dialysate flow is usually higher than blood flow, the influence on clearance is lower.

UF Rate, UF Volume. This error is usually less than 50–100 ml/h depending on the system used. Information about the UF accuracy is normally available from technical handbooks that come with dialysis machines.

References

1   Association for the Advancement of Medical Instrumentation: 2003-Hemodialysis Systems. ANSI/AAMI RD5–2005:www.aami.org
2   International Electrotechnical Commission: Medical electrical equipment. Part 2: Particular requirements for safety of haemodialysis equipment. Geneva, IEC 60601 2–16, 1998: www.iec.ch
3   System100. Brooklyn Park, CHF Solutions, Inc: www.chfsolutions.com
4   Polaschegg HD, Levin N: Hemodialysis machines and monitors; in Winchester J, Koch R, Lindsay R, Ronco C, Horl W (eds): Replacement of Renal Function by Dialysis, ed 5. Dordrecht, Boston, London, Kluwer Academic Publishers, 2004, pp 323–447.
5   International Organisation for Standardization: Medical devices-Application of risk management to medical devices. Geneva, ISO 14971:2000: www.iso.ch
6   International Organisation for Standardization: Concentrates for haemodialysis and related therapies. Geneva, ISO 13958:2002: www.iso.ch
7   Sternby JP, inventors. Gambro AB, assignee: Dialysis system. EP patent 0547025, June 12, 1996.
8   Polaschegg HD, inventors. Fresenius Medical Care Deutschland GmbH, assignee: Method for measuring the efficiency of mass and energy transfer in hemodialysis. US patent 6702774, March 9, 2004.

Hans-Dietrich Polaschegg, Dr. techn.
Medical Devises Consultant
Malerweg 12
AT–9231 Köstenberg (Austria)
Tel. +43 4274 4045, Fax +43 4274 4096, E-Mail pg@compuserve.com

Ronco C, Brendolan A, Levin NW (eds): Cardiovascular Disorders in Hemodialysis.
Contrib Nephrol. Basel, Karger, 2005, vol 149, pp 27–34

....

Mathematical Model to Characterize Internal Filtration

Gianfranco Beniamino Fiore[a], *Claudio Ronco*[b]

[a]Department of Bioengineering, Politecnico di Milano, Milano, and
[b]Department of Nephrology, St. Bortolo Hospital, Vicenza, Italy

Abstract

Convective-diffusive dialysis techniques have recently gained considerable favor. Indeed, convective fluxes through dialyzer membranes have been demonstrated to play a role in enhancing the clearance of middle-molecular-weight solutes. An interesting opportunity is given by exploiting the internal filtration (IF)/back filtration mechanism that occurs spontaneously in high-flux dialyzers, but is difficult to quantify. In view of overcoming this drawback, a semi-empirical, lumped-parameter mathematical model for characterization of IF phenomena was developed. The model considers a dialyzer as composed by N adjacent axial blocks. For each block, hydrodynamics in the blood and dialysate compartments are determined considering hydraulic resistance and calculating local filtration. Blood viscosity and oncotic pressure are calculated locally based on hematocrit and protein concentration. Resistance parameters were determined experimentally for the BS-UL (Toray Industries Inc., Tokyo, Japan) dialyzers. The set of equations describing the model, implemented into a software program, is solved using a numerical method. Simulations allow highlighting the role of device-, treatment- and patient-dependent parameters in affecting IF. Provided an extensive validation is carried out, the use of a mathematical model could be the key to make IF more understandable and its use reliable in clinical practice.

Principles Governing Internal Filtration

The driving force of the ultrafiltration (UF) flux that locally takes place through the membrane of a dialyzer is local transmembrane pressure difference (TMPD): $TMPD = p_b - p_d - \Pi$, where p_b and p_d are pressures in the blood and dialysate compartments, respectively, and Π is the oncotic pressure. If local TMPD >0, then UF flux is directed from blood towards dialysate

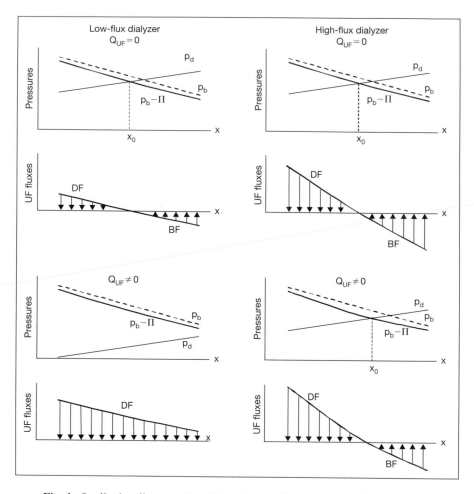

Fig. 1. Qualitative diagrams describing the ultrafiltration mechanisms along low-flux and high-flux dialyzers. Blood flow directed from left to right; dialysate flow from right to left. Upper graphs = At zero net ultrafiltration rate; lower graphs = at nonzero net ultrafiltration rate.

(direct filtration, DF); otherwise, if local TMPD <0, then UF flux is directed from dialysate towards blood (back filtration, BF).

When a dialyzer is used in a countercurrent arrangement and the net UF rate (Q_{UF}) is kept equal to zero, DF must occur in the proximal part, and BF in the distal part of the dialyzer (e.g. the upper graphs in fig. 1). Pressures in the two compartments decrease (as a result of hydraulic resistance) with opposite slopes. In order to yield $Q_{UF} = 0$, BF in the distal part must compensate for the

DF occurring in the proximal part, with the two pressure lines (the p_d line in the dialysate and the $p_b - \Pi$ line in blood compartment) intersecting at a certain axial position (x_0) around the device's middle section. The term internal filtration (IF) is associated to this situation to signify that, even with a null net UF rate, UF fluxes do take place inside the device.

Due to the countercurrent arrangement, the fluid pertaining to DF is discharged with the exhausted dialysate fluid, while the backfiltered fluid comes directly from ultrapure dialysate into the blood compartment. The IF/BF phenomenon may therefore represent a source of convective solute transport. The amplitude at which IF/BF occurs, however, depends on several factors: as an example, comparison of the IF/BF phenomenon with a low-flux or a high-flux device are qualitatively depicted in figure 1. At $Q_{UF} = 0$ (upper graphs), IF/BF takes place in both device types, but UF fluxes are much greater in the high-flux device due to elevated hydraulic permeability. When the devices are used with a nonzero net UF rate ($Q_{UF} \neq 0$), the dialysate pressure line must translate downward to provide for the necessary driving force to draw the desired Q_{UF}. Translation is consistent in low-flux devices (lower left graph), for which the IF/BF phenomenon vanishes, the only remaining source of convective transport being Q_{UF} (equaling the integral of DF fluxes). Translation is much less pronounced in a high-flux device (lower right graph), for which IF/BF remains. In this case, since net UF rate Q_{UF} is due to the difference between DF rate (Q_{DF} = the integral of DF fluxes) and BF rate (Q_{BF} = the integral of BF fluxes), it is yielded that: $Q_{DF} = Q_{UF} + Q_{BF}$, where Q_{DF} represents the total source of convective transport. In other words, in the presence of IF/BF, total convection is enhanced, with respect to the mere UF rate, by a portion that equals the BF rate taking place inside the device.

These considerations may lead to consider the potentialities of IF/BF as a means to promote convective transport in high-flux hemodialyzers, in order to enhance the clearance of middle-molecular-weight solutes (IF/BF is not expected to affect the removal of small molecules, for which diffusion still plays the primary role). This could be an interesting alternative to convective-diffusive dialysis techniques, such as on-line hemodiafiltration [1, 2], avoiding the use of replacement fluid for pre- or postdilution. Recent clinical findings confirm this possibility [3]. Still, a key issue to make the IF/BF effect useful in the clinical field is quantification of convective fluxes. The extent of IF/BF dependence on quantities related to the device design (geometry and membrane performance [4, 5]), to therapy conditions (blood flow, dialysate flow, and UF rate) or to blood characteristics [hematocrit (Hct), plasma protein concentration (C_P)] is rather complex and involves some nonlinear effects. With a view of shedding light on this matter, a semi-empirical mathematical model aimed at quantifying IF/BF is described in this work.

Mathematical Model Design and Implementation

The model was designed with the lumped-parameter approach, according to which hydraulic components are split into adequate number of units, each described through parameters that compactly address its hydrodynamic features. The fiber-bundle section of a hemodialyzer has been schematized by axially dividing it into N adjacent blocks (n = 100 has been used in simulations). The generic i-th block is characterized by its hydrodynamic resistance parameters in the blood side ($R_{b,i}$) and in the dialysate side ($R_{d,i}$). Since undisturbed laminar flow takes place in this section, a proportional relationship between pressure drop and flow rate (both on blood and on dialysate side) is expected: $\Delta p = R_i Q$, where the i-th block's resistance R_i is proportional to viscosity: $R_i = r\mu$. Blood viscosity may change along each fiber due to local concentration or dilution, thus resistance on the blood side is a function of position (each block features its own resistance $R_{b,i}$). The dependence of plasma viscosity μ_p upon total protein concentration C_p was described with the following equation [6]:

$$\frac{\mu_p}{\mu_w} = 1 + \left(\frac{\mu_{p,r}}{\mu_w} - 1 \right) \frac{C_p}{C_{p,r}} \tag{1}$$

where μ_w is water viscosity and the reference values for plasma viscosity and protein concentration are $\mu_{p,r} = 1.22$ cP and $C_{p,r} = 7$ g/dl, respectively.

Blood viscosity μ_b was locally calculated as a function of plasma viscosity and Hct according to the approximate equation [7]:

$$\mu_b = \mu_p (1 + 2.5 \text{ Hct}) \tag{2}$$

Local UF at the i-th block takes place depending on local TMPD:

$$Q_{UFi} = A_i L_P (p_{b,i} - p_{d,i} - \Pi_i) \tag{3}$$

where Q_{UFi} and A_i are the UF rate and membrane surface area pertaining to the i-th block, respectively, L_P is the membrane's hydraulic permeability, Π is the blood's oncotic pressure. Provided Π is evaluated locally and subtracted to hydraulic pressure difference across the membrane, local filtration may therefore be accounted for, in each axial block, with an additional resistance $R_{UFi} = 1/(A_i L_P)$, connecting the blood and dialysate side. The dependence of oncotic pressure on C_p was expressed according to Landis and Pappenheimer [8]:

$$\Pi = 2.1 C_p + 0.16 C_p^2 + 0.009 C_p^3 \tag{4}$$

with Π expressed in mm Hg and C_p in g/dl.

The resistance parameters $R_{b,i}$, $R_{d,i}$ were determined with in vitro tests for three high-flux polysulfone dialyzer types: BS-1.6 UL, BS-1.8 UL, BS-2.1 UL (Toray Industries Inc., Tokyo, Japan). For each device, pressure drop at different fluid flow rates (blood side: Q_b = 0 ml/min; 200–500 ml/min step 50 ml/min, dialysate side: Q_d = 0–1,500 ml/min step 250 ml/min) was measured. The blood or dialysate compartments were tested separately. An HT 110 Transonic® flowmeter was used to measure flow rates. Pressure was measured with Honeywell® 140PC sensors directly connected to the inlet/outlet blood/ dialysate chambers via holes in the devices' case, so as to remove the contribution to pressure drop due to elements external to the fiber-bundle region (connectors and distributors). Sucrose water solutions with calibrated viscosity (range 2–4 cP, measured with a Cannon-Fenske® viscometer) was used as a blood-analog fluid in the blood-side tests. Normal saline was used on the dialysate side. The membrane resistance to UF, R_{UFi}, was determined using the characteristics of the membranes supplied by the manufacturer.

The developed model was implemented into a software program in the LabVIEW® 6.1 environment for Windows®. The set of equations describing the model is numerically solved by an iterative routine essentially based on the Newton-Rhaphson method. The time needed for a typical problem to be solved is far less than one second with a common personal computer, so that the user may work with the program in quite an interactive way.

Analysis and Discussion of Model Simulation Results

At fixed patient values (i.e. Hct = 33% and C_p = 7 g/dl) and at Q_{UF} = 0, the effect exerted by changing blood or dialysate flow rates may be evaluated. The results for device BS-1.8 UL are shown in figure 2 (upper panels): IF/BF rate is more sensitive to blood flow rate changes (upper left graph) than to dialysate (upper right graph). Because of a higher hydraulic resistance, flow rate increments on blood side cause steeper changes in pressure drop (and therefore on IF/BF rate) than on dialysate side. Combined changes of both Q_b and Q_d would display superimposed effects.

At fixed patient values (Hct = 33%; C_p = 7 g/dl) and at given flow rate settings (Q_b = 300 ml/min; Q_d = 500 ml/min), the effect of increasing net UF rate (Q_{UF}) is shown in figure 2 (lower panel) for device BS-1.8 UL. Increases in Q_{UF} correspond to concomitant DF rate increments and BF rate decrements. Such changes are respectful of the mass flow equilibrium for the whole system, however, in this particular case, are not symmetric with respect to a horizontal line, rather the DF increase is larger than the BF decrease.

It is interesting to analyze the influence exerted by a patient-dependent parameter such as the inlet blood's protein concentration C_p with all other parameters (Hct, Q_b, Q_d and Q_{UF}) unchanged. In order to emphasize the model's behavior, C_p was varied even out of the normal range in simulations. The upper

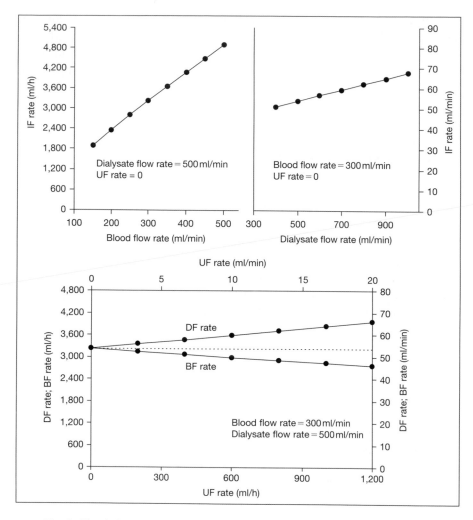

Fig. 2. Simulation results (BS-1.8 UL): effects of changing the dialysis machine parameters at fixed patient values (Hct = 33% and C_p = 7 g/dl).

panel of figure 3 shows that IF/BF rate ($Q_{IF/BF}$) tends to increase at increasing C_p, by virtue of the dependence of blood viscosity upon C_p; at C_p values higher than normal, increases are dampened. This particular behavior becomes clear when looking at the lower panel of figure 2, where the increase of the local values of C_p with respect to the inlet C_p value are plotted along the device's axial coordinate. The shape of curves is due to blood concentration taking place at proximal positions and blood dilution taking place at distal positions.

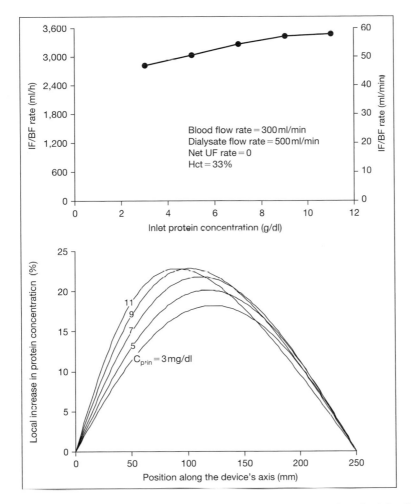

Fig. 3. Simulation results (BS-1.8 UL): effects of changing the inlet blood's protein concentration C_p (Hct = 33%, Q_b = 300 ml/min, Q_d = 500 ml/min, Q_{UF} = 0). Upper panel = IF/BF rate as a function of C_p; lower panel = percent local increase in C_p versus the device's axial position.

The maxima tend to increase at increasing inlet C_p because of a higher IF/BF rate. At large inlet C_p, however, this tendency is counterbalanced by the local increases in oncotic pressure. Also visible is a backward shift of the x_0 position (x_0 is where the curves reach their maxima) at increasing inlet C_p, due to the nonlinearity of the relationship between oncotic pressure and protein concentration (see eq. 4).

Conclusions

The aim of this work was to build up a simple, semi-empirical model and related software tool to characterize the IF of a particular dialyzer as a function of dialysis condition, including the main variability related to the patient's blood. The results obtained with the model may be used to evaluate the concrete possibility to perform diffusive-convective dialysis by virtue of the sole IF/BF phenomena. Several trials in the past might have failed to demonstrate the superiority of high-flux membranes because of the poor utilization of the convective potential. The use of mathematical modeling may therefore be helpful in the design process of dialyzers prone to enhancement of the IF/BF amount, such as high-flux polysulfone filters with augmented fiber length, that are now moving from the prototype to the commercial phase.

Provided that extensive validation of this approach is carried out, using mathematical modeling could thus be the key to make IF more understandable and its use reliable in the clinical practice.

References

1 Samtleben W, Dengler C, Reinhardt B, Nothdurft A, Lemke HD: Comparison of the new polyethersulfone high-flux membrane DIAPES® HF800 with conventional high-flux membranes during on-line haemodiafiltration. Nephrol Dial Transplant 2003;18:2382–2386.
2 Ward RA, Schmidt B, Hullin J, Hillebrand GF, Samtleben W: A comparison of on-line hemodiafiltration and high-flux hemodialysis: A prospective clinical study. J Am Soc Nephrol 2000;11:2344–2350.
3 Lucchi L, Fiore GB, Guadagni G, Perrone S, Malaguti V, Caruso F, Fumero R, Albertazzi A: Clinical evaluation of internal hemodiafiltration (iHDF): A diffusive-convective technique performed with internal filtration enhanced high-flux dialyzers. Int J Artif Organs 2004;27:414–419.
4 Mineshima M, Ishimori I, Ishida K, Hoshino T, Kaneko I, Sato Y, Agishi T, Tamamura N, Sakurai H, Masuda T, Hattori H: Effects of internal filtration on the solute removal efficiency of a dialyzer. ASAIO J 2000;46:456–460.
5 Ronco C, Brendolan A, Feriani M, Milan M, Conz P, Lupi A, Berto P, Bettini M, La Greco G: A new scintigraphic method to characterize ultrafiltration in hollow fiber dialyzers. Kidney Int 1992;41:1383–1393.
6 Pallone TL, Petersen J: A mathematical model of continuous arteriovenous hemofiltration predicts performance. ASAIO Trans 1987;33:304–308.
7 Lightfoot EN: Transport Phenomena and Living Systems. New York, Wiley & Sons, 1974.
8 Landis EM, Pappenheimer JR: Handbook of Physiology. Washington, DC, Am Physiol Soc, 1963.

Gianfranco B. Fiore, PhD
Dipartimento di Bioingegneria
Politecnico di Milano, Piazza Leonardo da Vinci, 32
IT–20133 Milano (Italy)
Tel. +39 02 2399 3361, Fax +39 02 2399 3360, E-Mail gianfranco.fiore@polimi.it

Ronco C, Brendolan A, Levin NW (eds): Cardiovascular Disorders in Hemodialysis.
Contrib Nephrol. Basel, Karger, 2005, vol 149, pp 35–41

....

Extracorporeal Sensing Techniques

Daniel Schneditz

Institute of Physiology, Medical University Graz, Graz, Austria

Abstract

Many physiologic variables have been measured in the extracorporeal circulation by experimental systems but only a few systems have reached technical maturity for everyday application. Variables relating to cardiovascular function, which today can be measured in the extracorporeal system, are pressure, temperature, and measures of blood composition such as hematocrit, hemoglobin, and total protein concentration. While the measurement of blood composition and temperature is well established, and while the use of extracorporeal pressure information awaits further analysis for robust application, recent interest focused on continuous measurement of plasma sodium concentration which is believed to be of major importance for optimal blood treatment. However, problems with a simple, reliable, and continuous measurement of plasma sodium for everyday use have not yet been resolved. As can be seen from the growing interest in isothermic or isonatremic treatment modes which turns away from constant and profiled treatment modes without feedback control, the treatment goal is now to provide stable conditions within the patient so as to minimize interference with intrinsic physiological control mechanisms. This, however, requires valid and reliable measurement of the specific patient variables of interest.

Introduction

The aim of this contribution is to present an overview of current techniques to measure physiological variables using the extracorporeal circulation and other components of the extracorporeal system. The overview further concentrates on techniques and variables related to cardiovascular function.

The extracorporeal circulation offers an almost perfect access to the measurement of physiologic variables. Arterial and venous blood lines take the form of catheters with transducers attached for the measurement of selected variables. In addition, bloodlines can be used for administration of markers whereby some of these markers such as sodium are delivered by the extracorporeal machine under controlled and reproducible conditions.

Concentration

A growing number of solutes is measured by on-line techniques but the concentration of sodium and the composition of blood determined by hematocrit, hemoglobin concentration, or total protein concentration remain the most relevant with regard to extracorporeal effects on cardiovascular function.

Local Measurement

Concentrations measured in the extracorporeal circulation or inside the machine in general provide local information which do not necessarily represent concentrations in other parts of the circulation. Examples can be given for the well-known effects caused by recirculation of treated blood in the access and in the cardiopulmonary loop of the cardiovascular system [1]. However, when the degree of recirculation is known, local concentrations can be corrected to provide physiologically meaningful arterial or mixed venous concentrations. On the other hand, when extracorporeal gradients are abolished, extracorporeal concentrations represent mixed venous concentrations.

Resolution

The resolution of extracorporeal measurements is usually much higher than what is required for clinical purposes. For many applications such as the calculation of relative blood volume changes, it is necessary to obtain differences and ratios of measured concentrations which, therefore, increases the demands on the reproducibility and stability of concentration measurements.

Sampling Rate

On-line techniques provide a continuous readout at sampling rates ranging from fractions of seconds to minutes. The dimension of time greatly extends the application of on-line sensing techniques as it then becomes possible to follow kinetic processes and to analyze transport phenomena between body compartments [2, 3].

Perturbation

The perturbation of a concentration, for example by injecting normal saline into the venous bloodline and measuring its concentration in the arterial bloodline, opens the whole field of indicator dilution technology to determine cardiovascular transport characteristics such as recirculation, access blood flow, and cardiac output [4, 5].

Sodium

The formula for sodium dialysance can be rearranged to express plasma water sodium concentration ($[Na^+]_{pw}$) in terms of dialysance (D), dialysate flow (Q_d), dialysate in- and outflow sodium concentrations ($[Na^+]_{din}$, $[Na^+]_{dout}$), and the Donnan coefficient (f^+) for univalent cations:

$$[Na^+]_{pw} = \frac{([Na^+]_{dout} - [Na^+]_{din}) \cdot \frac{Q_d}{D} + [Na^+]_{din}}{f^+} \qquad (1)$$

With the exception of f^+ all quantities on the right side of Eq. 1 can be measured in dialysate so that $[Na^+]_{pw}$ can be derived from dialysate measurements. What are the difficulties? First, f^+ is assumed for a given plasma protein concentration which will change during extracorporeal blood treatment. At a normal plasma protein concentration f^+ is assumed with 0.95, but with ultrafiltration-induced hemoconcentration f^+ may fall to much lower levels. The effect of hemoconcentration could be accounted for by parallel measurements of plasma volume changes. Secondly, sodium is not directly measured and electrical conductivity (λ) is used as a surrogate for $[Na^+]$. Since the measurement of dialysate λ is now used to determine dialysance as a surrogate for effective urea clearance in several dialysis machines, the same technique has been claimed to provide enough information to measure $[Na^+]_{pw}$ [6]. However, the measurement of λ is not 100% specific for sodium because of the contribution of other electrolytes. For example, an increase in λ in the dialysate outflow can be caused by an increase in $[Na^+]$, but it could also be related to an increase in potassium concentration, especially since potassium enters the dialysate during hemodialysis. Thus, the actual sodium gradient ($[Na^+]_{dout} - [Na^+]_{din}$) in Eq. 1 is not exactly known using current λ technology, not even when dialysate in- and outflow conductivities are matched to the same value. And thirdly, any unnoticed change in dialysance caused by a change in blood flow, recirculation, or dialyzer operation characteristics will lead to an apparent change in $[Na^+]_{dout}$ for non-zero dialysate sodium gradients ($[Na^+]_{dout} - [Na^+]_{din} \neq 0$).

Specific Application

There is an increasing consensus that hemodialysis should be conducted under isonatremic conditions, i.e., that sodium should be removed by convective flow without any change in $[Na^+]_{pw}$ [7]. Such a procedure requires dialysate composition to be adjusted to the individual level of $[Na^+]_{pw}$ which, given the caveats mentioned above, could be provided by the measurement of dialysate λ and feedback control of $[Na^+]_{pw}$.

Hemoconcentration

Optical, acoustical, mechanical, electrical, and rheological properties of blood which show a strong dependence on hematocrit, hemoglobin, and total protein concentration can be used for non-invasive measurement of blood composition [8]. Three devices based on optical and acoustical principles are now used in several machines to provide a continuous readout of concentrations suitable to track the degree of hemoconcentration during extracorporeal blood treatments. The CritLine (Hemametrics, Utah) and Hemoscan (Hospal-Gambro, Italy) devices analyze the intensity of light at different wavelengths transmitted across and/or reflected by the blood sample passing the measuring site. The blood volume monitor (BVM, Fresenius Medical Care, Germany) measures the time of flight of ultrasonic pulses transmitted across the blood sample. To reach the stability and resolution required for extended periods of time, measurements are done using dedicated measuring cells which are either part of special bloodlines (BVM, Hemoscan) or which can be inserted into the extracorporeal circulation, for example between the arterial line and the arterial port of the dialyzer (CritLine).

Specific Application

The core application of these techniques is to measure ultrafiltration-induced hemoconcentration essentially by determining the relative change in hematocrit, hemoglobin, or total protein concentration over time. These changes can then be used to estimate a relative change in blood volume. The basic assumptions made for blood volume monitoring have been summarized previously [8]. A major uncertainty in the calculation of volume changes with available techniques and persisting to this day, however, resides with the non-uniform and non-steady distribution of red blood cells in the circulation during extracorporeal blood treatment [9].

Temperature

Blood temperature in the extracorporeal circulation can be measured by thermistor probes attached to the outer surface of bloodlines using a control algorithm to minimize temperature gradients in the measuring head [10]. Even though temperature can be treated like a concentration, for example with regard to recirculation and recirculation measurements, it is important to recognize that heat exchange normally occurs across the whole extracorporeal circuit, including the bloodlines. The measured blood temperature, therefore, importantly depends on the location of the sensor placed on the bloodline. However, to determine the thermal effects of the whole extracorporeal circulation blood temperatures must be determined at the level of the vascular access. In quiet air, arterial and venous access temperatures can be estimated from arterial and venous bloodline measurement when room temperature, blood flow, and insulation characteristics as well as the distance of the measuring site from the access are known [11]. This is done in the blood temperature monitor (BTM, Fresenius Medical Care, Germany).

Specific Application

In the BTM the continuous measurement of arterial and venous line temperatures together with intermittent measurement of recirculation allows for a continuous determination of body temperature. Even though most extracorporeal treatments provide cooling, body temperatures tend to increase during extracorporeal treatments which is considered to contribute significantly to hemodynamic instability [12, 13]. It has therefore been suggested to measure body temperature and to continually adjust dialysate temperature and extracorporeal blood cooling so that body temperature remains constant for an isothermic treatment. Since individual differences in body temperature provide different degrees of cooling when using constant dialysate temperatures, it is important to measure body temperature and to adjust dialysate temperatures to individual needs.

Pressure

Dynamic pressures measured in the extracorporeal circulation under normal operating conditions are largely determined by the components of the extracorporeal circulation, such as blood pump, dialyzer, and access needles so that pressure effects originating in the access and in the circulation of the patient are masked. The relationships between pressure and blood flow in the extracorporeal circulation have been modeled with various degrees of complexity and have been reviewed elsewhere [14, 15]. But also static pressures measured without blood flow do not necessarily match intra-access pressures because of unfavorable and unknown catheter and transducer characteristics. While extracorporeal pressures are measured in all systems, this information has not yet been fully exploited for detailed physiologic and functional information.

Specific Application

Most applications with extracorporeal pressure measurements have focused on monitoring access function with the aim to identify peripheral accesses at risk for thrombosis because of access strictures [16]. The oscillations of intra-access pressure caused by the arterial pulse are transmitted to the extracorporeal circulation. However, under normal operating conditions the pressure measured in the arterial line also varies at a frequency comparable to the frequency of the arterial pulse because of the cyclic action of the blood pump. If the signals are adequately processed by spectral analysis the measurement of arterial line pressure can provide a continuous record of heart rate [17].

Conclusion

The variables discussed in this short overview which are non-invasively and automatically measured in the extracorporeal circulation refer to intensive properties of blood and of the cardiovascular system. Such intensive variables are of special interest for the optimal control of extracorporeal treatments. As physiologic control seem to be exclusively focused on stability of intensive variables such as sodium concentration, temperature, or pressure, the measurement and the control of these variables by the extracorporeal system apparently holds the clues to an optimal control of extracorporeal blood treatments.

References

1　Schneditz D: Recirculation, a seemingly simple concept. Nephrol Dial Transplant 1998;13: 2191–2193.
2　Schneditz D, Roob JM, Oswald M, Pogglitsch H, Moser M, Kenner T: Nature and rate of vascular refilling during hemodialysis and ultrafiltration. Kidney Int 1992;42:1425–1433.
3　Chamney PW, Johner C, Aldridge C, Krämer M, Valasco N, Tattersall JE, Aukaidey T, Gordon R, Greenwood RN: Fluid balance modelling in patients with kidney failure. J Med Eng Technol 1999;23:45–52.
4　Schneditz D, Probst W, Kubista H, Binswanger U: Kontinuierliche Blutvolumenmessung im extrakorporellen Kreislauf mit Ultraschall. Nieren- und Hochdruckkrankheiten 1991;20:649–652.
5　Krivitski NM: Theory and validation of access flow measurement by dilution technique during hemodialysis. Kidney Int 1995;48:244–250.
6　Petitclerc T: Festschrift for Professor Claude Jacobs. Recent developments in conductivity monitoring of haemodialysis session. Nephrol Dial Transplant 1999;14:2607–2613.
7　Locatelli F, Covic A, Chazot C, Leunissen K, Luno J, Yaqoob M: Optimal composition of the dialysate, with emphasis on its influence on blood pressure. Nephrol Dial Transplant 2004;19: 785–796.
8　Schneditz D, Levin NW: Non invasive blood volume monitoring during hemodialysis: Technical and physiological aspects. Semin Dial 1997;10:166–169.
9　Mitra S, Chamney P, Greenwood R, Farrington K: The relationship between systemic and whole-body hematocrit is not constant during ultrafiltration on hemodialysis. J Am Soc Nephrol 2004;15:463–469.

10 Krämer M, Steil H, Polaschegg HD: Optimization of a sensor head for blood temperature measurement during hemodialysis. Proc Annu Int Conf IEEE-EMBS 1992;14:1610–1611.

11 Schneditz D, Martin K, Krämer M, Kenner T, Skrabal F: Effect of controlled extracorporeal blood cooling on ultrafiltration induced blood volume changes during hemodialysis. J Am Soc Nephrol 1997;8:956–964.

12 Barendregt JN, Kooman JP, van der Sande FM, Buurma JH, Hameleers P, Kerkhofs AM, Leunissen KML: The effect of dialysate temperature on energy transfer during hemodialysis (HD). Kidney Int 1999;55:2598–2608.

13 Maggiore Q: Isothermic dialysis for hypotension-prone patients. Semin Dial 2002;15:187–190.

14 Pallone TL, Hyver SW, Petersen J: A model of the volumetrically-controlled hemodialysis circuit. Kidney Int 1992;41:1366–1373.

15 Polaschegg HD, Levin NW: Hemodialysis machines and monitors; in Hörl WH, Koch KM, Lindsay RM, et al. (eds): Replacement of Renal Function by Dialysis. Dordrecht/Boston/London, Kluwer Academic Publishers, 2004, pp 325–449.

16 Frinak S, Zasuwa G, Dunfee T, Besarab A, Yee J: Dynamic venous access pressure ratio test for hemodialysis access monitoring. Am J Kidney Dis 2002;40:760–768.

17 Moissl U, Wabel P, Leonhardt S, Isermann R, Krämer M: Continuous observation and analysis of heart rate during hemodialysis treatment. Med Biol Eng Comput 1999;37:S558–S559.

Daniel Schneditz, PhD
Institute of Physiology, Medical University Graz
Harrachgasse 21/5, AT–8010 Graz (Austria)
Tel. +43 (316) 380 4269, Fax +43 (316) 380 9630
E-Mail daniel.schneditz@meduni-graz.at

Ronco C, Brendolan A, Levin NW (eds): Cardiovascular Disorders in Hemodialysis.
Contrib Nephrol. Basel, Karger, 2005, vol 149, pp 42–50

....

Water Treatment for Hemodialysis: A 2005 Update

Gianni Cappelli, Federica Ravera, Marco Ricardi,
Marco Ballestri, Salvatore Perrone, Alberto Albertazzi

Nephrology Dialysis and Renal Transplantation Unit,
University Hospital of Modena, Modena, Italy

Abstract

Water for dialysis represents an additive risk factors to the chronic infammatory state documented in patients on ESRD. The possibility of sustaining proinflammatory cytokines trough microbial derived products, coming from dialysate or infused solutions, is enhanced by biofilm presence on piping and on water treatment system or monitor components. Spread use of reverse osmosis, loop distribution system and pre-treatment components tailored to local raw water characteristics have greatly contributed to a general improvement in final water quality. Notwithstanding these contributions literature still reports fatal accidents or significant percentage of dialysis units not complying to the water quality standards. Technological improvement lowers chemical contamination but microbial quality relays more on quality assurance programs than on technology. Optimal water quality represents part of the anti-inflammatory strategies we need to assure to our dialysis patients to improve outcome.

Water Quality as Anti-Inflammatory Treatment

Dialysate quality enters as a fundamental agent in the biocompatibility evaluation of hemodialysis treatment. Chemicals and microbiological contaminants from dialysate are responsible not only for acute toxic reactions, nowadays sharply reduced in literature reports, but also for some chronic clinical pictures linked to a chronic inflammatory reaction [1]. The introduction of technological improvements into water treatment system of most dialysis units has greatly improved the chemical quality but microbiological contamination still remains a problem. As a matter of fact, main chronic effects are caused by small microbial products derived from bacterial breakdown: endotoxins and cytokine-inducing substances [2]. Carpal tunnel syndrome and decreased erythropoietin response have been linked to microbiological contamination [3, 4]. Most recent data also link chronic inflammation from endotoxin exposure to malnutrition and progressive atherosclerotic cardiovascular disease (MIA syndrome: malnutrition, inflammation, atherosclerosis) with proinflammatory cytokines having a central role [5]. Serum levels of C-reactive protein appear to reflect generation of interleukin-1, interleukin-6 and tumor necrosis factor-α and to predict mortality. Causes for hypercytokinemia and increased C-reactive protein of ESRD are multifactorial and relative importance of contributory factors is not well understood [6]. Non-dialysis-related factors as well as the dialysis procedure itself interact with genetic factors as causative agents. The avoidance of endotoxins or smaller bacterial fractions passage through the dialyzer membrane significantly reduces inflammation indices, such as C-reactive protein and interleukin-6 [7]. Water quality, therefore, represents a part of various anti-inflammatory treatment strategies to improve outcome on these patients.

Water Quality in Dialysis: Beyond European Pharmacopoeia

Notwithstanding problem of water contamination, the use of dialysate has moved from a classic medium for diffusion to an intravenous drug prepared for infusion during on-line treatments or for backfiltration in high-flow modalities. These treatments have been easily realized through some technical improvements and control of contamination by setting limit values for most important toxic substances. Historically the first proposal for a set of standard in dialysis has been a draft from AAMI in 1970, subsequently issued as a final document in 1981 and finally approved as national standards in 1982 (ANSI/AAMI RD5). Recently this specific set of standards has been revised and a new edition (ANSI/AAMI RD62) has been issued in 2004 [8]. In Europe AAMI standards were used as reference by many dialysis units for a long time. During years some limits entered by law were included in different national pharmacopoeias. The 1st edition of European Pharmacopoeia (EP), issued in 1967, started to introduce monographies regulating some fluids used in dialysis. In 1986 the 2nd edition of EP-included standards levels for water for diluting concentrated hemodialysis solutions as well as analytical methods for testing. The 5th edition of EP is effective from January 1, 2005 and still reports almost similar chemical and microbiological contamination levels. To note that EP weakens standards importance by stating: 'monograph is given for information and guidance; it does not form a mandatory part of the Pharmacopoeia; the analytical methods

Table 1. Water for dialysis: Comparison of allowed limits for chemical contaminants in European Pharmacopoeia [9] and in AAMI Standards [8]

Contaminant ppm	European Pharmacopoeia (4th ed.)	AAMI RD62
Aluminium	0.01	0.01
Ammonium	0.2	
Antimony		0.005
Arsenic		0.005
Barium		0.1
Beryllium		0.0004
Cadmium		0.001
Calcium	2 (0.05 mmol/l)	2
Chloramines		0.1
Total chlorine	0.1	
Free chlorine		0.5
Chlorides	50	
Chromium		0.014
Copper		0.1
Cyanide		0.002
Fluorides	0.2	0.2
Heavy metals (Pb)	0.1	0.005
Magnesium	2 (0.07 mmol/l)	4
Mercury	0.001	0.0002
Nitrates	2	2
Potassium	2 (0.1 mmol/l)	8
Sodium	50 (2.2 mmol/l)	70
Selenium		0.09
Silver		0.005
Sulfates	50	100
Thallium		0.002
Zinc	0.1	0.1

described and the limits proposed are intended to be used for validating the procedure for obtaining the water' [9].

Apart from this note, EP presents major differences to AAMI standards, both in chemical (table 1) and microbiological parameters (table 2). EP lists a reduced number of chemical contaminants but a greater attention is focused on microbiological evaluation with a lower limit value not only for bacteria but also for endotoxins. Notwithstanding these reduced values the EP standards do not assure a perfect microbiological quality at the point of use: high volumes of infusion or backfiltration are at risk of exceeding pyrogenic dose for humans if 0.25 EU/ml of endotoxins in dialysate are observed. Nephrological

Table 2. Water for dialysis and dialysate: Comparison of allowed limits for microbial contaminants in European Pharmacopoeia [9], in AAMI standards [8] and in EDTA-ERA Guidelines [11]

	Water for dialysis			Ultrapure dialysate
	European Pharmacopoeia (4th ed.)	AAMI (RD62)	EDTA-ERA guidelines	EDTA-ERA guidelines
Bacteria, CFU/ml	100	200	100	0.1
Endotoxin, EU/ml	0.25	2.0	0.25	0.003

community has, therefore, introduced new limits for microbiological contamination and the definition of ultrapure dialysate (UPD) has been given to this solution [10]. These new references have been spread out through European Guidelines [11] or introduced in some countries as law directive for on-line treatments [12].

The Optimal Water Treatment System

The need to produce dialysis water having both a high degree of chemical and microbiological specifications has led to great improvements in water purification technology in dialysis units [13]. The most important are the diffusion of reverse osmosis (RO) and ultrafiltration as well as the improvement in distribution of pure water through loop systems. A water purification system has to be tailored to the local needs and these will mainly contribute to definition of the pre-treatment section. A multi-media depth filter can be used for primary clarification in row water, but usually a simple microfiltration (5 μm filters at entrance and 1 μm filter at exit from pretreatment) is enough to remove visible particles or colloids and to avoid damage to RO membranes. Pretreatment is based on softeners, where resins with an ion-exchange mechanism substitute hardness with salinity, and on activated carbon filters (also called granular-activated carbon) with high adsorptive capacity for small organic contaminants (<300 Da), chlorine and chloramines. Deionization has largely been superseded, as a final treatment, by RO. RO membranes reject 90–98% of monovalent ions (i.e., sodium) and 95–99% of divalent ions; they also remove larger organic items (>200 Da) and therefore result fundamental for removing dissolved inorganic and organic contaminants, bacteria, pyrogens,

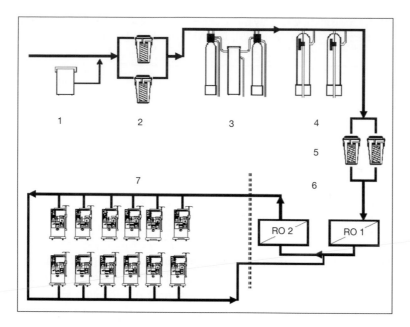

Fig. 1. Water treatment system for dialysis: state-of-the-art plant. 1 = Bleach (sodium hypochlorite) tank; 2 = filtration at 5 μm; 3 = two softeners with brine tank; 4 = two granular-activated carbon filters in series; 5 = filtration at 1 μm; 6 = two reverse osmosis in series; 7 = distribution loop in a low rugosity and disinfection resistant material.

and particulates. A double RO in series represents state-of-the-art in water purification technology (fig. 1); it allows an optimal water quality, both chemical and microbiological, independently from the need of a second RO if feedwater has a high dissolved solid levels. Distribution system technology has also improved, assuring sterile water distribution through the piping in a loop configuration. Cross-linked polyethylene, polyvinylidene fluoride or AISI 316L stainless steel are gradually substituting food-grade polyvinyl chloride due to low rugosity surface, resistance to chemical or heat disinfection procedures and absence of chemical or particulate leaching [14].

Ultrapure Water for All Patients

Since 1990 Mion et al. [15] introduced the use of ultrafilters to produce by 'cold sterilization' a sterile and non-pyrogenic dialysate. Subsequently various authors confirmed these results and reported beneficial effects in reducing dialysis-related clinical complications [1, 3, 16–18]. Bacterial and endotoxin

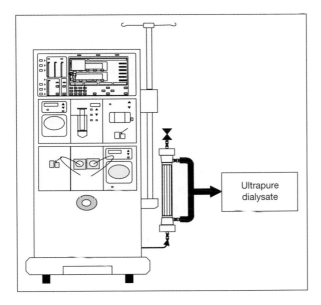

Fig. 2. Ultrafiltration as a dialysate treatment to result in an ultrapure solution.

value limits have been set up to differentiate standard from UPD and most commercial monitors now include the possibility to have an ultrafiltration treatment on produced dialysate to reach quality for UPD (fig. 2). Ultrafilter is included in the hydraulic circuit of dialysis monitor, periodically disinfected and regularly substituted according to manufacturer's specification. The inclusion of dialysis monitors in the medical device regulations assigns to companies the responsibility to assure quality of produced dialysate according to pre-defined procedures and requirements [19]. This procedures standardization has allowed development and safety of on-line techniques and high-flux treatments. The achievement of pyrogenic reaction eliminations [20], reduction of β_2M-amyloidosis [3], reduction of biochemical markers of chronic silent inflammation [6], sparing of erythropoietin [4], increased response to vaccines [21], improved nutritional status [22], and slowing loss of residual renal function in new dialysis patients [23], all confirms UPD as a valid anti-inflammatory treatment. The use of UPD should be diffused to all patients, independently of treatment modalities and dialyzer membrane. Recent European Guidelines move into this suggestion stating: 'UPD is absolutely required when it is used as substitution fluid for on-line haemofiltration or haemodiafiltration. In order to minimize inflammation, dialysis units should also work towards supplying UPD routinely for all dialysis modalities' [11].

Disinfection Procedures and Quality Assurance System

The challenge of water quality is essential for any treatment modality and for any dialysis units. Recent surveys from literature on dialysis water quality results show that still an high percentage of water samples, in many dialysis units, are not complying with standards of either microbiological (8–49%) or chemical (14%), both in United States [24] and Canada [25] or in Europe [26–29]. The presence of an updated water treatment system guarantees for water quality but cornerstones in assuring a persistent optimal quality are maintenance protocols, disinfection procedures and results evaluation through periodical testing: in a nutshell the implementation of a quality assurance system to validate water quality [30]. Disinfection protocols (type of disinfectant, concentration, time of exposure, frequency) for water treatment system and dialysis monitors need to be validated, as biofilm formation can modify with time the initial results [31]. All persons working in the dialysis unit should be involved in a continuous quality assurance process. All reports from technicians on maintenance data and from laboratory on compliance with standards parameters should be collected and stored for analysis. Side effects on patients, if related to water for dialysis, should also be registered to analyze clinical impact from water contaminants. Finally, to comply with quality assurance system, periodical audit should be scheduled; all data should be analyzed and discussed while proper corrective actions have to be programmed. Nowadays water quality in dialysis remains not simply a matter of technology but mainly a procedure process assuring optimal results, in the long term, to patients.

References

1   Brunet P, Berland Y: Water quality and complications of haemodialysis. Nephrol Dial Transplant 2000;15:578–580.
2   Bambauer R, Meyer S, Jung H, Goehl H, Nystrand R: Sterile versus non-sterile dialysis fluid in chronic hemodialysis treatment. ASAIO Trans 1990;36:M317–M320.
3   Baz M, Durand C, Ragon A, Jaber K, Andrieu D, Merzouk T, Purgus R, Olmer M, Reynier JP, Berland Y: Using ultrapure water in haemodialysis delays carpal tunnel syndrome. Int J Artif Organs 1991;14:681–685.
4   Richardson D: Clinical factors influencing sensitivity and response to epoietin. Nephrol Dial Transplant 2002;17(suppl 1):53–59.
5   Pecoits-Filho R, Lindholm B, Stenvinkel P: The malnutrition, inflammation, and atherosclerosis (MIA) syndrome – The heart of the matter. Nephrol Dial Transplant 2002;17(suppl 11):28–31.
6   Panichi V, Migliori M, De Pietro S, Taccola D, Andreini B, Metelli MR, Giovannini L, Palla R: The link of biocompatibility to cytokine production. Kidney Int 2000;76:S96–S103.
7   Schiffl H, Lang SM, Startakis D, Fischer R: Effects of ultrapure dialysis fluid on nutritional status and inflammatory parameters. Nephrol Dial Transpl 2001;16:1863–1869.

8   AAMI Standard and Recommended Practices: AAMI/RD62:2001. Water Treatment Equipment for Hemodialysis Applications. Association for the Advancement of Medical Instrumentation. Arlington, Virginia, USA, 2001.

9   European Pharmacopoeia 4th Ed. Haemodialysis Solutions, Concentrated, Water for Diluting. Monograph 1167:2002. Council of Europe, Strasbourg, 2002.

10   Ledebo I: Ultrapure dialysis fluid – Direct and indirect benefits in dialysis therapy. Blood Purif 2004;22(suppl 2):20–25.

11   European best practice guidelines for haemodialysis (part 1) Section IV: Dialysis fluid purity. Nephrol Dial Transplant 2002;17(suppl 7)45–62.

12   Ministere de l'Emploi et de la Solidarieté: Circulaire relative aux specifications techniques et a la securite de la pratique de l'hemofiltration et de l'hemodiafiltration en ligne dans etablissement de sante. Republique Francaise, Circulaire – DGS/DH/AFSSAPS n° 311 du 7 Juin 2000.

13   Cappelli G, Inguaggiato P: Water treatment for contemporary hemodialysis; in Horl WH, Koch KM, Lindsay RM, Ronco C, Winchester JF (eds): Replacement of Renal Function by Dialysis, ed 5. Dordrecht, Kluwer Academic Publisher, 2004, pp 491–503.

14   Cappelli G, Ballestri M, Facchini F, Carletti P, Lusvarghi E. Leaching and corrosion of polyvinyl chloride (PVC) tubes in a dialysis water distribution system. Int J Artif Organs 1995;18: 261–263.

15   Mion CM, Canaud B, Garred LJ, Stec F, Nguyen QV: Sterile and pyrogen-free bicarbonate dialysate: A necessity for hemodialysis today. Adv Nephrol Necker Hosp 1990;19:275–314.

16   Bambauer R, Walther J, Meyer S, East S, Shower M, Young WK, Gohl H, Vienken J: Bacteria and endotoxin-free dialysis fluid for use in chronic haemodialysis. Artif Organs 1994;18: 188–192.

17   Sitter T, Bergner A, Schiffl H: Dialysate related cytokine induction and response to recombinant human erythropoietin in haemodialysis patients. Nephrol Dial Transplant 2000;15:1207–1211.

18   Lonnemann G: Should ultra-pure dialysate be mandatory? Nephrol Dial Transplant 2000;15 (suppl 1):55–59.

19   Council Directive 93/42/EEC of 14 June 1993 concerning medical devices. Official Journal 169/1 of 1993–07–12.

20   Pegues DA, Oettinger CW, Bland LA, Oliver JC, Arduino MJ, Aguero SM, McAllister SK, Gordon SM, Favero MS, Jarvis WR: A prospective study of pyrogenic reactions in hemodialysis patients using bicarbonate dialysis fluids filtered to remove bacteria and endotoxin. J Am Soc Nephrol 1992;3:1002–1007.

21   Akrum R, Frohlich M, Gerritisen AF, Dalia MR, Chang PC: Ultrapure water for dialysate contributes to a lower activation state of peripheral blood mononuclear cells. Kidney Int 1999;55: 1158–1171.

22   Schiffl H, Lang SM, Stratakis D, Fisher R: Effects of ultrapure dialysis fluid on nutritional status and inflammatory parameters. Nephrol Dial Transplant 2001;16:1863–1869.

23   Schiffl H, Lang SM, Fisher R: Ultrapure dialysis fluid slows loss of residual renal function in new dialysis patients. Nephrol Dial Transplant 2002;17:1814–1818.

24   Klein E, Pass T, Harding GB, Wright R, Million C: Microbial and endotoxin conatmination in water and dialysate in the Central United States. Artif Organs 1990;14:85–94.

25   Laurence RA, Lapierre ST: Quality of hemodialysis water: A 7-year multicenter study. Am J Kidney Dis 1995;25:738–750.

26   Kulander L, Nisbeth U, Danielsson BG, Eriksson O: Occurrence of endotoxin in dialysis fluid from 39 dialysis units. J Hosp Infect 1993;24:29–37.

27   Bambauer R, Schauer M, Jung WK, Vienken J, Daum V: Contamination of dialysis water and dialysate, a survey of 30 centers. ASAIO Journal 1994;40:1012–1016.

28   Arvanitidou M, Spaia S, Katsinas C, Pangidis P, Constantinidis T, Katsouyannopoulos V, Vayonas G: Microbiological quality of water and dialysate in all haemodialysis centres of Greece. Nephrol Dial Transplant 1998;13:949–954.

29   Arvanitidou M, Spaia S, Askepidis N, Kanetidis D, Pazarloglou M, Katsouyannopoulos V, Vayonas G: Endotoxin concentration in treated water of all hemodialysis units in Greece and inquisition of influencing factors. J Nephrol 1999;12:32–37.

30   Cappelli G, Perrone S, Ciuffreda A: Water quality for on-line haemodiafiltration. Nephrol Dial Transplant 1998;13(suppl 5):12–16.
31   Cappelli G, Sereni L, Scialoja MG, Morselli M, Perrone S, Ciuffreda A, Bellesia M, Inguaggiato P, Albertazzi A, Tetta C: Effects of biofilm formation on haemodialysis monitor disinfection. Nephrol Dial Transplant 2003;18:2105–2111.

Gianni Cappelli, MD
Nephrology, Dialysis and Renal Transplantation Unit
Department of Medicine and Medical Specialties
University Hospital of Modena, Via Del Pozzo, 71
IT–41100 Modena (Italy)
Tel. +39 0594222481, Fax +39 0594222167, E-Mail cappelli@unimo.it

Ronco C, Brendolan A, Levin NW (eds): Cardiovascular Disorders in Hemodialysis.
Contrib Nephrol. Basel, Karger, 2005, vol 149, pp 51–57

··········· ··········

Online Convective Therapies: Results from a Hemofiltration Trial

A. Santoro[a], *E. Mancini*[a], *L. Bibiano*[b], *A. Specchio*[c], *A. Francioso*[d],
C. Robaudo[e], *M.A. Nicolini*[f], *G. Tampieri*[g], *A. Fracasso*[h], *M. Virgilio*[i],
W. Piazza[j], *M. Di Luca*[k], *G. Campolo*[l], *F. De Tomaso*[m], *A. Montanari*[n],
A. Gattiani[o], *F. Aucella*[p], *L. Fattori*[q], *R. Estivi*[r], *S. Costantini*[s]

[a]U.O. Nefrologia e Dialisi Malpighi, Policlinico S.Orsola-Malpighi, Bologna,
[b]Policlinico Umberto I, Ancona, [c]Ospedale T. Russo, Cerignola (Fg), [d]Ospedale S.
Croce, Fano (Ps), [e]Ospedale Dimi, Genoa, [f]Ospedale S.Andrea, La Spezia, [g]Ospedale
Civile, Lugo (Ra), [h]Ospedale Umberto I, Mestre (Ve), [i]Ospedale Civile, Molfetta (Ba),
[j]Fondazione Maugeri, Pavia, [k]Ospedale S. Salvatore, Pesaro, [l]Ospedale Misericordia e
Dolce, Prato (Fi), [m]Ospedale S. Maria degli Angeli, Putignano (Ba), [n]Ospedale S.
Maria delle Croci, Ravenna, [o]Ospedale degli Infermi, Rimini, [p]Casa Sollievo della
Sofferenza, S. Giovanni, Rotondo (Fg), [q]Ospedale Civile, Senigallia (An), [r]Ospedale
S. Salvatore, Tolentino (Mc), and [s]Ospedale Belcolle, Viterbo, Italy

Abstract

With the introduction of the on-line preparation of dialysis fluids, the hemofiltration
technique, which has never had a widespread diffusion in its old version with the infusion
bags, has gained a new interest. We planned a prospective, randomized, 3-year-long study
comparing survival and morbidity in ultrapure bicarbonate dialysis (BD) with on-line predilution hemofiltration (HF). Since comorbidity is one of the main factors limiting survival,
the study was addressed to patients with a severe degree of comorbidity. The paper presents
the preliminary results of the trial. Sixty-four patients were enrolled and randomized to either
BD (N = 32) or HF (N = 32). Mean age and dialysis vintage were comparable. Twenty
patients died during the study, 12 in BD and 8 in HF. The relative risk of death was 11%
higher in patients treated with BD compared to those in the HF group (p < 0.005). The number of hospitalisation events per single patient was lower, even though not significantly, in
HF compared to BD (1.94 ± 1.26 in HF vs 2.48 ± 1.98 in BD, p = NS). As concerns biochemistry, apart from beta-2-microglobulin, any other substantial difference was not found
during the study, though the small solute concentration was generally a little more elevated in
HF than in BD. Dialysis hypotension showed a trend to decrease in both the dialysis modalities up to near half of the trial, then, during the last year, it remained quite stable in HF, while,
on the contrary, it increased in the BD group. By the end of the protocol, patients in HF
showed a 2.5% incidence of acute dialysis hypotension, while patients in BD had 23%.

In conclusion, this medium-long run study showed that, in spite of a comparable level of small solute removal, survival was longer in patients treated with on-line HF than with BD. Morbidity was lower too, but without a statistical relevance. Hemodynamic stability, in the long run, became clearly better with convection. We should reconsider what is dialysis adequacy and which are its best clinical markers.

Introduction

Survival and quality of life of chronic hemodialysis patients is undoubtedly worse than in the general population. The ideal aim for a dialytic treatment, to mimic the natural kidney functions, has never been reached, and only bio-artificial organs, coupling artificial systems with cellular elements, could go in that direction.

Since the natural kidney operates by convection, hemofiltration (HF), a purely convective dialysis technique introduced by Lee Henderson [1] in 1967 presents as an ideal dialysis modality.

In the new format, with the online preparation of infusion fluids, HF has all the requisites to be compared with the traditional, purely diffusive dialysis techniques. This is the reason why we planned a study protocol comparing mortality and morbidity in the medium-long run, in two groups of patients, one treated with ultrapure bicarbonate dialysis (BD) and the other with online HF. Since comorbidity is one of the main factors limiting survival of dialysis patients [2], the study was addressed to those patients with a high number of comorbid conditions, to look for an outcome difference on the basis of the dialysis modality used. In this paper we present some preliminary results of the trial.

Patients and Methods

Study Design
The study was designed as a multicenter, randomized, prospective, 3-year-long, controlled trial with two-arm study (BD vs. HF). Nineteen Italian dialysis centers participated to the trial. It was approved by the local Ethical Committees; each patient gave his/her informed consent.

Patients
An overall number of 130 patients, on regular dialysis treatment for at least one year, were screened for eligibility. The comorbidity degree, defined as a Charlson Comorbidity Index [3] ≥ 3 was the inclusion criterion. Exclusion criteria were a residual urinary output greater than 500 ml/day, a malfunctioning vascular access, or body weight greater than 85 kg.

An overall number of 64 patients were enrolled and centrally randomized in a 1:1 ratio, to either BD (n = 32) or HF (n = 32).

Mean age (66.4 ± 10.3 years in BD, 67.9 ± 9.2 in HF) was comparable between the two groups of patients; the dialysis vintage was higher, but not significantly, in the HF group (60.2 ± 62 months in BD, 66.2 ± 61.7 in HF; p = NS).

Study Parameters

The overall mortality from any cause was the primary outcome of this study. The hospitalization rate was the main secondary outcome. The tolerance to treatment, in terms of both intra- and interdialysis symptoms, was an additional secondary outcome variable.

Hemodialysis Methods

To eliminate any confounding factors, the same dialysis machine (Gambro AK 100 Ultra), as well as the same dialysis membrane (polyamide) were used (Poliflux 8L, $1.7\,m^2$ and Poliflux 21S, $2.1\,m^2$, respectively in BD and HF). Sterile, nonpyrogenic substitution fluid, 37°C, was used in HF, with a target infusate volume aimed at 120% of the dry body weight, and ultrapure dialysate 500 ml/min, 37°C was used in BD.

Measurements

Pre- and postdialysis blood pressure and body weight, as well as the number of episodes of symptomatic hypotension and any other dialysis side effect was recorded. Hypotension was defined as a symptomatic fall of systolic arterial pressure by 20 mm Hg or more, requiring intervention of a nurse. Interdialysis symptoms (fatigue, hypotension, headache, muscular cramps, etc.) were investigated along a week every 4 months. Blood chemistry was tested every 4 months.

Statistics

The survival analysis and the hospitalization rate were based on the intention-to-treat principle, while the secondary analysis (biochemical data and side effects) was calculated on the as-treated-principle.

The Cox proportional hazard regression model was used to assess the effects of the two treatments in terms of relative risk of death. The Student t test and analysis of variance (ANOVA) were used for differences in quantitative variables changing in time. The SPSS statistical package was used for calculations.

Results

We present here the preliminary results of the statistical analysis, still not entirely completed.

Survival

Twenty patients died during the 3 years, 12 in BD and 8 in HF; the cardiovascular events were the leading cause of death (14/20, 70%).

Taking into account the interactive effect of both the age and type of treatment, patients treated with BD had a 11% increase in the relative risk of death compared to patients treated with HF (p < 0.005).

Any other clinical variable tested (duration of dialysis, diabetes status, myocardiopathy, baseline albumin) did not prove to have any significant effect on survival.

Morbidity

In the intention-to-treat approach, 36 patients out of 64, underwent hospitalization. The number of hospitalization events per single patient was lower, even though not significantly, in HF compared to BD (1.94 + 1.26 in HF vs. 2.48 + 1.98 in BD, p = NS).

Blood Chemistry

The comparison was carried out in those patients completing the study (10 in BD and 11 in HF). At baseline, the most important parameters of the biochemical control were completely similar in the two groups. Apart from β_2-microglobulin, no substantial difference was found during the study in these parameters, though the small solute concentration was generally a little more elevated in HF than in BD (table 1).

Intratreatment Hemodynamics

Dialysis hypotension showed a peculiar behavior along the study. In fact, in up to nearly half of the study there was a progressive trend to a reduction of the number of hypotension episodes, and at the 20th month the prevalence of dialysis-induced hypotension in the two groups was substantially comparable. However, during the last year of the protocol, the number of hypotension episodes remained quite stable in the HF group, and, on the contrary, it progressively increased in the BD group. By the end of the protocol, patients treated with HF showed a 2.5% incidence of acute dialysis hypotension, while patients treated with HF had 23% (fig. 1).

Discussion

Even though the results presented here are only preliminary, the main observation coming from this prospective, randomized study concerns the survival, which was found longer with online HF compared to BD. Old, but basic, clinical studies performed with conventional HF, with low convective volumes, suggested a better survival of older patients treated with postdilutional HF compared to dialysis [4, 5]. But our study is the first one evaluating survival with HF in the long run, and particularly in patients with a high comorbidity burden.

Table 1. Main predialysis biochemical parameters, baseline, at 12, 24, and 36 months, found in patients completing the study

	Bicarbonate dialysis				Online hemofiltration			
	Baseline	12 M	24 M	36 M	Baseline	12 M	24 M	36 M
Hemoglobin, g/dl	11.1 ± 1.1	11.3 ± 1.2	11.1 ± 1.2	11.6 ± 1.2	11.4 ± 1.7	11.9 ± 0.9	12.0 ± 0.8	11.6 ± 1.1
Hematocrit, %	33.3 ± 3.3	34.7 ± 4.2	34.2 ± 3.8	35.8 ± 3.4	34.9 ± 5.2	36.5 ± 3.3	37.0 ± 2.6	35.8 ± 3.5
Azotemia, mg/dl	160.5 ± 48.6	162.7 ± 57.4	132.3 ± 32.0	154.2 ± 57.0	153.2 ± 29.5	176.2 ± 33.9	160.4 ± 27.7	169.6 ± 40.8
Creatinine, mg/dl	9.2 ± 1.7	9.9 ± 2.5	9.1 ± 1.7	9.7 ± 1.9	10.0 ± 1.3	11.2 ± 1.5	10.6 ± 2.0	10.7 ± 1.8
Calcium, g/dl	9.8 ± 0.8	10.4 ± 1.2	10.3 ± 0.9	9.8 ± 1.0	9.8 ± 1.1	9.4 ± 0.6	9.5 ± 0.9	9.4 ± 0.7
Phosphate, mg/dl	5.1 ± 1.3	5.2 ± 1.2	4.9 ± 1.2	5.1 ± 1.5	5.1 ± 1.8	5.8 ± 2.4	5.6 ± 0.9	5.2 ± 1.4
Potassium, mEq/l	5.7 ± 0.6	5.8 ± 0.8	5.1 ± 0.7	5.6 ± 0.8	5.7 ± 0.6	5.6 ± 0.8	5.7 ± 0.8	5.4 ± 0.6
Bicarbonate, mEq/l	20.6 ± 2.8	22.2 ± 2.9	21.9 ± 2.8	21.2 ± 2.6	22.4 ± 1.8	21.8 ± 1.7	20.5 ± 3.1	21.8 ± 2.8
Kt/V	1.2 ± 0.1	1.00 ± 0.09	1.32 ± 0.15	1.23 ± 0.17	1.08 ± 0.09	0.94 ± 0.08	0.83 ± 0.11	0.83 ± 0.10
Albumin, g/dl	3.8 ± 0.3	3.6 ± 0.3	3.8 ± 0.3	3.7 ± 0.3	3.8 ± 0.6	3.8 ± 0.3	3.7 ± 0.5	3.8 ± 0.4
β_2-MG, mg/l	33.9 ± 9.3	35 ± 9.7	33.2 ± 6.1	36.9 ± 16.0	30.2 ± 11.2	23.9 ± 7.8	22.2 ± 8.7	23.9 ± 5.6

Significant differences: At 12 M: Calcium, p = 0.032; β_2-MG, p = 0.011; at 24 M: β_2-MG, p = 0.005; at 36 M: β_2-MG, p=0.028.
M = Months; β_2-MG = β_2-microglobulin.

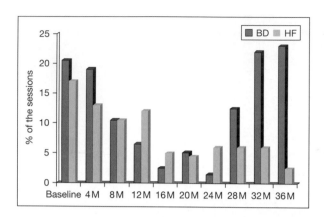

Fig. 1. Trend of the dialysis hypotension incidence, compared in the two dialysis modalities, expressed as the percentage of dialysis sessions complicated by acute hypotension. Data are reported at baseline, and every 4 months (M) until the end of the study.

The results, in terms of mortality were striking, and the relatively low number of patients gives even more robustness to the result. The probability of death was 11% greater in patients treated with BD; this result is particularly strong also because the patients treated with BD were slightly younger, even though not significantly, than patients in HF.

A better hemodynamic tolerance to fluid withdrawal and a better removal of medium-molecular-weight toxins are the only reasons hypothesizable today to explain the result on survival. The superiority of HF compared to hemodialysis in hemodynamic stability has been known since a long time [4–6]. Very recently, Altieri et al. [7] have confirmed this observation, also in comparison with a mixed, convective-diffusive technique, such as hemodiafiltration. In our study, we observed that in the first 2 years there was a progressive reduction in the incidence of hypotension with both the techniques, up to a condition in which, by the 20th month, there was no substantial difference between HF and BD. This phenomenon can be attributable in our opinion only to the *study effect*: a more rigid clinical surveillance of any patient included in a study protocol, especially taking into account that all patients were in critical conditions, due to their comorbidity burden. But, in the last year of the study, dialysis-induced hypotension showed a trend towards a progressive increase in BD, while in HF it remained at a low frequency. The impact of the dialysis-induced hypotension phenomenon on survival has been recently emphasized by Shoji et al. [8] who demonstrated that it is an independent factor affecting survival.

The comparison of the biochemical data during the study showed essentially, as expected, the greater removal of β_2-microglobulin; in the long run this may improve the quality of life by reducing the development of the amyloid bone disease, indirectly influencing survival [9]. On the contrary, the removal of the typical small solutes was not substantially different between the two techniques. This observation opens new horizons in the identification of ideal markers for a good dialysis treatment, and denies that urea should be considered as the hallmark for the depurative efficacy of a treatment.

In conclusion, in this prospective, randomized, 3-year-long study comparing ultrapure BD with online predilution HF, first analysis show that survival was better with HF. Morbidity too was lower with HF, compared with BD. Hemodynamic stability, which in the short-medium run does not seem different, in the long run instead proved clearly better with convection. These results were obtained in spite of a comparable level of small solute removal. Thus, the high volume convective dialysis modalities prove substantially similar to diffusive dialysis in terms of biochemical depuration, but better in terms of survival. What dialysis adequacy is made of, and which are its best markers, have to be reconsidered.

References

1   Henderson LW: Biophysics of ultrafiltration and hemofiltration; in Maher JF (ed). Replacement of renal function by dialysis. Dordrecht, Kluwer Academic, 1989, pp 300–326.
2   Vonesh EF, Snyder JJ, Foley RN, Collins AJ: The differential impact of risk factors on mortality in hemodialysis and peritoneal dialysis. Kidney Int 2004;66:2389–2401.
3   Van Manen JG, Korevaar JC, Dekker FW, Boeschoten EW, Bossuyt PM, Krediet RT: How to adjust for comorbidity in survival studies in ESRD patients: A comparison of different indices. Am J Kidney Dis 2002;40:82–89.
4   Baldamus CA, Ernst W, Lysaght MJ, Shaldon S, Koch KM: Hemodynamics in hemofiltration. Int J Artif Organs 1983;6:27–31.
5   Fox SD, Henderson LW: Cardiovascular response during hemodialysis and hemofiltration: Thermal, membrane and catecholamine influences. Blood Purif 1993;11:224–236.
6   Henderson LW: Hemodynamic instability during different forms of dialysis therapy: Do we really know why? Blood Purif 1996;14:395–404.
7   Altieri P, Sorba G, Bolasco P, Asproni E, Ledebo I, Cossu M, Ferrara R, Ganadu M, Cadinu F, Serra G, Cabiddu G, San G, Casu D, Passaghe M, Bolasco F, Pistis R, Ghisu T; Second Sardinian Multicentre Study: Predilution hemofiltration – The Second Sardinian Multicentre Study: Comparison between haemofiltration and haemodialysis during identical Kt/V and session times in a long-term cross-over study. Nephrol Dial Transplant 2001;16:1207–1213.
8   Shoji T, Tsubakhara Y, Fujii M, Imai E: Hemodialysis-associated hypotension as an independent risk factor for two-year mortality in hemodialysis patients. Kidney Int 2004;66:1212–1220.
9   Locatelli F, Marcelli D, Conte F, Limido A, Malberti F, Spotti D: Comparison of mortality in ESRD patients on convective and diffusive extracorporeal treatments. Kidney Int 1999;55:286–293.

Dr. Antonio Santoro
U.O. Nefrologia e Dialisi Malpighi
Policlinico S.Orsola-Malpighi, Va P.Palagi 9
IT- 40138 Bologna (Italy)
Tel. +39 051 6362430, Fax +39 051 6362511, E-Mail santoro@aosp.bo.it

Ronco C, Brendolan A, Levin NW (eds): Cardiovascular Disorders in Hemodialysis.
Contrib Nephrol. Basel, Karger, 2005, vol 149, pp 58–68

· ·

Selected Lessons Learned from the Dialysis Outcomes and Practice Patterns Study (DOPPS)

Ronald L. Pisoni[a], *Roger N. Greenwood*[b]

[a]University Renal Research and Education Association, Ann Arbor, Mich., USA;
[b]Lister Hospital, Stevenage, UK

Abstract

The Dialysis Outcomes and Practice Patterns Study (DOPPS) is a prospective, observational study of the relationships between hemodialysis (HD) patient outcomes and HD treatment practices. The DOPPS began in 1996 in the United States, expanding to France, Germany, Italy, Japan, Spain, and the United Kingdom in 1998–1999, and then to Australia, Belgium, Canada, New Zealand, and Sweden in 2002. More than 300 dialysis units have participated in the DOPPS since 1996, with mortality data collected from nearly 90,000 HD patients and detailed longitudinal data from nearly 30,000 HD patients. Large sample size and the large treatment practice variation observed in the DOPPS – given its international scope of participation – provide strong statistical power to investigate many different HD practices. Furthermore, the detailed patient data collected in the DOPPS allow relationships to account for differences in a large number of patient characteristics. More than 55 papers have been published from the DOPPS; here we provide a summary of selected DOPPS findings regarding nutrition, mineral metabolism, anemia management, vascular access, depression, and use of multivitamins and statins.

The Dialysis Outcomes and Practice Patterns Study (DOPPS) is a prospective observational study of the relationships between hemodialysis (HD) treatment practices and HD patient outcomes. A unique aspect of the DOPPS has been the large international participation within the study which allows a diverse range of practices to be observed. Data collection in the DOPPS first began in 1996 in the United States, and extended to France, Germany, Italy, Japan, Spain, and the United Kingdom in 1998–1999. Subsequently, the countries of Australia, Belgium, Canada, New Zealand, and Sweden joined the DOPPS as part of a second phase of DOPPS data collection in 2002. A third phase of DOPPS data collection including these 12 countries is commencing in 2005. Through to the end of 2004, mortality data thus far have been collected from >90,000 chronic HD patients in the DOPPS, with detailed longitudinal data collected from nearly 30,000 HD patients. This detailed patient data along with information regarding facility practices has allowed for many HD practices to be examined in the DOPPS with extensive adjustments for differences in patient characteristics.

More than 300 dialysis units have participated in each phase of DOPPS data collection. Dialysis units participating in the DOPPS have been randomly selected from a list of all dialysis units treating more than 24 in-centre HD patients within each country. This random selection is stratified to yield representative facility samples proportional to the types and geographical distribution of dialysis units within each country. Furthermore, patients for whom detailed data collection is performed are randomly selected from all chronic HD patients treated within each dialysis unit. Additional details of the DOPPS study design have been described extensively by Young et al. [1] and Pisoni et al. [2]. The representative nature of the DOPPS facility and patient sampling design has made it possible for the DOPPS to provide national statistics similar to those published by individual country registry reports. However, the main mission of the DOPPS has been to focus on treatment practices. Taking advantage of the large sample of patients participating in the DOPPS in conjunction with the wide variation in observed practices due to international participation, more than 50 papers have thus far been published from the DOPPS, many of which have tried to describe clinically relevant relationships between common HD practices and HD patient outcomes. Brief summaries of selected areas of investigation are provided below. Further information, citations, and web graphics for DOPPS publications are available at the DOPPS website: www.dopps.org, and extensive topical summaries also are provided in a recent American Journal of Kidney Diseases Supplement publication of DOPPS work [3].

Nutrition and Mineral Metabolism Practices

Maintaining a good nutritional state for HD patients is very important – and at the same time is very challenging when considering the dietary restrictions necessary for providing optimal care for HD patients. Pifer et al. [4] investigated the value of seven different nutritional measures for predicting mortality risk of HD patients in the US-DOPPS during 1996–2001. For this patient population, 11% of patients had severe malnutrition and 8% had moderate malnutrition as defined by a modified subjective global assessment (mSGA) score. Multivariate Cox survival analyses adjusted for patient demographics, time since ESRD onset, and 15 comorbidity classes, indicated that patients with severe malnutrition had a 33% higher risk of mortality ($p < 0.05$) compared with patients defined as being in the normal range for mSGA scores. Furthermore, lower baseline values for the nutritional markers of serum albumin, serum creatinine, body mass index, and lymphocyte count were each associated with a significantly higher mortality risk ($p < 0.05$), but not in a linear fashion for each measure. As expected, lower baseline neutrophil values were associated with a significantly lower mortality risk. Not all of these measures are thought to be strictly related to nutritional status, however, since serum albumin, neutrophil count, and lymphocyte count have also been shown to be associated with inflammation. In addition to the relationships seen between baseline nutritional measures and mortality, a large decline over a 6-month period in a patient's serum albumin, serum creatinine, or body mass index was also associated with a 35–80% higher mortality risk compared to patients having no change in these nutritional measures over a 6-month period [4]. Leavey et al. [5] showed that across different subgroups of HD patients (e.g., subgroups of age, race, diabetes status, smoking status, or illness severity), greater body mass index consistently was associated with lower mortality risk which is opposite to the relationship seen in the general population.

Although the above results indicate that maintaining good nutritional status is associated with improved outcomes, other studies from the DOPPS indicate that maintaining a healthy nutritional status for HD patients needs to consider a dietary care plan which also regulates dietary-related factors such as phosphorus, calcium, and a patient's interdialytic weight gain (IDWG). In a large study of >17,000 HD patients in 7 countries participating in the DOPPS, Young et al. [6] recently showed significantly higher all-cause and cardiovascular mortality risk for patients having serum calcium, phosphorus, or calcium/phosphorus product levels exceeding the National Kidney Foundation-Kidney Disease Outcomes Quality Initiative (NKF-K/DOQI) Guideline maximum target values. The risk was substantially higher for cardiovascular mortality than for all-cause mortality suggesting that one of the possible mechanisms for the effect of these mineral metabolism measures upon mortality could be through high calcium and phosphorus levels promoting vascular calcification leading to cardiovascular disease. Recently, Kimata et al. [7] have shown that high serum calcium levels above the NKF-K/DOQI guidelines have a much greater detriment on the survival of patients with prior coronary artery disease (CAD) than for patients without CAD. As an example, figure 1 demonstrates the large increase in cardiovascular mortality risk with serum phosphorus concentrations above the NKF-K/DOQI Guideline maximum target of 5.5 mg/dl [8].

These relationships between calcium and phosphorus with mortality risk for HD patients are particularly important since approximately 50% of HD

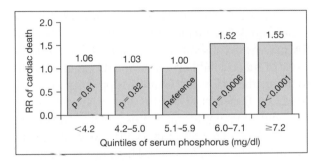

Fig. 1. Association of baseline serum phosphorus with relative risk (RR) of cardiac death in HD patients. The RR of cardiac death for HD patients according to quintiles of baseline serum phosphorus concentration was determined by Cox regression analysis for HD patients participating in the DOPPS during 1996–2001 from the countries of France, Germany, Italy, Japan, Spain, the United Kingdom, and the USA. The analysis was stratified by country, was restricted to patients on dialysis >90 days at time of study entry (n = 7,238), and was adjusted for age, gender, body mass index, years since onset of ESRD, 15 summary classes of comorbidity, baseline hemoglobin, serum albumin and serum calcium, dialysis dose, hospitalization during the 3 months prior to study entry, facility phosphate binder use, intravenous vitamin D use, and accounted for facility clustering effects [8].

patients have calcium or phosphorus values exceeding the NKF-K/DOQI and European Best Practices Guideline maximum target levels [6]. It is worth noting that the NKF-K/DOQI guidelines recommend an albumin-corrected serum calcium level of 8.4–9.5 mg/dl. However, the DOPPS results from Young et al. [6] do not demonstrate an adverse mortality risk for patients having albumin-corrected calcium values <8.4 mg/dl. Instead, patients having albumin-corrected serum calcium values <8.4 mg/dl display a lower mortality risk compared with patients having albumin-corrected serum calcium values of 8.4–9.5 mg/dl.

Another important aspect relevant to the nutritional care of HD patients is the amount of fluid gained between HD treatment sessions. Saran et al. [9] showed that patients having an IDWG >5.7% of postdialysis body weight had a 12% higher mortality risk (p = 0.05). Nearly 20% of HD patients displayed this level of IDWG across the 7 countries in DOPPS I. These analyses were adjusted for demographics, years with ESRD (to control for residual renal function), 15 summary comorbid conditions, depression, smoking status, and socioeconomic factors. The higher mortality risk with high IDWG may be due to excessive cardiovascular burden related to expanded extracellular volume.

Because of the concern that water-soluble vitamins may be depleted from HD patients due to three-times-a-week/HD sessions, a recent paper by Fissell et al. [10] looked at multivitamin use and related outcomes among HD patients in the 7 countries in DOPPS I. Multivitamins were found to be used by 72% of HD patients in the USA, 4–6% of patients in the UK, Japan, and Italy, 12–15% of patients in France and Germany, and 38% of patients in Spain. This higher use of multivitamins in the USA may be a consequence, in part, of early studies showing decreased serum levels of water-soluble vitamins in HD patients which led to initial recommendations in the USA for vitamin supplementation. A striking finding from this study by Fissell et al. [10] is that in a multivariate analysis stratified by continent and adjusting for numerous patient characteristics, patients taking a multivitamin had a 16% lower mortality risk compared to patients not taking a multivitamin (95% CI, 0.76–0.94). These results suggest a possibly large benefit for patients taking a multivitamin as part of their routine nutritional care and supports further investigation by randomized clinical trials in this area.

International Anemia Management

Anemia management in HD patients requires constant review since anemia is one of the common consequences of chronic kidney disease. A major advance in controlling anemia in HD patients was provided by the introduction of recombinant human erythropoietin (rHuEpo) in 1989 which has allowed for a large increase in the mean hemoglobin (Hgb) concentration for HD patients and a large reduction in the need for blood transfusions for this patient population. Recently, Pisoni et al. [11] described anemia management and related outcomes across 12 countries based on DOPPS data collected in 2002–2003. Mean Hgb concentrations for prevalent HD patients (table 1) ranged from 12 g/dl in Sweden, 11.6–11.7 g/dl in the USA, Spain, Canada, and Belgium, 11.1–11.5 g/dl in Australia/New Zealand, Germany, Italy, the UK, and France, and 10.1 g/dl in Japan, which has had a lower target mean Hgb than other countries. Use of rHuEpo for HD patients varied from 83 to 94% across these countries. Moreover, country mean rHuEpo doses ranged from nearly 5,600 units/week in Japan to 17,300 units/week in the USA with a significant trend seen between higher country mean Hgb concentration with higher country mean rHuEpo dose. In many countries, current practice guidelines recommend a target of ≥ 11 g/dl for Hgb in HD patients. Patient characteristics significantly related to a greater likelihood of achieving this target Hgb of 11 g/dl included: male gender, older age, not using a catheter for a vascular access, polycystic kidney disease, not having gastrointestinal bleeding, higher percentage of transferrin saturation (TSAT), lower ferritin and PTH concentrations, and higher concentrations of serum albumin, calcium, or phosphorus [11, 12]. From a facility practice pattern perspective, it was found that dialysis units with higher catheter use had significantly lower mean Hgb values.

Table 1. Mean hemoglobin (Hgb) concentrations for HD patients on dialysis >180 days and at time of starting dialysis, percentage of patients with Hgb <11 g/dl, and percentage erythropoietin (Epo) use for HD patients on dialysis >180 days and during the pre-ESRD period, by country: DOPPS II. Reprinted from Pisoni et al [11], © 2004, with permission from the National Kidney Foundation.

Country	Among patients on dialysis >180 days				Among patients new to ESRD, at start of dialysis*			
	n_1	Epo use (% of patients)	Mean Hgb (g/dl)	Hgb <11 g/dl (% of patients)	n_2	Epo use prior to ESRD (% of patients)	Mean Hgb (g/dl)	Hgb <11 g/dl (% of patients)
Sweden	466	94	12.0	23	168	65	10.7	55
United States	1,690	91	11.7	27	458	27	10.4	65
Spain	513	93	11.7	31	170	56	10.6	61
Belgium	442	94	11.6	29	213	33	10.3	66
Canada	479	91	11.6	29	150	43	10.1	70
Australia/New Zealand	423	86	11.5	36	108	50	10.1	70
Germany	459	86	11.4	35	142	46	10.5	61
Italy	447	87	11.3	38	167	59	10.2	68
United Kingdom	436	94	11.2	40	93	44	10.2	67
France	341	83	11.1	45	86	43	10.1	65
Japan	1,210	84	10.1	77	131	62	8.3	95

*Includes patients who were new to ESRD and entered DOPPS within 7 days of first-ever chronic dialysis treatment. Those receiving Epo prior to ESRD had a 0.35 g/dl higher Hgb at time of starting dialysis compared with patients not receiving Epo during the pre-ESRD period (p < 0.001).

Intravenous iron therapy for HD patients during a 4-month period in 2002–2003 varied from 38% of patients in Japan, 53–67% in Italy, the USA, UK, Canada, Australia/New Zealand, and France, and 75–89% of patients in Germany, Spain, Sweden, and Belgium [11]. In countries reporting TSAT or total iron-binding capacity (along with serum iron levels) for >75% of patients, the percentage of HD patients with a TSAT <20% ranged from 16% in Spain to 35% in Belgium. These results indicate that substantial iron store deficits exist in HD patients despite a large fraction of patients routinely receiving IV iron doses in many DOPPS countries. Furthermore, analyses adjusted for numerous patient characteristics failed to observe any significant relationship between patient TSAT levels and the percentage of facility patients receiving IV iron or the number of IV iron doses given during a 4-month period.

An indication of some of the possible benefits of improved anemia control has also been described in DOPPS investigations. Mortality risk was found to be 5% lower for every 1 g/dl higher concentration in baseline Hgb across the 12 countries in DOPPS in 2002–2003 [11] and in EuroDOPPS in 1998–2000 [13]. The mean patient follow-up time in these analyses after the baseline Hgb measurement was 1.4–1.7 years. Our recent work with B. Robinson et al. (unpublished) using a time-varying analysis found an even larger relationship between higher Hgb associated with lower mortality risk when the time interval is shortened to 3 or 6 months for observing the relationship between measured Hgb value and mortality risk. Locatelli et al. [13] and Pisoni et al. [11] also have observed that higher patient Hgb levels were significantly associated with lower hospitalization risk.

Statin Use and Outcomes in HD Patients

Cardiovascular disease is very common in HD patients and is the leading cause of death in this patient population. Consequently, management of cardiovascular disease is an important aspect of the care provided to HD patients. A recent DOPPS investigation by Mason et al. [14], indicated that 3-hydroxy-3-methylglutaryl coenzyme A reductase inhibitors (statins) were used by 11.8% of HD patients in France, Germany, Italy, Japan, Spain, the UK, and the USA in year 2000 [country range: 3.5–5% (Italy, Spain), 7–8% (Japan, UK), 12–17% (Germany, France, USA)]. Statin use by HD patients was associated with a significantly lower all-cause mortality risk (RR = 0.69, p < 0.001, for all HD patients; RR = 0.70, p < 0.001 for patients with CAD or patients with hypertension). In addition, statin use was associated with a lower risk of death from CAD (RR = 0.77, p = 0.03). Furthermore, every 10% greater use of statins within a dialysis unit was associated with a 5% lower overall mortality risk for patients treated in those dialysis units (p = 0.02). In summary, this large international investigation suggests that more frequent use of statins may lead to significant improvements in dialysis patient outcomes. It is hoped that ongoing randomized clinical trials such as the Die Deutche Diabetes Dialyse Studie, the Study to Evaluate the Use of Rosuvastatin in Subjects on Regular Hemodialysis, and the Study of Heart and Renal Protection will be able to provide additional insights regarding the benefit of statins for HD patients.

Under-Diagnosed Depression in HD Patients?

Recent work by Lopes et al. [15] indicated that *physician-diagnosed* depression in a prevalent cross-section of HD patients varied across the 12 DOPPS countries from 2% in Japan to 22% in the USA. However, 39% of HD patients in the USA to 62% in Italy had scores consistent with possible depression as measured using a patient self-reported depression-screening instrument (CES-D 10 item, with score ≥ 10 indicative of possible depression). Patients with a CES-D score >10 displayed a significantly elevated mortality risk (RR = 1.42, $p < 0.001$) and elevated risk of hospitalization (RR = 1.12, $p < 0.01$) compared with patients having a CES-D score of 0–4. Furthermore, for the group of possibly depressed patients having a CES-D score of 15–30, only 32% were diagnosed as being depressed by their physicians and only 21% were receiving anti-depressant medication. These results suggest possible under-diagnosis of depression in HD patients to a large degree in some practices, and that more frequent use of depression-screening tools may assist physicians in recognizing patients with possible depression for further evaluation and treatment.

Vascular Access Use and Outcomes

The DOPPS has published numerous papers describing large country differences in vascular access use for both prevalent and incident patients [16–21]. Some of the key findings from these investigations include: (1) The native arteriovenous fistula (AVF) displays the best access survival, fewest complications and procedures, and should be considered as the preferred access for HD patients. (2) Catheter use, at the patient level or the facility level, is associated with a substantially higher mortality risk, higher hospitalization risk, and poorer anemia control; unfortunately, recent trends suggest an increase in catheter use in many countries. (3) Dialysis units which have high AVF use are dedicated to AVF as the permanent vascular access of choice, and typically have a shorter time interval between referral until access placement; individuals at these units have developed the expertise to successfully perform the first cannulation of an AVF in a shorter period of time. (4) A vascular access surgeon's experience during training in vascular access placement is significantly related to vascular access use seen later for patients served by the surgeon when no longer in training [22]. (5) Certain commonly used drugs in HD patients have been found to be associated with survival of AVF and synthetic grafts [21].

Conclusion

We hope these brief summaries of selected topics provide a helpful synopsis of some of the work being done in the DOPPS. A recent paper by Port et al. [23] estimated estimated that a large number of patient life years could be gained by improving certain HD practices to bring more patients within current NKF-K/DOQI guidelines. Clearly, there have been major improvements in the care and management of HD patients during the last decade. Dissemination of results from the DOPPS can assist individual practitioners in benchmarking practices and outcomes in their own setting with those seen across many dialysis units and countries, and thereby help support ongoing global efforts to improve the lives of HD patients.

Acknowledgment

We wish to acknowledge the patients and the superb efforts of the dedicated staff members from the more than 300 dialysis units which have participated in the DOPPS. The DOPPS is coordinated by University Renal Research and Education Association (URREA) based in Ann Arbor, Mich., USA. We also wish to gratefully acknowledge Amgen, Inc and Kirin for their support for the DOPPS through scientific/educational grants with no restrictions upon publication from this study.

References

1   Young EW, Goodkin DA, Mapes DL, Port FK, Keen ML, Chen K, Maroni BL, Wolfe RA, Held PJ: The Dialysis Outcomes and Practice Patterns Study (DOPPS): An international hemodialysis study. Kidney Int 2000;57(suppl 74):S74–S81.
2   Pisoni RL, Gillespie BW, Dickinson DM, Chen K, Kutner M, Wolfe RA: The Dialysis Outcomes and Practice Patterns Study: Design, data elements, and methodology. Am J Kidney Dis 2004;44 (suppl 2):S7–S15.
3   Evidence for improving patient care and outcomes: The Dialysis Outcomes and Practice Patterns Study (DOPPS) and Kidney Disease Outcomes Quality Initiative (K/DOQI). Am J Kidney Dis 2004;44(suppl 2):S1–S67.

4   Pifer TB, McCullough KP, Port FK, Goodkin DA, Maroni BJ, Held PJ, Young EW: Mortality risk in hemodialysis patients and changes in nutritional indicators. DOPPS. Kidney Int 2002;62: 2238–2245.

5   Leavey SF, McCullough K, Hecking E, Goodkin D, Port FK, Young EW: Body mass index and mortality in 'healthier' as compared with 'sicker' haemodialysis patients: Results from the Dialysis Outcomes and Practice Patterns Study (DOPPS). Nephrol Dial Transplant 2001;16: 2386–2394.

6   Young EW, Albert JM, Satayathum S, Goodkin DA, Pisoni RL, Akiba T, Akizawa T, Kurokawa K, Bommer J, Piera L, Port FK: Predictors and consequences of altered mineral metabolism: The Dialysis Outcomes and Practice Patterns Study. Kidney Int 2005;67:1179–1187.

7   Kimata N, Albert JM, Akiba T, Akizawa T, Bommer J, Kerr PG, Pisoni RL, Port FK, Saran R: Association of serum calcium and phosphorus with cardiac mortality in hemodialysis patients: new results from the DOPPS. J Am Soc Nephrol 2004;15:3A–4A.

8   Pisoni R, Satayathum S, Young E, Akiba T, Akizawa T, Kurokawa K, Locatelli F, Combe C, Maroni B, Port F: Predictors of hyperphosphatemia and its association with cardiovascular deaths and hospitalization in chronic hemodialysis patients: International results from the DOPPS. Nephrol Dial Transplant 2003;18(suppl 4):678.

9   Saran R, Bragg-Gresham JL, Rayner HC, Goodkin DA, Keen ML, van Dijk PC, Kurokawa K, Piera L, Saito A, Fukuhara S, Young EW, Held PJ, Port FK: Nonadherence in hemodialysis: Associations with mortality, hospitalization, and practice patterns in the DOPPS. Kidney Int 2003;64:254–262.

10   Fissell RB, Bragg-Gresham JL, Gillespie BW, Goodkin DA, Bommer J, Saito A, Akiba T, Port FK, Young EW: International variation in vitamin prescription and association with mortality in the Dialysis Outcomes and Practice Patterns Study (DOPPS). Am J Kidney Dis 2004;44: 293–299.

11   Pisoni RL, Bragg-Gresham JL, Young EW, Akizawa T, Asano Y, Locatelli F, Bommer J, Cruz JM, Kerr PG, Mendelssohn DC, Held PJ, Port FK: Anemia management outcomes from 12 countries in the Dialysis Outcomes and Practice Patterns Study (DOPPS). Am J Kidney Dis 2004;44: 94–111.

12   Kimata N, Akiba T, Pisoni RL, Albert JM, Satayathum S, Cruz JM, Akizawa T, Andreucci VE, Young EW, Port FK: Mineral metabolism and haemoglobin concentration among haemodialysis patients in the Dialysis Outcomes and Practice Patterns Study (DOPPS). Nephrol Dial Transplant, Advance access published on February 22, 2005; doi: 10.1093/ndt/gfh732.

13   Locatelli F, Pisoni RL, Combe C, Bommer J, Andreucci VE, Piera L, Greenwood R, Feldman H, Port FK, Held PJ: Anaemia and associated morbidity and mortality among haemodialysis patients in five European countries: Results from the Dialysis Outcomes and Practice Patterns Study (DOPPS). Nephrol Dial Transplant 2004;19:121–132.

14   Mason NA, Bailie GR, Satayathum S, Bragg-Gresham JL, Akiba T, Akizawa T, Combe C, Rayner HC, Saito A, Gillespie BW, Young EW: HMG-coenzyme A reductase inhibitor use is associated with mortality reduction in hemodialysis patients. Am J Kidney Dis 2005;45:119–126.

15   Lopes AA, Albert JM, Young EW, Satayathum S, Pisoni RL, Andreucci VE, Mapes DL, Mason NA, Fukuhara S, Wikström B, Saito A, Port FK: Screening for depression in hemodialysis patients: Associations with diagnosis, treatment, and outcomes in the DOPPS. Kidney Int 2004;66: 2047–2053.

16   Pisoni RL, Young EW, Dykstra DM, Greenwood RN, Hecking E, Gillespie B, Wolfe RA, Goodkin DA, Held PJ: Vascular access use in Europe and the United States: Results from the DOPPS. Kidney Int 2002;61:305–316.

17   Combe CH, Pisoni RL, Port FK, Young EW, Canaud B, Mapes DL, Held PJ: Dialysis Outcomes and Practice Patterns Study: Données sur l'utilisation des cathéters veineux centraux en hémodialyse chronique. Nephrologie 2001;22:379–384.

18   Young EW, Dykstra DM, Goodkin DA, Mapes DL, Wolfe RA, Held PJ: Hemodialysis vascular access preferences and outcomes in the Dialysis Outcomes and Practice Patterns Study (DOPPS). Kidney Int 2002;61:2266–2271.

19   Rayner HC, Pisoni RL, Gillespie BM, Goodkin DA, Akiba T, Akizawa T, Saito A, Young EW, Port FK: Creation, cannulation and survival of arterio-venous fistulae: Data from the Dialysis Outcomes and Practice Patterns Study (DOPPS). Kidney Int 2003;63:323–330.

20 Saran R, Dykstra DM, Pisoni RL, Akiba T, Akizawa T, Canaud B, Chen K, Piera L, Saito A, Young EW: Timing of first cannulation and vascular access failure in haemodialysis: An analysis of practice patterns at dialysis facilities in the DOPPS. Nephrol Dial Transplant 2004;19:2334–2340.

21 Saran R, Dykstra DM, Wolfe RA, Gillespie BM, Held PJ, Young EW: Association between vascular access failure and the use of specific drugs: The Dialysis Outcomes and Practice Patterns Study (DOPPS). Am J Kidney Dis 2002;40:1255–1263.

22 Saran R, Elder SJ, Asano Y, Ethier J, Rayner HC, Saito A, Young EW, Goodkin DA, Pisoni RL. Training, experience, and attitudes of vascular access (VA) surgeons predict VA type: the DOPPS. J Am Soc Nephrol 2004;15:153A.

23 Port FK, Pisoni RL, Bragg-Gresham JL, Satayathum S, Young EW, Wolfe RA, Held PJ: DOPPS estimates of patient life years attributable to modifiable hemodialysis treatment practices in the United States. Blood Purif 2004;22:175–180.

Dr. Roger Greenwood
Lister Renal Unit
East and North Herts NHS Trust
Stevenage, Herts, SG1–4AB (UK)
Tel. +44 1438 781157, Fax +44 1438 781182, E-Mail rogergreenwood@nhs.net

Ronco C, Brendolan A, Levin NW (eds): Cardiovascular Disorders in Hemodialysis.
Contrib Nephrol. Basel, Karger, 2005, vol 149, pp 69–82

....

What Did We Learn from the HEMO Study? Implications of Secondary Analyses

Tom Greene

Department of Biostatistics and Epidemiology, Cleveland Clinic Foundation,
Cleveland, Ohio, USA

Abstract

Background: The HEMO Study was a randomized clinical trial designed to determine
whether increasing hemodialysis dose above current standards, or using high-flux membranes, would improve patient outcome. The primary results of the trial showed no statistically significant effects of either dialysis dose or membrane flux on the primary outcome of
mortality. **Methods:** This report examines the implications of secondary analyses involving
subgroups and secondary outcome measures for the overall interpretation of the trial.
Results and Conclusions: The secondary analyses of the HEMO Study do not alter the conclusions of the primary analysis: In the context of conventional three times per week
hemodialysis, neither the high-flux nor high-dose interventions substantially improved
patient outcome compared to low-flux and standard-dose levels. However, certain secondary
results from the trial are consistent with the hypothesis of subtle effects that may be magnified by more intensive therapies that extend beyond the limits of conventional three times
per week dialysis. This hypothesis will be addressed by a pair of new randomized trials
sponsored by the National Institute of Digestive and Kidney Disease (NIDDK), which will
compare six times per week daily and nocturnal therapies with conventional three times per
week dialysis.

Introduction

The HEMO Study was a randomized clinical trial designed to determine
whether increasing dialysis dose above current standards, or using high-flux
membranes, would improve patient outcome or not. With 1,846 randomized
patients, 5,237 patient-years of follow-up and 871 events for the primary outcome

of mortality, the HEMO Study is the largest randomized clinical trial performed to date among the maintenance hemodialysis population [1]. The study succeeded in achieving its operational objectives: pre-specified goals for recruitment, retention, and study power were exceeded, the targeted separation between treatment groups in measures of dose and flux was achieved, and the study investigators successfully randomized a broad cross-section of patients with a wide range of comorbidity levels [1, 2].

The primary results of the trial, published in December 2002, showed no statistically significant effects of either dialysis dose or membrane flux on the primary outcome of mortality or on composite outcomes defined by time to either death or selected first cause-specific hospitalizations [1]. The relative risk for the high dose compared to the standard dose was 0.96, with a 95% confidence interval (CI) of 0.84–1.10 and a p value of 0.53. The relative risk for the high flux compared to the low-flux intervention was 0.92; 95% CI (0.81–1.05), p = 0.23. Subsequent publications by the study investigators have examined the effects of the dose and flux interventions on other outcomes, and in various subgroups of the patient population [3–8]. The trial's results have also been critiqued by a number of researchers in the field [9–14]. While the fact that neither the dose nor flux interventions demonstrated an overall benefit is a disappointment for the renal community, the primary results of the study provide objective evidence to support previously designated standards, and have far reaching implications for the pathogenesis of uremia and the future direction of research in the treatment of dialysis patients [10].

From a methodological perspective, a randomized trial is a focused experiment addressing specific questions. This focus is essential to the rigorous evaluation of cause-and-effect relationships with a clearly defined of precision. However, this high degree of specificity can also lead to frustration, particularly after a negative study, from those interested in related questions which were not directly addressed by the trial design. Such questions are abundant for the HEMO trial, as researchers have questioned whether the results might have been different had the study been conducted in a different population based on different entry criteria, or had different methods been used to define the interventions. Moreover, the primary intent-to-treat analysis addresses only the average benefit in the full study population, leaving open the possibility of benefits in subgroups of patients. It is also of interest whether the interventions may have had beneficial effects on other outcomes besides the primary outcome of mortality.

It is possible to address some of the questions that have been raised based on evidence from secondary analyses conducted within the trial. In doing so, it is essential to maintain awareness of the limitations of such secondary results – most importantly, the possibility of 'false positive' conclusions when carrying out a large number of statistical analyses. This report reviews some of the key implications of the primary and secondary analyses of the HEMO Study, while taking into account the inherent limitations of secondary analyses.

Summary of Study Design and Interventions

The HEMO Study design has been described elsewhere [15–17]. Briefly, following a pilot study in 1993, the full-scale trial was performed between March 1995 and December 2001. The 1,846 randomized patients were recruited by 15 clinical centers from 72 dialysis facilities in the United States, and assigned according to a 2×2 factorial design with equal allocation to either a high dose (target urea equilibrated Kt/V, eKt/V, of 1.45) or standard dose (target eKt/V of 1.05) of dialysis and to either a high-flux (mean β_2-microglobulin clearance >20 ml/min) or low-flux (mean β_2-microglobulin clearance <10 ml/min) membrane.

Key entry criteria included a residual kidney urea clearance of not more than 1.5 ml/min per 35 liters of urea distribution volume, achievement of an eKt/V >1.3 within two of three consecutive monitored dialysis sessions in which the high-dose goal was targeted, serum albumin >2.6 g/dl by nephelometry, and an age of 18–80 years. Within the constraints of the entry criteria and a high proportion of urban centers with a preponderance of African-American patients, the study investigators were encouraged to recruit a broad spectrum of patients, including those with high levels of comorbidity. As a result of these efforts, the proportions of randomized patients with diabetes (45%), cardiac disease (80%) and other comorbid conditions were similar to those in the general United States dialysis population [1, 2].

Centrally generated dialysis prescriptions and monthly monitoring of the delivered dialysis led to a mean difference in eKt/V between the dose groups of 0.37 eKt/V units, which was 92.5% of the study's targeted separation of 0.40 eKt/V units. Mean follow-up eKt/V was <1.25 for 93% of standard-dose patients and >1.35 for 92% of high-dose patients. To increase generalizability, different dialyzers and reuse methods were permitted in the trial as long as the criteria defining either the low-or high-flux interventions were satisfied [1, 17]. In all, 8 low-flux and 17 high-flux dialyzers were approved for use in the study. The mean (SD) achieved follow-up β_2-microglobulin clearance was 33.8 (11.4) ml/min in the high-flux group and 3.4 (7.2) ml/min in the low-flux group. The mean β_2-microglobulin clearance was less than 5 ml/min for all low-flux dialyzers, and greater than 30 ml/min for all high-flux dialyzer-reuse combinations with the exception the CT-190 when reused with renalin (25 ml/min).

The potential follow-up time from randomization to the end of the trial averaged 4.48 years (range 0.9–6.6 years, depending on when the patient was randomized). Reflecting the high mortality of dialysis patients, the actual mean follow-up time was 2.84 years. The observed mortality rate of 16.6% was lower than that reported for the entire United States hemodialysis population during the period of the study [13], but was higher than that projected in the study's power calculations and similar to the death rate of 17.6% of hemodialysis patients in the United States within the age limits and racial distribution of the HEMO Study [2].

Implications of Subgroup Results

There are two main limitations of subgroup analyses in randomized trials. The first is the statistical problem of interpreting multiple hypothesis tests: if a large number of subgroup analyses are conducted, some may appear to reach statistical significance by chance alone, even the absence of a true effect [18]. This problem is aggravated if subgroup analyses are conducted post hoc, after examination of the data, since it is then difficult to determine the number of distinct analyses that has been performed to determine the appropriate statistical adjustments. Recognizing this problem, the HEMO Study investigators pre-specified in the protocol that subgroup analyses would be conducted for 7 baseline factors: (1) age, (2) race, (3) gender, (4) diabetes, (5) duration of dialysis prior to the trial, (6) comorbidity level by the index of coexisting disease (ICED Score), and (7) serum albumin. The second limitation of subgroup analyses comes from sample size constraints; due to resource limitations, randomized trials, including the HEMO Study, are typically powered for the full study cohort, and have the capacity to detect only large effects within subgroups.

Within the constraints of these limitations, we examine the implications of subgroup analyses in the HEMO Study subgroup for several key issues related to the interpretation of the trial's results. Tables 1 and 2 recapitulate the results of the HEMO Study subgroup analyses for the 7 pre-specified factors, plus 3 post-hoc factors relevant to questions that have arisen since the start of the trial: (8) flux of the dialyzer membrane used by the patient prior to the trial, (9) dialysis dose prior to the trial, and (10) use of renalin as a reprocessing agent. The interactions of the interventions with follow-up time, with the treatment effects estimated separately before and after 2 years of follow-up, are also summarized.

Comorbidity
Four of the seven pre-specified subgroup factors are directly associated with higher comorbidity: greater age, diabetes, lower albumin, and a higher

Table 1. Interactions of dose intervention with baseline factors

Factor	Subgroup	Relative risk	95% CI	Interaction p value[1]
Age[2]	≤58 years	0.95	(0.74–1.22)	0.92
	>58 years	0.97	(0.82–1.13)	
Sex[2]	Male	1.16	(0.94–1.43)	0.014
	Female	0.81	(0.68–0.97)	
Race[2]	Nonblack	1.13	(0.91–1.39)	0.06
	Black	0.87	(0.73–1.03)	
Diabetes[2]	Nondiabetic	0.90	(0.74–1.09)	0.35
	Diabetic	1.02	(0.85–1.22)	
Duration of dialysis[2]	≤3.7 years	1.03	(0.88–1.22)	0.12
	>3.7 years	0.83	(0.66–1.04)	
ICED[2]	≤2	0.92	(0.70–1.22)	0.96
	≥3	0.93	(0.76–1.15)	
Baseline albumin[2]	≤3.6 g/dl	0.89	(0.75–1.06)	0.16
	>3.6 g/dl	1.08	(0.88–1.32)	
Baseline flux	Low flux	0.91	(0.73–1.12)	0.51
	High flux	1.00	(0.84–1.18)	
Baseline eKt/V	<1.42	0.96	(0.79–1.15)	0.96
	≥1.42	0.96	(0.80–1.16)	
Renalin use	Nonrenalin	0.91	(0.77–1.09)	0.42
	Renalin	1.02	(0.83–1.26)	
Follow-up time	Year 0–2	0.96	(0.80–1.16)	0.96
	After year 2	0.96	(0.79–1.16)	

[1] p values for interactions provided on a comparison basis, without adjustment for multiple tests.
[2] Pre-specified factor for subgroup analysis.
Interactions tests for continuous factors based on dichotomization at their mean values.

ICED score. None of these factors were associated with trends for benefits of either the dose or flux interventions. Due to the constraints on the statistical power of subgroup analyses, these negative results do not rule out limited benefits of the interventions in patients with higher comorbidity. However, the limits of the 95% CI for the treatment effects are generally inconsistent with large benefits of the interventions within these subgroups. Further, these results indicate that it is unlikely that the trial would have yielded different results had a greater proportion of high-comorbidity patients been randomized, as had been hypothesized by some reviewers [9, 12].

Table 2. Interactions of flux intervention with baseline factors

Factor	Subgroup	Relative risk	95% CI	Interaction p value[1]
Age[2]	≤58 years	0.98	(0.76–1.26)	0.69
	>58 years	0.92	(0.79–1.08)	
Sex[2]	Male	1.03	(0.84–1.26)	0.27
	Female	0.88	(0.74–1.06)	
Race[2]	Nonblack	1.04	(0.84–1.28)	0.24
	Black	0.88	(0.74–1.04)	
Diabetes[2]	Nondiabetic	0.95	(0.78–1.15)	0.87
	Diabetic	0.93	(0.77–1.11)	
Duration of dialysis[2]	≤3.7 years	1.05	(0.89–1.24)	0.005
	>3.7 years	0.68	(0.53–0.86)	
ICED[2]	≤2	0.95	(0.72–1.24)	0.94
	≥3	0.93	(0.76–1.14)	
Baseline albumin[2]	≤3.6 g/dl	0.91	(0.76–1.09)	0.65
	>3.6 g/dl	0.97	(0.79–1.19)	
Baseline flux	Low flux	0.94	(0.76–1.16)	0.99
	High flux	0.94	(0.79–1.11)	
Baseline eKt/V	<1.42	1.00	(0.83–1.20)	0.36
	≥1.42	0.88	(0.73–1.06)	
Renalin use	Nonrenalin	0.95	(0.80–1.13)	0.88
	Renalin	0.93	(0.75–1.14)	
Follow-up time	Year 0–2	0.88	(0.73–1.05)	0.29
	After year 2	1.01	(0.83–1.22)	

[1]p values for interactions provided on a comparison basis, without adjustment for multiple tests.
[2]Pre-specified factor for subgroup analysis.
Interactions tests for continuous factors based on dichotomization at their mean values.

Prevalent versus Incident Patients

Some critics have argued that the HEMO Study design was flawed by the enrollment of prevalent patients rather than restriction of enrollment to incident patients [9, 12]. Three study design considerations led to the decision to enroll prevalent patients: (1) increased generalizability, allowing the trial's results to apply to prevalent as well as incident patients, (2) concern that restriction to incident patients would lead to higher average levels of residual renal function among trial participants, reducing the proportional effect of the dose and flux interventions on total solute clearance, and (3) the infeasibility of enrolling enough incident patients from the participating dialysis units to obtain the target sample size required for adequate statistical power.

Although it is plausible to hypothesize that certain types of prior dialysis therapy might have attenuated effects of the interventions in prevalent patients, this hypothesis is not supported by the data from the trial. Characteristics of dialysis at entry, including dialysis dose and membrane flux, were not associated with the effects of the interventions. Further, as shown in tables 1 and 2, the relative risks of the high-dose and high-flux interventions both exceeded 1 in the subgroup of patients with less than 3.7 years of prior dialysis. Similar results are obtained if the analysis is restricted to patients with less than 1 year (n = 490 patients) of prior dialysis. Finally, since the effects of the interventions were similar before and after 2 years of follow-up, the data do not support the hypothesis that carry over effects from prior dialysis therapy obscured long-term benefits of the interventions in the months following randomization.

Body Size
A number of arguments have been advanced to suggest that it may not be optimal to standardize dialysis dose by the volume of urea distribution V, as was done in the HEMO Study [19, 20]. The most extreme of these perspectives holds that the minimum threshold for an optimal dialysis dose should be based on the total treatment clearance, $K \times T$ (or Kt), with no standardization for body size at all [19]. If so, this would imply a higher minimum threshold for eKt/V for smaller patients than for larger patients. This, in turn, corresponds to the hypothesis that the dose effect in the HEMO Study should have been larger in smaller patients than in larger patients. The data from the trial do not directly support this hypothesis: an exploration of potential interactions the dose intervention with a wide array of size-related indices was undertaken, but no clear evidence of a larger dose effect for smaller levels of any of these indices was identified [5]. On the other hand, as shown in table 1, a possible interaction of the dose intervention with gender was identified (p = 0.014), with a trend for a benefit of high dose in women (who had a mean anthropometric volume of 31.3 liters), but not in men (mean anthropometric volume = 39.6 liters). Although joint analyses of both gender and body size indices indicated that the gender interaction could not be explained by any of the size indices [5], due to statistical power limitations of subgroup analyses we cannot rule out the possibility of an undetected benefit of the high dose in smaller patients.

Interactions Involving Gender and Prior Years of Dialysis
Perhaps, the most tantalizing data to come out of the HEMO Study are interactions suggesting a benefit of the high-dose intervention in women (interaction p = 0.014) and of the high-flux intervention in patients with more than 3.7 prior years of dialysis (interaction p = 0.005) [1, 3, 5]. Since 7 interaction

tests were prespecified for each intervention, the key question is whether one might expect 2 interaction tests of this level of significance by chance or not. The bootstrap method [21] was used to estimate the correlations among the 14 interaction tests of the two interventions with the 7 pre-specified subgroup factors in the HEMO Study. Based on these correlations, in the absence of any true subgroup effects, the probability of attaining at least one p value of 0.014 or smaller for the 7 interaction tests with the dose intervention is 0.092, and the corresponding probability of observing at least one interaction p value smaller than 0.005 for the flux intervention is 0.034. The probability of obtaining two p values smaller than 0.014 among all 14 interaction tests is 0.017. That is, if a series of randomized trials were conducted, each with 14 interaction tests with correlations similar to those of the HEMO Study, then by chance alone approximately 1 trial in 58 would report 2 interactions as strong as the two in observed in the HEMO Study.

These calculations indicate that it would be unusual, although not beyond the realm of possibility, to observe the interactions reported in the HEMO Study in the absence of any true subgroup effects. The HEMO Study investigators have interpreted these subgroup results with caution for other reasons besides the multiple testing problem. Since the gender interaction could not be explained by gender differences in size parameters, at the present time there does not appear to be a compelling physical basis for an interaction of dose with gender [5]. Additionally, the strength of the dose by gender interaction depends in part on an unexpected trend in men favoring the standard-dose over the high-dose intervention. The interaction of the flux intervention with years of dialysis turned out to be partially dependent on the cutoff of 3.7 years for the analysis; when years of dialysis is treated as a continuous variable, the p value for the interaction is increased from 0.005 to 0.040 [3]. This is not to say that these subgroup results should be dismissed; rather, given the noted limitations, the HEMO Study investigators have interpreted these subgroup analyses as generating important hypotheses for investigation in future studies [1]. In fact, following up on the HEMO Study results, Port et al. [22] have confirmed the presence of similar interactions of dose with gender in large observational databases in the United States and other countries, with stronger effects of higher dose observed in women than in men. Although observational relationships are subject to the risk of bias due to confounding, the main risk of error in the observational analyses (a greater confounding in the dose-mortality relationship in women than in men) is independent from the potential error in the HEMO Study (a spurious false-positive finding due to multiple subgroup tests). Thus, although further investigation is still required, the striking agreement of the results pertaining to the gender interaction from the randomized trial and observational data strengthens the case for a benefit of a higher dialysis dose in women.

Alternative Methods of Expressing Dialysis Dose

Some researchers have questioned whether the dose group comparison in the HEMO Study might have been obscured by the use of equilibrated Kt/V as defined by a specific urea kinetic model to define the dose intervention [13]. For example, might the results of the trial have been different if the dose intervention had been based on some other measure, such as the $K \times T$ product, or the treatment time itself? The short answer to this question is 'Probably not'. Experience from the trial suggests that the dose levels obtained in the study were close to the lowest that could be feasibly achieved in the Standard-Dose group and the highest that could be achieved in the High-Dose group, irrespective of what measure of dose was used to calibrate the interventions. The actual separation that was achieved in eKt/V, treatment time, and in the $K \times T$ product are displayed in figure 1, overall and stratified by the level of anthropometric volume. The separation in other measures based on urea concentration, such as the urea reduction ratio (URR) or single pool Kt/V, was similar to that of eKt/V.

The above conclusion that the dose intervention approximated the maximum feasible separation, irrespective of the dose measure, is based on two practical observations. First, during the trial national standards and policies of dialysis chains stipulated a minimum URR of 65% and a minimum single pool Kt/V of 1.2. Due to pressures on participating facilities to exceed these standards, one of the largest logistical challenges of the trial was preventing an attenuation of separation between the dose groups by an upward drift of achieved dose in the Standard-Dose group. Since the protocol stipulated the use of high-efficiency dialysis unless this interfered with fluid removal [1, 17], the Standard-Dose arm in effect targeted the lowest dose that allowed for adequate fluid removal and met national standards. This is true irrespective of the specific measure used to calibrate the intervention. For example, had the dose intervention been based on $K \times T$, any attempt to reduce $K \times T$ in a particular patient would also have reduced spKt/V and URR, but spKt/V and URR were already at the lowest level acceptable to the dialysis units to assure satisfaction of national standards.

Conversely, in the High-Dose arm the achieved dose levels represented the highest that could be achieved under a conventional three times per week schedule with treatment time under 4.5 h for patients with urea volume of around 40 liters or greater. The problem of convincing patients and their physicians to adhere to the high-dose target in spite of treatment times often exceeding 4 h was the second major logistical challenge of the study. Thus, for patients with urea volumes of around 40 liters or greater, the high-dose target in the HEMO Study required the highest feasible levels of clearance

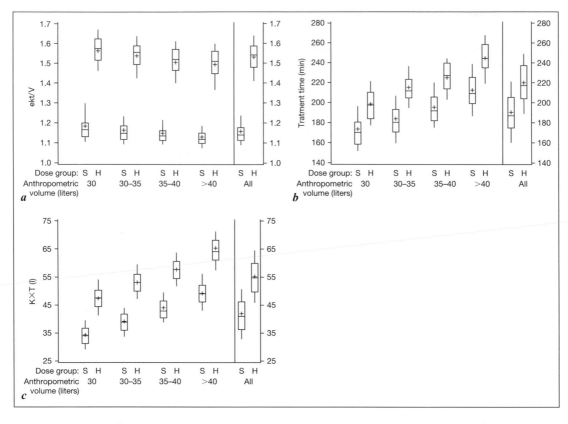

Fig. 1. Shown is the separation in the mean levels of achieved eKt/V (*a*), treatment time (*b*), and in the K × T product (*c*) between the dose groups during follow-up. Results were first averaged over all follow-up kinetic modeling sessions for each patient, and then summarized with stratification by the level of baseline anthropometric volume (Watson formula) and overall. The box plots show the 10th, 25th, 50th, 75th, and 90th percentiles, with plus signs indicating the mean value.

and treatment time, and hence the highest feasible value for any of the dose measures which have been proposed. It is possible that greater separation could have been achieved in the smaller patients by ratcheting up treatment time in the smaller patients to match that of the larger patients. However, aside from the practical difficulties of greatly increasing total average treatment time for patients and participating dialysis units, this would have resulted in a design in which the separation between dose groups was greater for the smaller patients than the larger patients, complicating the interpretation of the study results.

Effects on Secondary Outcomes

While mortality is the appropriate primary outcome in a study of dialysis patients, effects of an intervention on other factors, including hospitalization, nutritional status, and quality of life, are also of substantial importance. The effects of the treatment interventions on 22 secondary outcomes have now been described in publications from the trial and are summarized in table 3. When interpreting these results it is important to note that because the HEMO Study was powered for mortality, the study had sufficient power to detect tiny treatment effects on most continuous outcomes (such as the nutritional status and quality of life measures in the table) which were measured multiple times throughout the follow-up. Thus, it is possible for effects on these factors to be statistically significant without demonstrating clinical significance.

Of 46 tests of the treatment effects presented in table 3, 7 reached nominal significance at the 5% level, without adjustment for multiple testing. Of these, the 2 which have drawn the most attention are the effects of the flux intervention on cardiac death, with a relative risk of 0.80, 95% CI (0.65–0.99), p = 0.042, and the composite of cardiac hospitalization and cardiac death [relative risk = 0.87, 95% CI (0.76–1.00), p = 0.045] [3]. Among the remaining 5 nominally significant results, the effect of the dose intervention on equilibrated normalized protein catabolic rate was interpreted as likely an artifact of the model used to compute this quantity and the remaining effects as of insufficient magnitude to represent clinical significance [7].

Thus, with the possible exception of the effect of flux on the cardiac end points, no clinically important effects of either intervention have been identified for any of the main secondary end points in the entire cohort. Further, while the flux effects on cardiac end points are of potential importance, the borderline significance of these effects precludes firm conclusions in the context of multiple analyses of secondary end points.

Conclusion

In conclusion, the secondary analyses of the HEMO Study do not alter the interpretation of the primary analysis; namely, that in the context of conventional three times per week hemodialysis, neither the high-dose nor the high-flux interventions substantially improve patient outcome compared to low-flux and standard-dose levels consistent with current standards. The secondary analyses from the trial indicate that this conclusion is unlikely to have been different had the trial been restricted to patients with higher comorbidity levels

Table 3. HEMO Study results for primary and secondary end points

Outcome	Primary HEMO Study references	High dose vs. standard dose	High flux vs. low flux
Mortality	1, 6	ns	ns
1st cardiac hospitalization or death	1, 6	ns	ns
1st infection hospitalization or death	1, 4	ns	ns
1st 15% decline in serum albumin or death	1, 7	ns	ns
All non-access hospitalizations	1	ns	ns
Cardiac death	6	ns	RR = 0.80, $(0.65 - 0.99)$[2]
1st cardiac hospitalization or cardiac death	6	ns	RR = 0.87, $(0.76 - 1.00)$[2]
Infection death	4	ns	ns
1st infection hospitalization or infection death	4	ns	ns
Serum albumin (g/dl)	7	ns	ns
Post-dialysis weight (kg)	7	ns	ns
Equilibrated protein catabolic rate (g/kg/day)	7	$\Delta = 0.019 \pm 0.07$[1]	ns
Adjusted energy intake (g/kg/day)	7	ns	ns
Adjusted protein intake (kcal/kg/day)	7	ns	ns
Calf circumference (cm)	7	ns	$\Delta = 0.31 \pm 0.13$[1]
Upper arm circumference (cm)	7	ns	$\Delta = 0.35 \pm 0.16$[1]
Appetite assessment (5 pt scale)	7	ns	ns
SF-36 PCS	8	$\Delta = 1.23 \pm 0.46$[1]	ns
SF-36 MCS	8	ns	ns
SF-36 physical functioning	8	ns	ns
SF-36 vitality	8	ns	ns
KDQOL symptoms	8	ns	ns
KDQOL sleep index	8	ns	$\Delta = 2.25 \pm 0.95$[1]

Treatment effects with uncorrected p values of 0.05 or smaller are highlighted. Additional details on the outcomes and analysis methods are provided in the listed HEMO Study references.

[1]Indicated are mean treatment effects ($\pm$SE) over the first 3 years of follow-up, expressed as mean differences in changes from baseline to follow-up between the dose or flux groups. Positive values indicate larger values for the high-dose or high-flux groups compared to the standard-dose or low-flux groups.

[2]Indicated are the relative risks (and 95% CI) for the high-flux group compared to the low-flux group.

ns = Not significant.

or to incident patients, or if a different measure had been used to define the dose intervention. The absence of clinically important effects on continuous measures of nutritional status and quality of life corroborates the absence of treatment effects on the primary analysis of mortality.

On the other hand, the negative results of the HEMO Study do not rule out benefits of very substantial increases in dialysis dose or treatment time that may be obtained from interventions that extend beyond the limits of the conventional three times per week interventions tested in the trial. In particular, in recent years considerable interest has been expressed in the use of hemodiafiltration to substantially boost removal of middle molecules, and of increasing the frequency and/or total duration of dialysis treatments with so-called daily dialysis or nocturnal dialysis. The daily and nocturnal interventions have the potential to dramatically increase solute clearance and improve volume homeostasis compared to the interventions of the HEMO Study. It is possible that the hints in the HEMO Study of beneficial effects of the dose and flux interventions in certain subgroups or on certain secondary end points may reflect subtle benefits which could be magnified by these more intensive interventions. This hypothesis will be addressed by a pair of prospective randomized trials now being sponsored by the National Institute of Digestive and Kidney Disease (NIDDK), which will compare intermediate physiological and quality of life end points between six times per week daily and nocturnal therapies to conventional three times per week dialysis [14].

References

1   Eknoyan G, Beck GJ, Cheung AK, Daugirdas JT, Greene T, Kusek JW, Allon M, Bailey J, Delmez JA, Depner TA, Dwyer JT, Levey AS, Levin NW, Milford E, Ornt DB, Rocco MV, Schulman G, Schwab SJ, Teehan BP, Toto R for the HEMO Study Group: Effect of dialysis dose and membrane flux on mortality and morbidity in maintenance hemodialysis patients: Primary results of the HEMO Study. N Engl J Med 2002;347:2010–2019.
2   Rocco MV, Cheung AK, Greene T, Eknoyan G for the Hemodialysis (HEMO) Study Group: The HEMO Study: Applicability and generalizability. Nephrol Dial Transplant 2005, in press.
3   Cheung AK, Levin NW, Greene T, Agodoa L, Bailey J, Beck G, Clark W, Levey AS, Leypoldt JK, Ornt DB, Rocco MV, Schulman G, Schwab SJ, Teehan B, Eknoyan G for the HEMO Study Group: Effects of high-flux hemodialysis on clinical outcomes: Results of the HEMO Study. J Am Soc Nephrol 2003;14:3251–3263.
4   Allon M, Depner TA, Radeva M, Bailey J, Beddhu S, Butterly D, Coyne D, Gassman J, Kaufman A, Kaysen G, Lewis J, Schwab SJ for the HEMO Study Group: Impact of dialysis dose and membrane on infection-related hospitalization and death: Results of the HEMO Study. J Am Soc Nephrol 2003;14:1863–1870.
5   Depner T, Daugirdas J, Greene T, Allon M, Beck G, Chumlea C, Delmez J, Gotch F, Kusek J, Levin N, Macon E, Milford E, Owen W, Star R, Toto R, Eknoyan G for the HEMO Study Group: Dialysis dose and effect of gender and body size on outcome in the HEMO Study. Kidney Int 2004;65: 1386–1394.

6 Cheung AK, Berkoben M, Heyka R, Kaufman A, Lewis J, Ornt D, Rocco M, Sarnak M, Toto R, Windus D, Yan G, Levey AS for the HEMO Study Group: Cardiac diseases in maintenance hemodialysis patients: Results of the HEMO study. Kidney Int 2004;65:2380–2389.

7 Rocco MV, Dwyer JT, Larive B, Greene T, Cockram DB, Chumlea WC, Kuse JW, Leung J, Burrowes JD, McLeroy SL, Poole D, Uhlin L for the HEMO Study Group: The Effect of dialysis dose and membrane flux on nutritional parameters in hemodialysis patients: Results of the HEMO Study. Kidney Int 2004;65:2321–2334.

8 Unruh M, Benz R, Greene T, Yan G, Beddhu S, DeVita M, Dwyer JT, Kimmel PL, Kusek JW, Martin A, Rehm-McGillicuddy J, Teehan BP, Meyer KB: Effect of hemodialysis dose and membrane flux on health-related quality of life in the HEMO Study. Kidney Int 2004;66:1–12.

9 Locatelli F: Effect of dialysis dose and membrane flux in maintenance hemodialysis. (letter; comment). N Engl J Med 2003;348:1491–1494.

10 Depner TA, Gotch FA, Port FK, Wolfe RA, Lindsay RM, Blake PG, Locatelli F: How will the results of the HEMO study impact dialysis practice? Semin Dial 2003;16:8–21.

11 Levin N, Greenwood R: Reflections on the HEMO study: The American viewpoint. Nephrol Dial Transplant 2003;18:1059–1060.

12 Locatelli F: Dose of dialysis, convection and haemodialysis patients outcome – What the HEMO study doesn't tell us: The European viewpoint. Nephrol Dial Transplant 2003;18:1061–1065.

13 Himmelfarb J: The HEMO Study – Where do we go from here? Nephrol Dial Transplant 2003;18: 587–591.

14 Briggs JP: Evidence-based medicine in the dialysis unit: A few lessons from the USRDS and the NCDS and HEMO trials. Semin Dial 2004;17:136–141.

15 Eknoyan G, Levey AS, Beck GJ, Agodoa LY, Daugirdas JT, Kusek JW, Levin NW, Schulman G for the HEMO Study Group: The Hemodialysis (HEMO) Study: Rationale for selection of interventions. Semin Dial 1996;9:24–33.

16 Daugirdas JT, Depner TA, Gotch FA, Greene T, Keshaviah P, Levin NW, Schulman G for the HEMO Study Group: Comparison of methods to predict equilibrated Kt/V in the HEMO pilot study. Kidney Int 1997;52:1395–1405.

17 Greene T, Beck GJ, Gassman JJ, Gotch FA, Kusek JW, Levey AS, Levin NW, Schulman G, Eknoyan G for the HEMO Study Group: Design and statistical issues of the Hemodialysis (HEMO) Study. Control Clin Trials 2000;21:502–525.

18 Piantadosi S: Clinical Trials: A Methodologic Perspective: Exploratory or Hypothesis-Generating Analyses. 1997, chapt 12.7, pp 320–324. New York, Wiley.

19 Lowrie E, Chertow G, Lew N, Lazarus M, Owen WF: The urea [clearance × dialysis time] product (Kt) as an outcome-based measure of hemodialysis dose. Kidney Int 1999;56:729–737.

20 Lindholm B, Waniewski J, Werynski A: KT/V: The denominator dilemma. Pol Merkuriusz Lek 2003;15:311–315.

21 Westfall PH, Young SS: Resampling-based multiple testing: Examples and methods for p-value adjustment. New York, Wiley, 1993.

22 Port FK, Wolfe RA, Hulbert-Shearon TE, McCullough KP, Ashby VB, Held PJ: High dialysis dose is associated with lower mortality among women but not among men. Am J Kidney Dis 2004;43: 1014–1023.

Tom Greene
Department of Biostatistics and Epidemiology, Wb-4
Cleveland Clinic Foundation, 9500 Euclid Avenue
Cleveland Ohio, 44195 (USA)
Tel. +1 216 444 9933, Fax +1 216 445 2781, E-Mail tgreene@bio.ri.ccf.org

Ronco C, Brendolan A, Levin NW (eds): Cardiovascular Disorders in Hemodialysis.
Contrib Nephrol. Basel, Karger, 2005, vol 149, pp 83–89

....

What Are We Expecting to Learn from the MPO Study?

Francesco Locatelli, Pietro Pozzoni, Salvatore Di Filippo

Department of Nephrology and Dialysis, A. Manzoni Hospital, Lecco, Italy

Abstract

High-flux membranes represented a major improvement in dialysis technique, but evidences supporting their clinical superiority over conventional low-flux dialysis are still inconclusive. Although several studies, most of which were observational, showed an association between high-flux dialysis and lower morbidity and mortality, the Hemodialysis (HEMO) study, the first large-scale randomized clinical trial specifically aimed at testing the effect of membrane permeability on patients' outcome, failed to demonstrate a statistical significant benefit of high-flux membranes on all-cause mortality. Although disappointing, these results should however be interpreted in light of some important limitations of the HEMO study, first of all the inclusion of both incident and prevalent hemodialysis patients, the exclusion of sicker patients and the allowance of dialyzer reuse. In this context, much is expected from the Membrane Permeability Outcome (MPO) study, a randomized clinical trial investigating the effect of high-flux membranes in a large population of incident hemodialysis patients across Europe. Inclusion of only incident patients, absence of severe exclusion criteria and no dialyzer reuse are all distinguishing features of this study. Analyses of the baseline data of the MPO study confirm the high burden of cardiovascular disease among incident dialysis patients, although comparison with the Dialysis Outcomes and Practice Patterns Study data provides further evidence of a positive selection of patients in clinical trials.

Introduction

Despite several technological advances achieved in hemodialysis over the last decades, life expectancy of dialysis patients is still significantly poorer when compared to nondialysis subjects with similar demographics. Although most of such excess of mortality results from a disproportionate burden of cardiovascular disease [1, 2], a number of factors directly related to hemodialysis, including inadequate removal of uremic toxins, incomplete correction of chronic uremia-related disorders, or short- and long-term adverse effects of the dialysis procedure, may still severely affect survival, morbidity and quality of life of patients. High-flux membranes are characterized by enhanced removal of middle- and large-molecular-weight solutes. As such molecules are claimed to be involved in the genesis of many complications of chronic hemodialysis, high-flux hemodialysis has been suggested to have important clinical advantages over standard hemodialysis. However, evidences supporting the clinical superiority of high-flux hemodialysis are still inconclusive, but the analysis is made difficult by the relatively low number of properly designed clinical studies. To this purpose, in the past years two large-scale, randomized, controlled clinical trials, the Hemodialysis (HEMO) study and the Membrane Permeability Outcome (MPO) study, have been specifically designed with the aim of better assessing the effect of high-flux membranes on morbidity and mortality of hemodialysis patients.

Benefits of High-Flux Dialysis: Evidences before the HEMO Study

After the introduction of high-flux dialysis into clinical practice, several studies showed its association with an improvement in many aspects of chronic hemodialysis-related morbidity, preservation of residual renal function and, above all, long-term survival.

High-flux dialysis has been claimed to potentially improve the cardiovascular stability of patients on extracorporeal treatment, because of a different effect of convective fluid removal on peripheral resistance and an increased removal of peptides with possible hypotensive action, but, despite these theoretical advantages, results obtained in clinical studies have been rather disappointing, showing no significant difference in the occurrence of intradialytic hypotensive episodes among patients treated with membranes with different permeability characteristics; although, the number of hypotensive episodes was much lower than expected [3].

β_2-microglobulin plasma levels have been shown to significantly decrease in patients treated with high-flux dialysis as compared to those assigned to low-flux treatment [3, 4], and some studies found that clinical manifestations of β_2-microglobulin-related amyloidosis are less likely to occur in patients undergoing high-flux hemodialysis [5, 6]. Some uncertainty, however, still remains as to whether the reduced β_2-microglobulin plasma levels associated with the use of high-flux membranes are the result of increased membrane permeability per se or, rather, of different biocompatibility; although, both flux characteristics and biocompatibility of the membrane have been found to be independent determinants of clinical manifestations of β_2-microglobulin-derived of amyloidosis [7].

Despite some evidences of an association between high-flux membranes and high dietary protein intake [8], the same finding was not confirmed in a randomized clinical trial, showing the absence of any correlation between membrane biocompatibility and flux and protein catabolic rate [3]. Furthermore, other studies [4, 9] failed to demonstrate any major influence of convective treatments on the time evolution of nutritional markers in hemodialysis patients, but these studies were either characterized by a too short follow-up period [4] or dealt with relatively well-nourished patients [9] to possibly detect a significant effect.

High-flux dialysis has also been suggested to result in improved anemia control, but this hypothesis, although supported by some preliminary studies, was not confirmed in a multicentre, controlled, randomized trial, where no significant differences in hemoglobin levels were found between patients treated with a high-flux membrane (BK-F polymethylmethacrylate) and those assigned to a cellulose dialyzer [4]. Shortness of follow-up and positive selection of relatively healthy patients could, however, have decreased the power of the study, thus contributing to its failure to detect a positive effect.

Evidences of a putative superiority of high-flux membranes in better preserving residual renal function of hemodialysis patients, possibly in light of the increased removal of large-molecular-weight uremic substances with potential nephrotoxic action, have been also weak, and moreover supported by clinical studies unable to discriminate the effects of membrane flux characteristics from those of biocompatibility [10]. Therefore, the hypothetical beneficial effects of high-flux membranes per se on the preservation of residual renal function in hemodialysis patients still needs to be demonstrated by properly designed clinical trials.

As to the most important question concerning the impact of high-flux hemodialysis on patients' survival, several observational, retrospective studies found a significant association with reduced mortality, with a percent reduction in the relative risk of death ranging from 5 to 76% [6, 11–14]. Interestingly, Port et al. [14] evidenced an independent contribution of membrane permeability to the observed improvement in hemodialysis patients' survival. Only a randomized clinical trial by Locatelli et al. [3] did not reveal any statistically significant difference in survival related to membrane biocompatibility or flux, but this study was not primarily designed to do so. The enthusiasm about high-flux dialysis, which arose from these promising findings, became milder after the results of the HEMO study, the first large-scale randomized clinical trial specifically aimed at testing the effect of membrane permeability on hemodialysis patients' mortality, were published.

What the HEMO Study Did and Did Not Tell Us About High-Flux Dialysis

The HEMO study was conducted between 1995 and 2001 in the United States, enrolled a total of 1,846 hemodialysis patients, and its results were published at the end of 2002 [15]. The main aim of this study consisted in evaluating the effect of standard versus high dialysis dose and of low- versus high-flux membranes on morbidity and mortality of hemodialysis patients, randomized to different membranes and different dialysis doses according to a two-by-two factorial design. Similarly biocompatible membranes, defined as low- or high-flux according to β_2-microglobulin clearance, were used in the two membrane permeability arms of the study, so that eventual differences in patients' outcome would have been attributable to an independent effect of membrane flux characteristics per se. Despite the large sample of patients, the HEMO study failed to demonstrate a statistically significant benefit of high-flux membranes on all-cause mortality: the adjusted relative risk of all-cause death was 8% lower (p = 0.24) in patients randomized to the high-flux group [15]. Only when the analysis was limited to cardiac death or to patients who had been on dialysis for more than 3.7 years at the time of enrollment was better survival found in the high-flux group (adjusted relative risk 0.78, p = 0.03, and 0.68, p = 0.001, respectively) [15], but these results should be interpreted with caution, having been obtained by secondary analyses. Although clearly disappointing, the results of the HEMO study must be interpreted in the light of some important study drawbacks. First of all, the study did not consider patients with low plasma albumin and high body weight, thus actually excluding those patients who were at particularly high risk for morbidity and mortality, namely, those who could have likely benefited more from high-flux dialysis. Mean age at inclusion was relatively low (58 years) and many long survivors were included (the average time on dialysis at baseline was 3.7 years), further suggesting the existence of some form of positive selection in the enrollment of patients. A carry-over effect could also not be excluded, as 60% of the patients had been treated with high-flux dialysis before enrollment, so that at least some patients with a previous long treatment on high-flux dialysis were shifted to low-flux dialysis. Finally, the reuse practice allowed by protocol may have confounded the results of the study, as the performance characteristics of the membranes may be influenced by the reprocessing procedure. For all these reasons, doubts on the general applicability of these results are emerging, and the results of the MPO study are therefore awaited with much interest, particularly as far as the prevention of cardiovascular mortality is concerned, before definitive conclusions as to the clinical benefits of high-flux membranes may be drawn.

Present and Future Lessons from the MPO Study

The MPO study [16] is a still on-going prospective, randomized, clinical trial, which enrolled 744 patients between 1998 and 2003 across nine European countries. The overall duration of the study will be 7.5 years, comprehensive of a 4.5-year recruitment period and a 3- to 7.5-year follow-up, so that the final results of the study will be available by 2006. The MPO study has been specifically designed to evaluate the long-term effects of membrane permeability on multiple clinical outcomes, including mortality, morbidity, vascular access survival and nutritional status. The main differences with the HEMO study deal with the characteristics of the included patients, as only incident patients (defined as patients being on dialysis for no longer than 2 months) have been considered, in order to rule out any possible confounding effect of previous treatment schedules, thus allowing the evaluation of the independent effect of flux on outcome. As such, the MPO study is the first randomized trial on the effect of high-flux dialysis on mortality thus far performed in incident hemodialysis patients. Moreover, no exclusion criteria, apart from dialysis duration and a minimum dialysis dose, have been applied when selecting the patients, in order to avoid the possibility that the exclusion of high-risk patients could interfere with the final results of the study. Finally, no reuse of dialyzers has been allowed, thus excluding this further potentially important confounding variable. For all these reasons, the results of the MPO study are expected to provide basic information on the actual impact of high-flux dialysis on the prognosis of hemodialysis patients.

At present, the analysis of the baseline characteristics of the patients included in the MPO study also allowed to give a further insight to the clinical characteristics of the incident dialysis population. It has been widely shown by several large-scale observational studies and national registries that the cardiovascular conditions of dialysis patients are already severely compromised when they start dialysis, as documented by the significant proportion of incident dialysis patients with clinical manifestations of cardiovascular disease [1, 17]. Analyses, although partial, of the baseline characteristics of the MPO study patient sample confirmed the considerable burden of cardiovascular disease at the time of starting dialysis; in nearly one half of the incident hemodialysis patients enrolled in the study, the prevalence of congestive heart failure, coronary artery disease, cerebrovascular disease and peripheral vascular disease was found to be 3, 10 and 12%, respectively [18]. Interestingly enough, the percentages of patients affected by cardiovascular disease were significantly different between patients enrolled in the North and in the South of Europe [19]. These results are however significantly different from those observed in a sample of more than 1,000 European incident dialysis patients randomly selected from the Euro-DOPPS (Dialysis Outcomes and Practice Patterns Study) population, in whom the prevalence of cardiovascular disease was found to be much higher, whereas no differences were observed as to the prevalence of diabetes and the type of vascular access [18]. These differences are not surprising and find an explanation in the fact that, unlike the MPO study in which, in spite of the largest acceptance criteria, a strict compliance with current practice guidelines was requested by the protocol for the patients to be enrolled, the population included in the Euro-DOPPS more faithfully reflects the hemodialysis population seen in everyday clinical practice. Findings similar to those observed in the MPO study have been obtained in another multicentre, prospective trial performed on incident hemodialysis patients, the Netherlands Cooperative Study on the Adequacy of Dialysis (NECOSAD) [20], further confirming the existence of a 'trial effect', namely, that the population included in a clinical trial is always the result of a positive selection.

Conclusions

In the past years, several observational studies showed that high-flux dialysis was associated with reduced morbidity and mortality of hemodialysis patients; however, the results of prospective randomized controlled trials, above all the HEMO study, have been somehow disappointing. In this context, much is expected from the results of the MPO study, the first large-scale randomized clinical trial investigating the effect of high-flux membranes on patient outcome, including overall and cardiovascular mortality, in a population of incident hemodialysis patients. Inclusion of only incident patients, absence of severe exclusion criteria and no dialyzer reuse are all distinguishing features of the study. The analyses of the baseline data of the MPO study confirm the high burden of cardiovascular disease among incident dialysis patients, although the observation that this is lower than that reported in observational studies, such as the Euro-DOPPS, provides further evidence of a positive selection of patients enrolled in clinical trials.

References

1   US Renal Data System: USRDS 2004 Annual Data Report. Bethesda, National Institute of Diabetes and Digestive and Kidney Diseases, 2004.
2   Rayner HC, Pisoni RL, Bommer J, Canaud B, Hecking E, Locatelli F, Piera L, Brugg-Gresham JL, Feldman HI, Goodkin DA, Gillespie B, Wolf RA, Held PJ, Port FK: Mortality and hospitalization in haemodialysis patients in five European countries: Results from the Dialysis Outcomes and Practice Patterns Study (DOPPS). Nephrol Dial Transplant 2004;19:108–120.
3   Locatelli F, Mastrangelo F, Redaelli B, Ronco C, Marcelli D, La Greca G, Orlandini G: Effects of different membranes and dialysis technologies on patient treatment tolerance and nutritional parameters. The Italian Cooperative Dialysis Study Group. Kidney Int 1996;50:1293–1302.

4   Locatelli F, Andrulli S, Pecchini F, Pedrini L, Agliata S, Lucchi L, Farina M, La Milia V, Grassi C, Borghi M, Redaelli B, Conte F, Ratto G, Cabiddu G, Grossi C, Modenese R: Effect of high-flux dialysis on the anemia of haemodialysis patients. Nephrol Dial Transplant 2000;15:1399–1409.
5   Locatelli F, Marcelli D, Conte F, Limido A, Malberti F, Spotti D: Comparison of mortality in ESRD patients on convective and diffusive extracorporeal treatments. The Registro Lombardo Dialisi e Trapianto. Kidney Int 1999;55:286–293.
6   Koda Y, Nishi S, Miyazaki S, Haginoshita S, Sakurabayashi T, Suzuki M, Sakai S, Yuasa Y, Hirasawa Y, Nishi T: Switch from conventional to high-flux membrane reduces the risk of carpal tunnel syndrome and mortality of hemodialysis patients. Kidney Int 1997;52:1096–1101.
7   Schiffl H, Fischer R, Lang SM, Mangel E: Clinical manifestations of AB-amyloidosis: Effects of biocompatibility and flux. Nephrol Dial Transplant 2000;16:840–845.
8   Lindsay RM, Spanner E: A hypothesis: The protein catabolic rate is dependent upon the type and amount of treatment in dialyzed uremic patients. Am J Kidney Dis 1989;13:382–389.
9   Wizemann V, Lotz C, Techert F, Uthoff S: On-line hemodiafiltration versus low-flux haemodialysis: A prospective randomized study. Nephrol Dial Transplant 2000;15(suppl 1):S43–S48.
10   Hartmann J, Fricke H, Schiffl H: Biocompatible membranes preserve residual renal function in patients undergoing regular hemodialysis. Am J Kidney Dis 1997;30:366–373.
11   Hornberger JC, Chernew M, Petersen J, Garber AM: A multi-variate analysis of mortality and hospital admissions with high-flux dialysis. J Am Soc Nephrol 1992;3:1227–1237.
12   Leypoldt JK, Cheung AK, Carroll CE, Stannard DC, Pereira BJ, Agodoa LY, Port FK: Effect of dialysis membranes and middle molecule removal on chronic hemodialysis patient survival. Am J Kidney Dis 1999;33:349–355.
13   Woods HF, Nandakumar M: Improved outcome for haemodialysis patients treated with high-flux membranes. Nephrol Dial Transplant 2000;15(suppl 1):S36–S42.
14   Port FK, Wolfe RA, Hulbert-Shearon TE, Daugirdas JT, Agodoa LY, Jones C, Orzol SM, Held PJ: Mortality risk by hemodialyzer reuse practice and dialyzer membrane characteristics: Results from the USRDS dialysis morbidity and mortality study. Am J Kidney Dis 2001;37:276–286.
15   Eknoyan G, Beck GJ, Cheung AK, Daugirdas JT, Greene T, Kusek JW, Allon M, Bailey J, Delmez JA, Depner TA, Dwyer JT, Levey AS, Levin NW, Milford E, Ornt DB, Rocco MV, Schulman G, Schwab SJ, Teehan BP, Toto R, Hemodialysis (HEMO) Study Group: Effect of dialysis dose and membrane flux in maintenance hemodialysis. N Engl J Med 2002;347:2010–2019.
16   Locatelli F, Hannedouche T, Jacobson S, La Greca G, Loureiro A, Martin-Malo A, Papadimitriou M, Vanholder R: The effect of membrane permeability on ESRD: Design of a prospective randomised multicentre trial. J Nephrol 1999;12:85–88.
17   Locatelli F, Marcelli D, Conte F, Del Vecchio L, Limido A, Malberti F, Spotti D, Sforzini S for the Registro Lombardo Dialisi e Trapianto: Patient selection affects end-stage renal disease outcome comparisons. Kidney Int 2000;57(suppl 74):S94–S99.
18   Locatelli F, Port FK, Pisoni RL, Martin-Malo A, Papadimitriou M, Vanholder R, Hannedouche T, Jacobson SH, Ronco C, Loureiro A, Czekalski S, Wizemann V, MPO Study Group: Patient and treatment characteristics in the MPO Study: Validation against the DOPPS population. Nephrol Dial Transplant 2003;18(suppl 4):A196.
19   Locatelli F, Hannedouche T, Jacobson SH, La Greca G, Loureiro A, Martin-Malo A, Papadimitriou M, Vanholder R, MPO Study Group: Risk factors for cardiovascular diseases in patients starting dialysis in Northern and Southern Europe: Results from the MPO Study. Nephrol Dial Transplant 2002;17(suppl 1):A11.
20   Jager KJ, Merkus MP, Boeschoten EW, Dekker FW, Stevens P, Krediet RT: Dialysis in the Netherlands: The clinical condition of new patients put into a European perspective. NECOSAD Study Group. Netherlands Cooperative Study on the Adequacy of Dialysis. Nephrol Dial Transplant 1999;14:2438–2444.

Prof. Francesco Locatelli
Department of Nephrology and Dialysis, A. Manzoni Hospital
Via dell'Eremo 9/11, IT–23900 Lecco (Italy)
Tel. +39 0341 489850, Fax +39 0341 489860, E-Mail nefrologia@ospedale.lecco.it

Ronco C, Brendolan A, Levin NW (eds): Cardiovascular Disorders in Hemodialysis.
Contrib Nephrol. Basel, Karger, 2005, vol 149, pp 90–99

A New Initiative in Nephrology: 'Kidney Disease: Improving Global Outcomes'

Norbert Lameire[a], *Garabed Eknoyan*[b], *Rashad Barsoum*[c],
Kai-Uwe Eckardt[d], *Adeera Levin*[e], *Nathan Levin*[f],
Francesco Locatelli[g], *Alison MacLeod*[h], *Raymond Vanholder*[a],
Rowan Walker[i], *Haiyan Wang*[k]

[a]University Hospital of Ghent, Ghent, Belgium; [b]Baylor College of Medicine,
Houston, Texas; [c]Cairo Kidney Center, Cairo University, Cairo, Egypt;
[d]University of Erlangen-Nuremberg, Erlangen, Germany; [e]University of British
Columbia, Vancouver, British Columbia, Canada; [f]Renal Research Institute,
New York, New York; [g]A. Manzoni Hospital, Lecco, Italy; [h]University of Aberdeen
Medical School, Aberdeen, United Kingdom; [i]Royal Melbourne Hospital,
Melbourne, Australia; and [k]Peking University, The First Hospital, Beijing,
Peoples Republic of China

Abstract

The burden of kidney disease: Improving global outcomes.

Chronic kidney disease (CKD) is a worldwide public health problem with an increasing incidence and prevalence of patients requiring replacement therapy. There is an even higher prevalence of patients in earlier stages of CKD, with adverse outcomes such as kidney failure, cardiovascular disease, and premature death. Patients at earlier stages of CKD can be detected through laboratory testing and their treatment is effective in slowing the progression to kidney failure and reducing cardiovascular events.

The evidence-based care of these patients are universal and independent of their geographic location. This paper describes the need to develop a uniform and global public health approach to the worldwide epidermic of CKD. It is to this end that a new initiative 'Kidney Disease: Improving Global Outcomes' has been established.

Some current and future activities of this initiative are described. They include among others modification of the classification of CKD, the development of guidelines on hepatitis C, the organisation of consensus conferences like on Renal Osteodystrophy, and the creation of a website allowing the comparison of the five main English language clinical practice guidelines in kidney disease worldwide.

Chronic kidney disease (CKD) is a worldwide medical and public health problem that is assuming epidemic proportions. Appreciation of the problem began when data accrued by the US Renal Data System (USRDS) revealed an increasing incidence of terminal kidney failure. Based on the international comparisons made in the most recent USRDS registry [1], the number of individuals suffering from end-stage renal disease (ESRD) in need of renal replacement therapy continues to increase worldwide, with the highest incident rates persisting in Taiwan, the United States, and Japan (254–365 per million population; pmp). Qatar and the Basque region have the highest pediatric incident rates, at 33.7 and 22.7 pmp, and for patients aged 20 and older rates are highest in Qatar, the United States, and Taiwan. In Europe, the highest rate is in Germany (175 pmp in 2000). In Eastern Europe, where renal replacement therapy programs are still developing, the reported rates are lower than in Western Europe [2].

Overall prevalence of renal replacement therapy worldwide was 240 pmp in 2001, whilst it was highest in North America (1,400 pmp), Japan (1,830 pmp), and Europe (490 pmp), followed by Latin America (310 pmp) and the Middle East (150 pmp) [3]. By contrast, the lowest rates of 22.3–79.3 are reported from the Phillipines, Bangladesh, and Russia. The dissimilarities in these rates likely reflect economic limitations available to support ESRD programs.

Factors contributing to the increase particularly in the developed world include: greater referral/acceptance, increased incidence of ESRD due to demographic changes (i.e. longer life expectancy) and, due to changes in risk for CKD due to the increased incidence of diseases such as diabetes and hypertension, and a reduction in competing risks such as cardiovascular mortality [4].

Some of the increase in acceptance rates can be ascribed to greater awareness of ESRD, a lower threshold for referral or acceptance of older and sicker patients with ESRD to nephrologists. This has partly been due to technical advances in dialysis therapy allowing safe long-term care of such patients.

In the developed world, as the overall incidence of ESRD has increased, the proportional contribution to the total of each individual cause has changed. Thus, whilst the incidence of glomerulonephritis may not have fallen, the percentage of patients with this, and many other primary kidney diseases, is falling, as other causes, particularly systemic diseases such as diabetes, become more common. For example, in Canada glomerulonephritis fell from 20 to under 15% between 1992 and 2000, and there were also small falls in pyelonephritis and polycystic kidney disease. In contrast diabetic ESRD rose from 24 to 32% and there were smaller percentage rises in renovascular and hypertensive kidney disease [5]. The trend in European countries is the same. As these proportional changes are in the context of rising acceptance rates, in absolute terms they are even greater. Patterns vary by age; in particular, renal disease of uncertain cause and renovascular disease are more common in the elderly. Thus, as more and more old patients with kidney failure are accepted for treatment, the changing patterns reflect the different age distributions of causes of ESRD.

What has been distressing is the poor survival of this growing segment of the population. While heart and cerebrovascular diseases and cancer are the major causes of mortality resulting from chronic diseases, diseases of the kidney have now also assumed epidemic proportions as well, and are among the leading causes of death in the industrialized world. This is in part due to the number of patients with CKD who progress to kidney failure requiring dialysis or transplantation, whose prognosis is comparable to those with metastatic cancer. In addition, and at least as important, it is now evident that in those whose kidney disease may never progress to dialysis dependence the presence of kidney injury, often signaled by proteinuria, and evidence of decreased kidney function, detected by estimating the glomerular filtration rate (GFR), are associated with increased risk of death from heart and cerebrovascular disease [6, 7]. As a result, the presence of kidney disease is now listed as an independent risk factor for cardiovascular disease in the most recent report of the Joint National Committee on Prevention, Detection and Treatment of High Blood pressure (JNC VII) and in a position statement of the American Heart Association [8, 9].

In contrast to the dramatic decline in mortality from heart and cerebrovascular disease [10], the mortality of kidney disease remains unacceptably high and appears to be increasing, at least in the United States. While similar worldwide detailed figures are lacking, wherever investigated or projected the data are supportive of an increasing prevalence of deaths attributed to kidney disease well before ESRD sets in [3, 8, 11–15]. Strikingly, cardiovascular mortality remains elevated in ESRD patients and is about 10 to 20 times higher than those in the general population, and accounts for more than 50% of deaths in the first year of dialysis [16, 17].

Several factors account for the increasing numbers and poor outcomes of ESRD; the three most notable are age, underlying comorbid conditions, and state of health at the initiation of dialysis. More than half the patients who begin treatment for ESRD are older than 65 years.

This reflects the ageing population in the industrialized world and the fact that most kidney diseases are insidious in onset, progress at variable rates, and take several years to reach kidney failure, when replacement therapy with dialysis or transplantation becomes necessary. Although age is a nonmodifiable risk factor, the years it takes to reach kidney failure provides for opportunities to retard the progression of kidney disease, to treat underlying comorbidities, and to prevent the systemic complications that develop in the course of gradual loss of kidney function [18, 19]. Up to this time, these opportunities have been missed.

Patients who begin dialysis currently have on average four comorbid conditions, with a preponderance of atherosclerotic cardiovascular disease and diabetes mellitus [20].

In patients with CKD, the prevalence of left ventricular hypertrophy shows an inverse relationship with the level of kidney function, affecting 30.8% of those patients with a GFR of 25–50 ml/min and 45.2% of those with a GFR of less than 25 ml/min [21]. On the basis of these data, it has been estimated that in the course of CKD, the risk for left ventricular hypertrophy increases by 3% for each decline of 5 ml/min in GFR.

The connecting link in this association is hypertension, which contributes to the progression of kidney disease and is a major risk factor for cardiovascular disease.

The third principal determinant of the high mortality rate in patients with ESRD is the state of their health at the initiation of dialysis. Much of the morbidity and mortality on dialysis is due to complications that develop during the course of progressive loss of kidney function.

Several studies have shown a relationship between the severity of these complications and the outcomes on dialysis.

Apart from the most important of these cardiovascular complications disease, there also is an inverse association between the level of kidney function and the severity of anemia, malnutrition, bone disease, and neuropathy. In diabetes, the combination of early anemia [22, 23], cardiovascular disease and neuropathy accounts for the relatively very high comorbidity already present in the early stages of the kidney disease of these patients. Each of these conditions affect their outcome and quality of life, not only after the initiation of dialysis but also during the course of progressive kidney disease. In addition, the progression of kidney disease to kidney failure is usually associated with escalating psychosocial stress, which results from the increasing burden of the complications of kidney disease, the complexity of the required treatments, the likelihood of limitations in functioning and well-being, and the expected shortened life span. Failure to attend appropriately to these complications accounts for the increased number of comorbidities with which ESRD patients begin dialysis [20]. Hence, the clear need to call to action all health professionals in order to address the worldwide epidemic of kidney disease.

The best recent approach to resolve problems of inadequate CKD and dialysis care has been the development of guidelines. It has now been shown that rigorously developed evidence-based clinical practice guidelines (CPGs), when implemented, can reduce variability of care, improve patient outcomes and ameliorate deficiencies in health care delivery [24]. The practical specificity of well-defined guideline statements, which facilitates their translation into clinical practice, differentiates CPGs from other important evidence-based approaches (meta-analyses and systematic reviews), which distill and analyze the evidence in the literature but do not make necessary practical recommendations for clinical practice. There is no evidence that the passive dissemination of these publications, even when linked to consensus-derived recommendations, results in changes in clinical practice [25]. On the other hand, the actionable recommendations of CPGs, when implemented, make them the best tool now available to close the gap between actual practice and evidence-based best practice [24]. Indeed, the implementation of rigorously developed guidelines has been said to 'lead to even greater improvements in patient care than the introduction of some new technologies' [25].

The National Kidney Foundation launched the Dialysis Outcomes Quality Initiative (DOQI) in 1995. The initial set of CPGs developed by DOQI on hemodialysis adequacy, peritoneal dialysis adequacy, vascular access, and anemia were issued in 1997 and have been updated in 2000. Their favorable impact on the quality of care delivered to dialysis patients has been documented [26].

In the process of developing DOQI guidelines, it became evident that in order to actually improve outcomes it was essential to improve the health status of patients who are initiated on dialysis, and that therein existed an even greater opportunity to improve outcomes for all individuals with CKD. To reflect this more ambitious phase, in which guideline development would encompass the entire spectrum of CKD, reference to 'dialysis' in DOQI was changed to 'disease' and a new initiative termed Kidney Disease Outcomes Quality Initiative (K/DOQI) was launched in 1999.

In Europe, the European Best Practice Guidelines on anemia have been first published in 1999 [27] and revised in 2004 [28], followed by guidelines on hemodialysis [29], and renal transplantation [30, 31]. Guidelines have also been developed in other countries. Notable among those are the guidelines developed by the Canadian Society of Nephrology [32], the United Kingdom Renal Association [33], and the Australian and New Zealand Society of Nephrology [34]. Others translated DOQI guidelines, which are now available in over 12 languages, and adopted selected components of them for local implementation.

The best guideline implementation strategy is the linkage of guideline recommended targets to actual data gathered on site and its feedback to providers, particularly when this is linked to continuous quality improvement programs and their evaluation through clinical performance measures. It is the first component of this goal, data gathered on site, which prompted the launch of another initiative, the Dialysis Outcomes and Practice Patterns Study (DOPPS) in 1996. During the mid 1990s, when DOQI was initiated, DOPPS was conceived with the goal of increasing the longevity of hemodialysis patients by evaluating practice patterns in dialysis facilities. The DOPPS investigates a large number of practice patterns to detect new evidence for modifiable practice patterns that are associated with improved outcomes.

A recent analysis by DOPPS calculated the percentage of patients outside published DOQI hemodialysis guidelines and their associated mortality risk. The number of life years that could be gained from adherence to four of these guidelines and two other modifiable practices, i.e. dialysis dose, phosphate control, improved anemia, partial correction of serum albumin, reduced interdialytic weight gain and less use of catheters for vascular access were estimated [35]. DOPPS data on these practices and guidelines were extrapolated to the US hemodialysis population for a 5-year projected period. The adjusted sum of the patient life years gained attributable to all six practice patterns was 143,617; a more conservative estimate, modeling life years potentially gained by bringing half of all patients outside targets within them, is 69,367. The magnitude of potential savings in life years is thus impressive and should encourage greater adherence to guidelines and practices that are associated with significantly better survival [35].

An important next step in the approach to CKD occurred when it became apparent to those developing guidelines that there was a need for a more uniform and global approach to the process. The rationale for a global initiative is simple and self-evident. Essentially, there is a recognized and increasing prevalence of kidney disease worldwide. As such, there is a clear need to develop a public health approach to the global epidemic of kidney disease, coupled with strategic initiatives that can improve the care of patients with kidney disease worldwide. The complications and problems encountered by those afflicted with kidney disease are universal. The science and evidence-based care of these complications and problems are also universal and thus independent of geographic location or national borders. It is important to increase the efficiency of utilizing available expertise and resources in improving global outcomes of kidney disease. There is definite room for improving international cooperation in the development, dissemination, and implementation of CPG to achieve these goals [36].

It was on this basis that an initial exploratory and consultative meeting of a group of individuals active in the field was convened on July 14, 2002, in Copenhagen, Denmark during the Annual ERA/EDTA meeting. At the meeting, enthusiastic support for undertaking a global initiative was expressed and the decision made that a Global Coordinating Board would be formed to explore the issue further. The history of this global organization has recently been described [36]. The name under which the initiative is now incorporated is 'Kidney Disease: Improving Global Outcomes' (KDIGO). Its mission statement is to: 'Improve the care and outcome of kidney disease patients worldwide through promoting coordination, collaboration and integration of initiatives to develop and implement CPG.' At several meetings of the KDIGO Executive Committee and Board of Directors action plans were developed and should be realized over the next 2–3 years. These plans include (1) the development of a common methodologic approach, including a uniform rating system of the strength of the evidence and the recommendations of guidelines; (2) the adoption of a common evaluation and classification of CKD to facilitate a unified nomenclature worldwide; (3) the establishment of an electronic interactive web-based clearinghouse of currently available CPGs, including implementation tools and performance measures to provide direct comparison of recommended targets in different guidelines together with the rationale for their differences; (4) the collection and evaluation of current implementation tools to use them in facilitating implementation strategies at national or regional levels; (5) the development of educational plans and a structure for countries without guidelines to help the adoption of selected guidelines most suitable for their regional needs; (6) the evaluation of current and planned guidelines in different countries with the intent of integrating the next phase of new or updated guidelines, on a voluntary basis, into a common process and their release as global guidelines; (7) the use of controversy conferences to reconcile existing guidelines, establish what is known, decide what can be done with what is known, and determine what needs to be known; (8) integration of the future updating and development of guidelines by KDIGO; (9) the initiation of the development of one set of new CPG; and (10) the presentation and dissemination of results achieved at various national and international meetings.

The KDIGO website http://www.kdigo.org is a new major portal which allows one to compare and contrast the five main English language CPG in kidney disease worldwide: the Kidney Disease Outcomes Quality Initiative (K/DOQI), the Caring for Australians with Renal Impairment (CARI), the European Best Practice Guidelines (EBPG); the United Kingdom Renal Association guidelines; and the Canadian Society of Nephrology guidelines. In the present version of the site guidelines for target hemoglobin, mineral metabolism targets, target blood pressure, and Kt/V can be compared. In the future, more comparative tables will be added as well as links to DOPPS, USRDS and other data bases. Gradually additional evidence-based guidelines from other parts of the world will be listed and compared.

Among the different working groups that have been composed, the Evidence Rating Group under the guidance of Alison McLeod and Katrin Uhlig provides advise on grading the quality of evidence and the strength of recommendations for guidelines to be developed by KDIGO. The evidence rating group first reviewed evidence grading systems in use by established entities that have issued guidelines in the domain of kidney disease. It then outlined steps to be followed in the process of developing a CPG, since the application of a transparent grading system is contingent on a clearly specified process which ensures that relevant questions are asked. A draft report of the evidence rating group has been formulated; this report will soon be submitted for final publication.

Another important activity was the organization of a first International Controversies Conference on 'Definition, Evaluation and Classification of Chronic Kidney Disease in Adults' in Amsterdam on November 16–17, 2004. The topics covered included the defintion and classification of CKD, estimation of GFR, and measurement of proteinuria. This conference involved more than 50 experts and clinicans and was co-chaired by Drs. Andrew Levey and Kai-Uwe Eckardt. The recommendations made during this conference will be published in the very near future.

Another controversies conference on the 'Definition, Evaluation and Classification of Renal Osteodystrophy' will be held in the fall of 2005.

Another workgroup reviewed the available literature on the most effective implementation strategies of guideline initiatives. Guidelines do not implement themselves and must finally lead to a change in attitude and behavior of the clinician. Substantial evidence suggests that a change of behavior is possible, but this change generally requires comprehensive approaches at different levels (doctor, team practice, hospital, wider environment), tailored to specific settings and target groups. Plans for change should be based on characteristics of the evidence or guideline itself and barriers and facilitators to change. How these can be applied to nephrology will be the subject of the report of this workgroup.

The ultimate success of all these initiatives will depend not only on the scientific rigor with which they are developed, but also on their reception by the worldwide nephrology community. To this end it will need the goodwill and support of local organizations and thought leaders. Ultimately, the goal of KDIGO is to ensure the best outcomes possible for all individuals with kidney disease. Through the support and collaboration of all concerned organizations and individuals, the recognition that there are common and universal principles of physiology, science, and clinical practice should help improve outcomes of kidney disease worldwide. It is certainly possible to make the future better than the present.

References

1   USRDS Annual Report. 2004.
2   Rutkowski B: Changing pattern of end-stage renal disease in central and eastern Europe. Nephrol Dial Transplant 2000;15:156–160.
3   Moeller S, Gioberge S, Brown G: ESRD patients in 2001: Global overview of patients, treatment modalities and development trends. Nephrol Dial Transplant 2002;17:2071–2076.
4   Van Dijk PC, Jager KJ, de Charro F, Collart F, Cornet R, Dekker FW, Gronhagen-Riska C, Kramar R, Leivestad T, Simpson K, Briggs JD: Renal replacement therapy in Europe: The results of

a collaborative effort by the ERA-EDTA registry and six national or regional registries. Nephrol Dial Transplant 2001;16:1120–1129.

5   Canadian Organ Replacement Register 2000 Report. 2000.

6   Collins AJ, Li S, Gilbertson DT, Liu J, Chen SC, Herzog CA: Chronic kidney disease and cardiovascular disease in the Medicare population. Kidney Int Suppl 2003;87:S24–S31.

7   Levin A: Cardiac disease in chronic kidney disease: Current understandings and opportunities for change. Blood Purif 2004;22:21–27.

8   Chobanian AV, Bakris GL, Black HR, Cushman WC, Green LA, Izzo JL Jr, Jones DW, Materson BJ, Oparil S, Wright JT Jr, Roccella EJ: The Seventh Report of the Joint National Committee on Prevention, Detection, Evaluation, and Treatment of High Blood Pressure: The JNC 7 report. JAMA 2003;289:2560–2572.

9   Sarnak MJ, Levey AS, Schoolwerth AC, Coresh J, Culleton B, Hamm LL, McCullough PA, Kasiske BL, Kelepouris E, Klag MJ, Parfrey P, Pfeffer M, Raij L, Spinosa DJ, Wilson PW: Kidney disease as a risk factor for development of cardiovascular disease: A statement from the American Heart Association Councils on Kidney in Cardiovascular Disease, High Blood Pressure Research, Clinical Cardiology, and Epidemiology and Prevention. Hypertension 2003;42:1050–1065.

10 National Vital Statistics Report. Vol 50,15. 2002.

11 Bommer J: Prevalence and socio-economic aspects of chronic kidney disease. Nephrol Dial Transplant 2002;17(suppl 11):8–12.

12 Campbell RC, Ruggenenti P, Remuzzi G: Halting the progression of chronic nephropathy. J Am Soc Nephrol 2002;13(suppl 3):S190–S195.

13 Cass A: Kidney disease: Are you at risk? Med J Aust 2002;176:515–516.

14 Lysaght MJ: Maintenance dialysis population dynamics: Current trends and long-term implications. J Am Soc Nephrol 2002;13(suppl 1):S37–S40.

15 Wang JG, Staessen JA, Fagard RH, Birkenhager WH, Gong L, Liu L: Prognostic significance of serum creatinine and uric acid in older Chinese patients with isolated systolic hypertension. Hypertension 2001;37:1069–1074.

16 Cheung AK, Sarnak MJ, Yan G, Dwyer JT, Heyka RJ, Rocco MV, Teehan BP, Levey AS: Atherosclerotic cardiovascular disease risks in chronic hemodialysis patients. Kidney Int 2000;58:353–362.

17 Foley RN, Parfrey PS, Sarnak MJ: Clinical epidemiology of cardiovascular disease in chronic renal disease. Am J Kidney Dis 1998;32:S112–S119.

18 Levey AS, Beto JA, Coronado BE, Eknoyan G, Foley RN, Kasiske BL, Klag MJ, Mailloux LU, Manske CL, Meyer KB, Parfrey PS, Pfeffer MA, Wenger NK, Wilson PW, Wright JT Jr: Controlling the epidemic of cardiovascular disease in chronic renal disease: What do we know? What do we need to learn? Where do we go from here? National Kidney Foundation Task Force on Cardiovascular Disease. Am J Kidney Dis 1998;32:853–906.

19 Sarnak MJ, Levey AS: Cardiovascular disease and chronic renal disease: A new paradigm. Am J Kidney Dis 2000;35:S117–S131.

20 Schmitz PG: Progressive renal insufficiency. Office strategies to prevent or slow progression of kidney disease. Postgrad Med 2000;108:145–148, 151–154.

21 Levin A, Singer J, Thompson CR, Ross H, Lewis M: Prevalent left ventricular hypertrophy in the predialysis population: Identifying opportunities for intervention. Am J Kidney Dis 1996;27:347–354.

22 Lameire N: The anaemia of silent diabetic nephropathy-prevalence, physiopathology, and management. Acta Clin Belg 2003;58:159–168.

23 Lameire N: Diabetes and diabetic nephropathy – A worldwide problem. Acta Diabetol 2004;41(suppl 1):S3–S5.

24 Steinberg EP: Improving the quality of care – Can we practice what we preach? N Engl J Med 2003;348:2681–2683.

25 Weingarten S: Using practice guideline compendiums to provide better preventive care. Ann Intern Med 1999;130:454–458.

26 Collins AJ, Roberts TL, St Peter WL, Chen SC, Ebben J, Constantini E: United States Renal Data System assessment of the impact of the National Kidney Foundation – Dialysis Outcomes Quality Initiative guidelines. Am J Kidney Dis 2002;39:784–795.

27   Cameron JS: European best practice guidelines for the management of anaemia in patients with chronic renal failure. Nephrol Dial Transplant 1999;14(suppl 2):61–65.
28   Revised European Best Practice Guidelines for the Management of Anaemia in Patients with Chronic Renal Failure. Nephrol Dial Transplant 2004;19:1–47.
29   Kessler M and expert group on haemodialysis. European Best Practice Guidelines on Haemodialysis. Nephrol Dial Transplant 2002;17(suppl 7): 1–111.
30   Berthoux F and the EBPG expert group on renal transplantation. European Best Practice Guidelines for Renal Transplantation (part I). Nephrol Dial Transplant 2000;15(suppl 7):1–85.
31   Berthoux F and the EBPG expert group on renal transplantation. European Best Practice Guidelines on renal transplantation (part II). Nephrol Dial Transplant 2002;17:1–67.
32   Canadian Clinical Practice Guidelines. (retrieved January 2005).
33   The Renal Association Standards. (retrieved January 2005).
34   CARI Guidelines. 2002 (retrieved January 2005)
35   Port FK, Pisoni RL, Bragg-Gresham JL, Satayathum SS, Young EW, Wolfe RA, Held PJ: DOPPS estimates of patient life years attributable to modifiable hemodialysis practices in the United States. Blood Purif 2004;22:175–180.
36   Eknoyan G, Lameire N, Barsoum R, Eckardt KU, Levin A, Levin N, Locatelli F, MacLeod A, Vanholder R, Walker R, Wang H: The burden of kidney disease: Improving global outcomes. Kidney Int 2004;66:1310–1314.

Prof. Dr. N. Lameire
Renal Division
University Hospital Ghent
De Pintelaan 185
B-9000 Ghent (Belgium)
Tel. +32 9 240 4524, Fax +32 9 240 4509, E-Mail norbert.lameire@ugent.be

Ronco C, Brendolan A, Levin NW (eds): Cardiovascular Disorders in Hemodialysis.
Contrib Nephrol. Basel, Karger, 2005, vol 149, pp 100–106

....

Is There a Magic in Long Nocturnal Dialysis?

Bernard Charra

Centre de Rein Artificiel de Tassin, Tassin, France

Abstract

Long 3×8 h/week hemodialysis (HD) has been used without modification in Tassin since 35 years with very satisfactory morbidity and mortality results. It can be performed in the day or overnight. The observed good outcome is mainly due to lower cardiovascular morbidity and mortality than usually reported in HD. This, in turn, is due to the good control of blood pressure (BP) and of serum phosphate level. The control of BP results from the strict extracellular volume normalization using an adequate ultrafiltration and a low salt diet. High doses of small and middle molecules lead to a satisfactory nutrition, correction of anemia, control of serum phosphate and potassium with minimal needs for medications. The treatment is cost-effective. It provides an optimal dialysis i.e. it corrects as perfectly as possible each abnormality of renal failure. Overnight dialysis is the most logical way of delivering long HD with the lowest possible hindrance on patient's life. Due to the change in case mix a decreasing number of patients are apt or willing to go on overnight dialysis; education to autonomy is more difficult, but the benefits are still there.

Introduction

Long 3×8–12 h/week hemodialysis (HD) on 1-m^2 flat plate cuprophan dialysers was in the late 70's the empirically most superior form of dialysis, the 'gold standard' [1]. Technical advances, changing scientific views on uremia pathophysiology and social and economical pressure (better use of the scarce dialysis stations) have led to the apparition and development of shorter HD sessions. In Tassin the 3×8 h/week dialysis has remained the unique treatment method for almost all patients whether in the unit or at home for three decades. The follow-up of long, slow dialysis representing over 6,500 patient-years of experience is summarized here.

Patients and Methods

Three sessions of 7–8 h/week HD are performed overnight during sleep or in the daytime according to the patient's convenience and preference. Home dialysis was used in a large proportion of the population up to the 80s; then 50% of the patients were treated at home. The multiplication of dialysis units, increased transplantation rate and worsening patients' case mix have led to a steady decrease in the proportion of patients treated at home. Today only 5% of Tassin patients continue to be treated at home.

Until 1995 the technical setting was very poorly 'biocompatible' and included cuprophan® membrane, acetate buffer and plain softened water. A moderate blood flow (220–250 ml/min) had been used throughout the experience. Since 1996 bicarbonate has substituted acetate. In 1998 low-flux polysulfone dialyzers replaced cellulosic ones. Finally in January 2001 a reverse osmosis treatment, completed by an ultrafiltration system, was set up. Since 1992 a 'short' HD program has been proposed as a complement to long HD. It uses a 3 × 5 or 6 h/week schedule with large-area-size dialyzers (1.7–2.5 m), and a 300-ml/min blood flow. This shortened schedule is presently used by 35% of Tassin patients.

Whatever the dialysis session duration, the dose provided is large. The mean delivered urea Kt/V (2nd generation Daugirdas method) is 2.0 per session [2]. The mean normalized PCR is over 1.2. The mean protein and calorie intakes are 1.2 g/kg and 32 kcal/kg/day, respectively. The patients are requested to restrict salt. No salt is added to the food and processed food is avoided. The average Tassin patients' sodium chloride intake is 5 g per day. The mean interdialytic weight gain is 1.8 kg (2.5% of dry weight). The dialysate sodium is set at 138 mmol/l. The patients are not requested to restrain from drinking. All antihypertensive medications (antiHT) are stopped in all patients within the first 2 months of HD. It is a crucial point that during the initial few weeks of dialysis each patient undergoes a systematic antiHT treatment withdrawal in conjunction with the lowering of his/her extracellular volume (ECV) to achieve 'dry weight' and normotension [3].

As in many units worldwide Tassin incident population has changed drastically with the calendar years. The mean age at start increased regularly from 36.1 years in 1968 to 64.7 years in 2004. During the same period diabetes and nephrosclerosis prevalence in the incident population crept up from 5 to 53% and the proportion of patients with cardiovascular (CV) comorbidity (myocardial infarction, angina, cerebrovascular accident, transient ischemic attack and peripheral vascular disease) increased from 6 to 61% of incident patients. As a consequence, the number of patients able to dialyze overnight with little or no nurse supervision has significantly decreased.

Results

Mortality

Due to the increase in risk factors, the crude mortality has steadily increased along the calendar years. The mean half-life of the cohort of patients starting HD between 1968 and 1975 (mean age 39 years, low comorbidity) was 18 years, the most recent cohort (mean age 63 years, high comorbidity) was only 5.6 years [2]. To get a realistic view of the mortality linked to the HD treatment method per se, one must take into account the changing demographic and comorbid patterns of the population. To achieve this, the patients' risk level is stratified according to identifiable individual risks. Standardized Mortality Ratio (SMR) adjusts for age, race, sex and cause of renal failure using the United States Renal Data System (USRDS) standard mortality table as the reference [4]. The average observed mortality in Tassin is 45% of the expected value according to US standards for similar risk patients. It has remained quite stable around this value over the last 15 calendar years in spite of the worsening case mix.

Comparison of Tassin mortality to the only available long-term French series of 4- to 5-h HD reported by Degoulet et al. [5] shows that long HD mortality was lower (52.4 vs. 99 deaths per 1,000 patient-years, $p < 0.001$). There was no difference in specific (infection, cancer, or others) causes of mortality between the two series. But CV mortality was much lower on long HD (19.8 vs. 44.6 CV deaths/1,000 patient-years, $p < 0.001$).

We analyzed the respective survival of two Tassin long HD subgroups of patients according to their predialysis integrated mean arterial pressure (MAP). The subgroup of patients with the lowest MAP (n = 382 patients; mean predialysis MAP = 89 mm Hg) had a significantly lower mortality ($p = 0.003$) than the subgroup with a slightly more elevated MAP (n = 383 patients; mean predialysis MAP = 107 mm Hg). The difference in survival was essentially due to a lower CV mortality in the lower MAP subgroup: 12.7 versus 28.1 CV deaths per 1,000 patient-years ($p < 0.01$).

The Cox proportional hazard model analyzing the same patients' survival shows that age, cause of renal failure and CV history are the most powerful predictors of mortality. These factors are not amenable to medical action as treatment-related factors. Among treatment-related factors, urea Kt/V is not a significant predictor of survival. On the other hand, the time-dependent middle molecule index calculated using Babb's method [6] is quite significantly correlated to survival (the higher the middle molecule removal rate, the longer the survival). In our experience the strongest predictors of mortality are serum albumin and predialysis MAP.

Morbidity

An essential feature of long HD is that it regularly achieves a good control of blood pressure (BP). The mean casual predialysis BP (128/79 mm Hg) is normal. Besides, ambulatory BP monitoring values are also within normal range at least for daytime (121/72 mm Hg) and circadian values (119/71 mm Hg). However the nighttime values (118/67 mm Hg) are slightly more elevated than normal (106/64 mm Hg) due to the lack of nocturnal dip in 50% of the patients. Intradialytic hypotensive episodes are scarcer on longer than on shorter dialysis: 47 events per 1,000 sessions on 8-hour dialysis versus 129 events on 5-hour dialysis (p < 0.005) in our own unit. The relationship between ECV and BP is illustrated by the initial months of long HD treatment. The initial sharp drop of ECV as expressed by the weight change contrasts with the more progressive pre-dialysis MAP decrease over months. This lag time between changes in ECV and BP [7] is a very important feature. It explains why the sustained decrease of the postdialysis weight is not immediately followed by BP normalization. After 2 months of dialysis BP continues to decrease but weight typically increases. This gain in weight does not reflect an increase in ECV but a gain in lean and fat body mass due to the improved appetite and anabolism occurring at the start of maintenance dialysis.

Tassin long HD population's hematocrit is 36% with erythropoietin being used for 65% of the patients. The average serum albumin (measured between two sessions) is 40.6 g/l. It remains stable after 20 years of dialysis. Serum PO_4 control is well controlled (mean overall population predialysis serum PO_4 is 1.34 ± 0.33 mmol/l). Only 30% of the long HD patients need to use a PO_4 binder.

Effects of Switching the Same Group of Patients from Short to Long HD and Conversely

One hundred and twenty-four patients were transiently dialyzed in Tassin while waiting for kidney transplantation in Lyon. All had been treated for 6 months or more on a 5-hour (or less) HD schedule in another unit. Half of them received an antiHT. After 3 months of 8-hour HD, their average postdialysis weight was reduced by 0.5 kg, their predialysis MAP was back to almost normal (mean = 101 mm Hg) and antiHT were stopped in all but one patient. Thereafter the predialysis MAP continued to decrease gently but, due to anabolism, the patients' weight increased progressively to plateau after one year [8]. At the same time the mean predialysis hematocrit level increased from 24 to 29% and the predialysis urea increased by 10% and creatinine by 25%.

Conversely, 49 Tassin 8-hour HD patients were switched to a 5-hour schedule. All had been dialyzed 3×8 h since at least 6 months. All were normotensive without antiHT. The dialyzer area and blood flow were increased to maintain an unchanged urea Kt/V after the patients were switched to the shorter schedule. After one year the delivered Kt/V per session had almost not changed (1.86 to 1.77) but the predialysis MAP had risen significantly by 10 mm Hg in spite of a mean 2.5 kg postdialysis weight reduction and the introduction of antiHT in 4 patients [8]. On the other hand, the mean hematocrit decreased from 31.5 to 27.5% and predialysis urea and creatinine serum levels decreased by 8 and 19%, respectively. Shortening HD time without decreasing the dose was therefore associated with an impaired BP and nutrition.

Discussion

Is Survival Better on Long HD?

The Tassin patients' case-mix has progressively worsened as in all developed countries' dialysis programs. In 2003 the mean age at the start of HD was 64 years, 58% of patients had diabetes or renal vascular disease and a significant CV story was found in 60% of incident patients. The slightly favorable patients' selection bias in Tassin does not account for the large mortality discrepancy with US results. One cannot exclude some 'center effect' in the achieved results. But other units, where a long dialysis is in use [9, 10], achieve the same high survival and low morbidity, as well as the same BP control without need for antiHT. Long HD allows for excellent ECV and BP control, satisfactory control of nutrition, anemia and phosphatemia.

What Is the Effect of a Large Kt/V?

Is survival better on long dialysis because the delivered urea Kt/V is higher than in the usual short HD? Longitudinal data – in well-controlled groups of patients – show that when the delivered spKt/V is deliberately increased mortality decreases [11, 12].

On the other hand, Lowrie et al. [13] recently pointed out that Kt is a better outcome-based measure of HD dose than Kt/V. Besides, the clearance term K depends on blood flow, recirculation and urea rebound. These factors limit the dialysis system performance and efficiency. Due to this, in usual operational conditions of dialysis it is almost impossible to substantially increase the urea Kt/V (over 1.8 or so) without increasing the session time t.

Other Time-Related Issues

The effect of increasing HD time is not limited to increasing the epuration of small molecules. It all the more proportionately increases the clearance of larger solutes such as 'middle molecules' (MM) and solutes which, due to their compartmental distribution, behave as MM (e.g. PO_4). The clearance of these solutes depends more on time (duration or frequency of the sessions) than on blood or dialysate flow. In uremia several MM have been identified that strongly influence vascular calcifications, nutrition, immunology and infection, BP and vascular disease [14]. A longer session time reduces the HD 'unphysiology' [15] due to the acute fluctuations in the volume and composition of fluid compartments. A longer dialysis is a slower, gentler dialysis with less intradialytic morbidity. This can be performed overnight during the sleeping hours, sparing time for active life.

The good ECV and BP control achieved are the most clinically relevant features of long HD. The development of shortened dialysis has led to an increased incidence of hypertension and CV morbidity. Shortened session time and higher ultrafiltration rates lead to a vicious circle [16] amplifying the BP variations and driving to both intradialytic hypotension and interdialytic hypertension. Conversely, a long session time allows for a better ECV and BP control without antiHT. The experience modifying the dialysis schedule in the same groups of patients reported here confirms the key role played by the session time.

The best way to deliver long HD without excessive hindrance on patients' everyday life is to perform it overnight. The change in patients case mix in the previous decades has led to a decreasing interest for home and overnight dialysis. One may question if nocturnal dialysis can fit this 'new' patient population. In fact this is possible if one takes into account the new patient's profile. The autonomy of the patient is not a pre-requisite anymore. In the center, and to a lesser degree in the self-dialysis, the patient does not need to take care of his blood access puncture or of his dialysis at all. But only clinically stable dialysis patients should be admitted into a 'community' overnight program so to keep it quiet and silent enough to allow the patients to sleeping. In 2005 our overnight patients' profile differs whether they dialyze at home (young age but old HD vintage) or in center (old age and usually rather recent HD vintage). They all receive the same treatment, but while all home or self-care patients self-stick, only 20% do so in the center. Clinical results observed in overnight dialysis do not differ from those in daytime dialysis.

The artificial kidney function cannot be any more reduced to urea Kt/V than the native kidney function can be restricted to urea clearance rate. An optimal dialysis needs several conditions, not just one. Any of these conditions, if lacking, suffices to wreck the whole maintenance dialysis ship. Time can not only provide a good dose of dialysis, in terms of small and even more MM, but can also provide a satisfactory nutrition, ECV and BP control. This last point is essential, as CV morbidity is by far the first cause of death in dialysis patients. So altogether, time appears to be a key factor for achieving an optimal dialysis treatment.

There is no magic in long dialysis. It is the sum of an adequate dialysis, high dose of small and MM, good ECV and BP control and adequate nutrition without hyperphosphatemia. But the respective effects of low-salt diet, low use of medications, high doctor's dose, unsophistication and easiness of the method should also be accounted for. Delivering a longer dialysis overnight adds comfort and leaves more time for everyday life.

References

1   Barber S, Appleton DR, Kerr DNS: Adequate dialysis. Nephron 1975;14:209–227.
2   Charra B, Calemard E, Laurent G: Importance of treatment time and blood pressure control in achieving long-term survival on dialysis. Am J Nephrol 1996;16:35–44.

3   Charra B: How important is volume excess in the etiology of hypertension in dialysis patients? Semin Dial 1999;12:297–299.
4   Wolfe RA, Gaylin DS, Port FK, Held PJ, Wood CL: Using USRDS generated mortality tables to compare local ESRD mortality rates to national rates. Kidney Int 1992;42:991–996.
5   Degoulet P, Legrain M, Reach I, Aimé F, Devriès C, Rojas P, Jacobs C: Mortality risk factors in patients treated by chronic hemodialysis. Nephron 1982;31:103–110.
6   Babb AG, Strand MJ, Uvelli DA, Scribner BH: The dialysis index: A practical guide to dialysis treatment. Dial Transplant 1977;6:9–12.
7   Charra B, Bergström J, Scribner BH: Blood pressure control in dialysis patients. The importance of the lag phenomenon. Am J Kidney Dis 1998;32:720–724.
8   Charra B, Laurent G: Long hemodialysis: The key to survival? in Brown EA, Parfrey PS (eds): Complications of Long-Term Dialysis. Oxford/New York/Tokyo, Oxford University Press, 1999, pp 228–256.
9   Ackrill P, Goldsmith DJA, Covic AA, Venning MC, Ralston AJ: Outcome of long hours self-care haemodialysis in a single unit from 1969 through 1994. Nephrology 1997;3(suppl 1):S536.
10 Buttimore AL, Lynn KL, Bailey RR, Robson RA, Little P: 25 years of universal home dialysis at Christchurch hospital. Nephrology 1997;3(suppl 1):S559.
11 Parker TF, Husni L, Huang W, Lew M, Lowrie EG, and the Dallas Nephrology Associates: Survival of hemodialysis patients in United States is improved with a greater quantity of dialysis. Am J Kidney Dis 1994;23:670–680.
12 Collins AJ, Ma J, Umen A, Keshaviah PR: Urea index and other predictors of hemodialysis patient survival. Am J Kidney Dis1994;23:272–282.
13 Lowrie EG, Chertow GM, Lew NL, Lazarus JM, Owen WF: The urea (clearance × dialysis time) product (Kt) as an outcome-based measure of hemodialysis dose. Kidney Int 1999;56:729–737.
14 Vanholder R: Middle molecules as uremic toxins: Still a viable hypothesis? Semin Dial 1998;7:65–68.
15 Kjellstrand KM, Evans RL, Petersen RJ, Shideman JR, von Hartitzsch B, Buselmeier TJ: The 'unphysiology' of dialysis: A major cause of dialysis side effects? Kidney Int 1975;10(suppl 7):S30–S34.
16 Letteri JM: Behind the scenes of the NKF-DOQI guidelines. Hemodialysis adequacy. Dial Transplant 1997;26:827–831.

Bernard Charra
Centre de rein artificiel de Tassin
42 Avenue du 8-Mai-1945
FR–69160 Tassin (France)
Tel. +33 472 323130, Fax +33 478 345940, E-Mail bcharra@aol.com

Ronco C, Brendolan A, Levin NW (eds): Cardiovascular Disorders in Hemodialysis.
Contrib Nephrol. Basel, Karger, 2005, vol 149, pp 107–114

....

Mid-Dilution: The Perfect Balance between Convection and Diffusion

A. Santoro[a], *P.A. Conz*[b], *V. De Cristofaro*[c], *I. Acquistapace*[c], *R. Gaggi*[a], *E. Ferramosca*[a], *J.L. Renaux*[d], *E. Rizzioli*[b], *M.L. Wratten*[d]

[a]U.O. Nefrologia e Dialisi Malpighi, Policlinico S. Orsola-Malpighi, Bologna,
[b]Monselice, [c]Sondrio, [d]Bellco Spa, Mirandola, Italy

Abstract

Although hemodiafiltration (HDF) offers the advantage of increased convective clearance for middle molecules, there is still controversy as to whether reinfusion should occur pre- or postfilter. Mid-dilution hemodiafiltration (MD HDF) is a new HDF technique that uses a special dialyzer, MD190, which allows both pre- and postreinfusion. While externally the dialyzer looks similar to conventional hemodialyzers, the internal fibers are divided into two bundles by a special annular header that first lets the blood pass through the peripheral bundle in 'postdilution', mix with the reinfusion fluid at the opposite end of the dialyzer and then proceed (after 'predilution') to the dialyzer blood exit. The dialyzer is able to support substantially higher reinfusion rates (10–12 l/h). We have compared the removal characteristics of several small solutes and larger middle-molecular-weight toxins by examining instantaneous clearance at 45 min, the dialysis reduction ratio and total mass removal (by spilling) in a three-center prospective cross-over study. Twenty patients were randomized to a treatment sequence of one-week high-flux bicarbonate hemodialysis (HD) followed by MD HDF, or vice versa. The parameters evaluated included urea, creatinine, β_2-microglobulin, angiogenin, leptin, retinol-binding protein, and the effects on sodium, potassium, bicarbonate and calcium. Blood flow rates ranged between 300–450 ml/min (mean 359 ± 44 HD, 367 ± 35 MD HDF). The mean reinfusion for MD HDF was 166 ± 17 ml/min. MD HDF had a significantly better instantaneous clearance for urea (328 ± 28 vs 277 ± 40); creatinine (292 ± 32 vs. 212 ± 66); phosphate (324 ± 38 vs. 242 ± 63); β_2-microglobulin (249 ± 27 vs. 100 ± 24); angiogenin (173 ± 27 vs. 28 ± 32); and leptin (202 ± 29 vs. 63 ± 43). Treatments were well tolerated with no adverse reactions occurring during any of the treatments. The MD HDF filter's unique configuration is designed to deliver high-efficiency HDF with a significant improvement in small and middle molecule removal. MD HDF supports substantially higher ultrafiltration rates, and as such, results in a higher removal of middle-molecular-weight toxins.

Introduction

Hemodiafiltration (HDF) continues to grow in popularity as evidence continues to support the benefits of the convective removal of uremic toxins [1]. Many studies have shown that HDF offers good cardiovascular stability, an improvement in renal anemia, a reduction in bone/joint pain and amyloidosis [2]. In the early stages of dialysis, the primary focus of most nephrologists was on removal of small solutes such as urea. Although the 'middle molecule hypothesis' was introduced in the late 1970s, few techniques were able to offer therapeutic strategies to reduce these toxins [3, 4]. Over the last two decades we have started to appreciate the importance of middle molecules and their role in associated comorbidities [5].

Early experiences with HDF were also limited to the availability and type of substitution fluid. Sterile bags of substitution filter were associated with higher costs and problems with storage and handling. Today, online HDF can be performed by the machine preparation of ultrapure dialysate/substitution fluid. This requires a strict compliance with a global program of water quality, disinfection and redundant ultrafiltration – however, it also offers the opportunity to use increased (and unlimited) ultrafiltration without the associated problems of sterile substitution bags.

Nephrologists now faced new issues like 'Which are the most important toxins to remove?'; 'What are the best methods for removing the toxins?' and 'Is there a danger of removing too much?' While new technologies try to address these questions, there is still disagreement as to the best approach for the convective removal of uremic toxins [6]. HDF removes uremic toxins by both convection and diffusion [7]. As ultrafiltration results in a net loss of plasma water, substitution fluid must be given to the patient. Substitution fluid given before the dialyzer is termed 'predilution', whereas fluid infused after the dialyzer is termed 'postdilution'. In addition, there are many variations such as 'mixed HDF', 'push-pull HDF', acetate-free biofiltration and HDF with endogenous reinfusion.

Proponents of predilution HDF sustain that it is possible to use higher ultrafiltration rates, and thereby increase convective removal. However, predilution HDF has the disadvantage of diluting small solutes before the dialyzer and potentially decreasing small solute clearance by reducing the concentration gradient between the blood and the dialysate.

On the other hand, postdilution HDF has the disadvantage of hemoconcentration within the dialyzer, and therefore has limited ultrafiltration. Postdilution HDF is also associated with higher transmembrane pressures, albumin leakage and problems related to ultrafiltration failure and fiber clotting. An innovative solution to the dilemma between pre- and postdilution is mixed HDF. This

Table 1. Molecular weight of the uremic toxins measured in the study

Molecule	Molecular weight
Urea	60
Phosphate	96
Creatinine	113
β_2-Microglobulin	12 kDa
Angiogenin	14 kDa
Leptin	16 kDa
Retinol-binding protein	20 kDa

technique, utilizes a dedicated machine and allows simultaneous performance of the two infusion modalities [8].

Mid-dilution hemodiafiltration (MD HDF) has been developed to address these many of the above-mentioned technical concerns [9, 10]. It uses a proprietary dialyzer, MD190, which permits the reinfusion of the substitution fluid at a mid-way point of the dialyzer (fig. 1). The internal fibers of the dialyzer are divided into two bundles by an annular header that first lets the blood pass through the peripheral bundle in 'postdilution', mix with the reinfusion fluid at the opposite end of the dialyzer and then proceed (after 'predilution') to the dialyzer blood exit. The dialyzer is able to support substantially higher reinfusion rates (10–12 l/h) than postdilution HDF. The objective of our study was to evaluate the performance of the MD190 dialyzer using MD HDF for the removal of urea, creatinine, inorganic phosphate, β_2-microglobulin, angiogenin, leptin, and retinol-binding proteins (table 1).

Materials and Methods

Twenty patients were randomized to a one-week treatment sequence of MD HDF followed by one week of high-flux hemodialysis (HF HD) (or vice versa). Instantaneous clearance, reduction rate and mass transfer in the dialysate were evaluated in the mid-week session. Blood flows (Q_B) were 359 ± 44 for HF HD and 367 ± 35 for MD HDF. The ultrafiltration was 10 ± 4 ml/min for HD and 166 ± 17 ml/min for MD HDF and included the amount of ultrafiltration to offset the patient's interdialytic weight gain. QD was set at 800 ml/min for both techniques. The effective QD entering the filter during MD HDF was the set QD (800 ml/min) – the substitution rate.

MD HDF was performed with an Olpūr MD190 dialyzer (Bellco, 1.9 m², DIAPES γ sterilized). HF HD was performed with a BLS 819 (Bellco, 1.9 m² DIAPES, γ sterilized). Treatment efficacy was determined by measuring the instantaneous clearance at 45 min (for urea, creatinine, phosphate, β_2-microglobulin, leptin and angiogenin). Reduction ratios were

Table 2. Comparison of total mass removal between HF HD and MD HDF

Molecule	HF HD (mass, mg)	MD HDF (mass, mg)	p
Urea	37,723 ± 9,259	36,375 ± 11,493	ns
Creatinine	2,095 ± 556	2,152 ± 951	ns
β_2-Microglobulin	154 ± 70	187 ± 73	<0.05
Angiogenin	7.14 ± 1.71	9.11 ± 1.95	<0.001
Retinol-binding protein	905 ± 1,000	836 ± 542	ns

ns = Not significant.

determined by the equation:

$$RR = (1 - C_{post}/C_{pre}) \times 100 \tag{1}$$

Proteins (β_2-microglobulin, leptin and angiogenin) were corrected for hemoconcentration for C_{post} [11].

Plasma concentrations were determined from the blood withdrawn from the arterial and venous bloodlines of the extracorporeal circuit. β_2-Microglobulin was measured by the nephelometric method (Beckman-Coulter, Fullerton, Calif., USA); angiogenin was measured with an enzyme immunoassy kit (R&D System, Minneapolis, Minn., USA) and leptin was measured with an enzyme immunoassay kit (Cayman Chemical Company, Ann Arbor, Mich., USA).

Results

All the patients completed the study and no adverse events were reported. Instantaneous clearance at 45 min was significantly higher with MD HDF than HF HD for urea (328 ± 28 vs. 277 ± 40; $p < 0.05$), creatinine (292 ± 32 vs. 212 ± 66; $p < 0.05$), phosphate (324 ± 38 vs. 242 ± 63; $p < 0.05$) (fig. 2), β_2-microglobulin (249 ± 27 vs. 100 ± 24; $p < 0.05$), angiogenin (173 ± 27 vs. 28 ± 32; $p < 0.05$) and leptin (202 ± 29 vs. 63 ± 43; $p < 0.05$) (fig. 3).

The overall reduction ratio was not significantly different between any of the measured parameters with the exception of β_2-microglobulin ($p < 0.05$) (figs 4 and 5), although there was a significant increase in the total mass removed in the dialysate for angiogenin and β_2-microglobulin (table 2).

Discussion

There is still a substantial debate regarding dialysis adequacy and how this can be best achieved. The average age of the dialysis patients is increasing, their

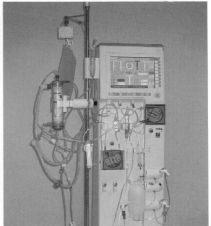

Fig. 1. Photograph of MD190 filter.

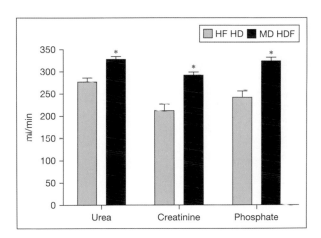

Fig. 2. Instantaneous clearance of urea, creatinine and phosphate at 45 min. * p < 0.05.

overall survival is also increasing [12] and many patients are now receiving renal replacement therapies for longer periods of time. All of these factors lead to a greater risk of dialysis comorbidities, such as amyloid deposition on bones and joints. Today the concept of dialysis adequacy is much more than just 'urea removal' and includes the removal of middle molecules as well as biocompatibility.

The USA-based HEMO study concluded that there was a nonsignificant trend towards a beneficial effect on mortality of HF HD compared to low-flux dialysis. Locatelli et al. [13] also showed a nonsignificant trend towards better

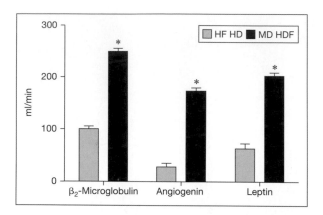

Fig. 3. Instantaneous clearance of β_2-microglobulin, angiogenin and leptin at 45 min. * $p < 0.05$.

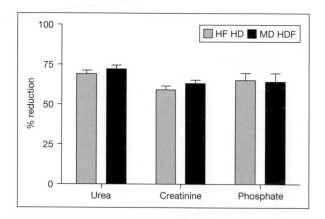

Fig. 4. Reduction ratio of small molecules: urea, creatinine and phosphate.

survival and significant delay in carpal tunnel surgery for HDF versus HD in the Lombard Registry (Italy). One criticism leveled at these studies is that it is hard to compare convection and diffusion as 'black' and 'white' entities. The internal filtration of HF HD is much lower than the expected convection of HDF. Ward et al. [14] also compared online HDF (21 l) to HF HD in a prospective one-year clinical study and found that although online HDF provides better solute removal over a wide molecular weight range, there were no significant differences in pretreatment plasma concentrations.

Our study compared HF HD with MD HDF (40 l reinfusion) using the same blood flows, membrane (DIAPES) and area (1.9 m²). As expected, we observed a significantly better instantaneous clearance of middle molecules

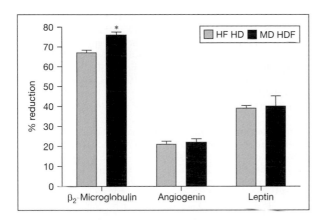

Fig. 5. Reduction ratio of middle molecules: β_2-microglobulin, angiogenin and leptin.
* $p < 0.05$.

(β_2-microglobulin, angiogenin and leptin) as compared with HF HD due to the increased convection. Although we observed an increase in the mass transfer of β_2-microglobulin and angiogenin in the dialysate (collected by spilling), we only observed a significant difference in the reduction ratio for β_2-microglobulin. This may be due to differences in pool refilling. The removal of small solutes and electrolyte balance was similar to HF HD.

As compared with other forms of HDF, MD HDF appeared to be safe, well tolerated and easy to use. The choice of renal replacement therapy depends on many factors including economic constraints, machine capability, water quality and the individual patient's clinical needs. The MD190 dialyzer does not require specific software or hardware modifications for the reinfusion, and can be used on any machine equipped for online HDF.

Similar to all online HDF techniques, MD HDF requires strict compliance with water quality guidelines, machine disinfection and the proper use of redundant ultrafiltration for cold sterilization of the dialysate/reinfusion fluid.

References

1   Beerenhout CH, Kooman JP, Luik AJ, Jeuken-Mertens, von der Sands FM, Leunissen KML: Optimizing renal replacement therapy – A case for online filtration therapies? Nephrol Dial Transplant 2002;17:2065–2070.
2   Maduell F, Del Pozo C, Garcia H, Sanchez L, Hdez-Jaras J, Albero MD, Calvo C, Torrengrosa I, Navarro V: Change from conventional haemodiafiltration to online haemodiafiltration. Nephrol Dial Transplant 1999;14:1202–1207.
3   Babb A, Ahmad S, Bergstrom J, Scribner BH: The middle molecule hypothesis in perspective. Am J Kidney Dis 1981;1:46–50.

4 Milutinovic J, Babb AL, Eschbach JW, Follette WC, Graefe U, Strand MJ, Scribner BH: Uremic neuropathy: Evidence of middle molecule toxicity. Artif Organs 1978;2:45–51.
5 Vanholder R, De Smet R, Glorieux G, Argiles A, Baurmeister U, Burnet P, Clark W, Cohen G, De Deyn PP, Deppisch R, Descamps Latscha B, Henle T, Jorres A, Lemke HD, Massy ZA, Passlick-Deetjen J, Rodriguez M, Stegmayer B, Stenvinkel P, Tetta C, Wanner C, Zidek W; European Uremic Toxin Work Group (EUTox): Review on uremic toxins: Classification, concentration, and interindividual variability. Kidney Int 2003;63:1934–1943.
6 Masakane I: Selection of dilutional method for on-line HDF, pre- or post-dilution. Blood Purif 2004;22:49–54.
7 Leypoldt J: Solute fluxes in different treatment modalities. Nephrol Dial Transplant 2000;15:3–9.
8 Pedrini LA, De Cristofaro V: On-line mixed hemodiafiltration with a feedback for ultrafiltration control: Effect on middle-molecule removal. Kidney Int 2003;64:1505–1513.
9 Krieter D, Falkenhain S, Chalabi L, Collins G, Lemke HD, Canaud B: Clinical cross-over comparison of mid-dilution hemodiafiltration using a novel dialyzer concept and post-dilution hemodiafiltration. Kidney Int 2005;67:349–356.
10 Krieter D, Collins G, Summerton J, Spence E, Moragues HL, Canaud B: Mid-dilution on-line haemodiafiltration in a standard dialyser configuration. Nephrol Dial Transplant 2005;20:155–160.
11 Bergstrom J, Wehle B: No change in corrected beta 2-microglobulin concentration after cuprophane haemodialysis. Lancet 1987;1:628–629.
12 Marcelli D, Stannard D, Conte F, Held PJ, Locatelli F, Port FK: ESRD patient mortality with adjustment for comorbid conditions in Lombardy (Italy) versus the United States. Kidney Int 1996;50:1013–1018.
13 Locatelli F, Marcelli D, Conte F, Limido A, Malberti F, Spotti D: Comparison of mortality in ESRD patients on convective and diffusive extracorporeal treatments. Kidney Int 1999;55:286–293.
14 Ward R, Schmidt B, Hullin J, Hillebrand GF, Samtleben W: A comparison of on-line hemodiafiltration and high-flux hemodialysis: A prospective clinical study. J Am Soc Nephrol 2000;11:2344–2350.

Antonio Santoro, MD
U.O. Nefrologia e Dialisi Malpighi, Policlinico S.Orsola-Malpighi, Via P.Palagi 9
IT–40138 Bologna (Italy)
Tel. +39 051 6362430, Fax +39 051 6362511, E-Mail santoro@aosp.bo.it

Ronco C, Brendolan A, Levin NW (eds): Cardiovascular Disorders in Hemodialysis.
Contrib Nephrol. Basel, Karger, 2005, vol 149, pp 115–120

....

Sequential Hemofiltration-Hemodiafiltration Technique: All in One?

M. Amato[a], *A. Brendolan*[b], *G. Campolo*[a], *D. Petras*[b], *M. Bonello*[b],
C. Crepaldi[b], *C. Ronco*[b]

[a]Nephrology Department, Prato Hospital, Prato, and [b]Department of Nephrology,
St. Bortolo Hospital, Vicenza, Italy

Abstract

Sequential dialysis techniques (i.e. pure ultrafiltration followed by dialysis) have been used in the past, due to their capability to remove large volumes of fluids without inducing hemodynamic instability. The disadvantages of the inadequate dialysis and the lack of technology lead to the decline such methods. Hemofiltration (HF) and hemodiafiltration (HDF) are recently being utilized in a greater proportion thanks to the on line fluid preparation systems. Each process (HF and HDF) has its own benefits in the removal of small, medium and high-molecular weight substances and in the hemodynamic stability. Sequential hemofiltration/hemodiafiltration (SHF/HDF), may combine the benefits and eliminate the disadvantages of each method. Furthermore they can be easily applied nowadays, due to the development of new high technological hemodialysis machines. In order to evaluate the feasibility and the effects of SHF/HDF we studied 7 chronic hemodialysis patients (6 months of treatment with SHF/HDF switched to 6 months of SHDF/HF), using the same machine (AK200 ULTRA), with on line fluid preparation system and the same type of dialyzer (Polyflux 210). The feasibility of such techniques (SHF/HDF or vice versa) resulted excellent. All sessions left the patients in a condition of well-being making fulltime work. No difference was observed between the different period of treatment, but a reduction in pre value was observed in calcium-phosphorous product, C-reactive protein and β_2-microglobulin, at the end of the sequential techniques. SHF/HDF therapy is a very promising technique. Further studies are needed to better explore the potential of such a therapeutic approach in the quality of life, the hemodialysis adequacy and the hemodynamic stability of our patients.

Introduction

Sequential techniques (i.e. ultrafiltration alone followed by isovolemic dialysis) allow for removal of large volumes of fluid without inducing hemodynamic instability. The downside of this approach is that adequacy can be impaired, due to the relatively shorter time of dialysis in the overall duration of the session. In the past sequential ultrafiltration dialysis was used with satisfactory results in patients with fluid overload and hemodynamic instability. Such procedures were abandoned due to adequate reasons including the lack of a simple and dedicated technology [1–3].

Hemofiltration (HF) has been suggested as an optimal method to achieve stable hemodynamic conditions during dialysis and for increased removal of medium- and high-molecular-weight substances [4]. Hemodiafiltration (HDF) has recently been largely utilized as the method that combines diffusive, convective and absorptive removal in the same exchange module. This seems to represent an ideal compromise and it carries high efficiency especially now that on-line fluid preparation systems are available [5, 6] and fluid exchange can be increased.

Sequential hemofiltration-hemodiafiltration (SHF-HDF) or vice versa could be a useful combined tool to exploit the best of each treatment modality. To perform SHF-HDF and to get the maximum efficiency in both phases, we need high substitution fluid volumes and large biocompatible membranes with elevated permeability characteristics. The new dialysis machines equipped with on-line fluid supply make possible the production of high quantities of ultra-pure pyrogen-free liquid and thus they represent the ideal platform to perform such therapies. New features in the AK200 machine (Gambro, Sweden) allow for an easy passage from HF to HDF and vice versa, making sequential therapies much simpler and easier to perform today.

Our aim has been to test the clinical feasibility of new sequential techniques in a small group of dialyzed patients using the hydrophilic-hydrophobic polyflux membrane (polyamide) and the AK200 machine.

Subjects and Methods

The clinical and demographical characteristics of 7 chronic hemodialysis patients involved in the study are shown in table 1. All subjects gave their informed consent to the study protocol design.

The main purpose of our study was to evaluate the feasibility of sequential techniques and their effects after 6 months of treatment (SHF-HDF switched to 6 months of SHDF-HF). Meanwhile we reported possible technical problems, if any, observed by nurses on duty.

To avoid any potential bias due to different kind of extracorporeal treatment, all patients were treated with high-efficiency hemodialysis during a 3-month period immediately before starting the protocol; a cross-over design was planned, based on the assumption that chronic hemodialysis subjects would not change their dietary or other similar habits during the observation periods.

Table 1. Clinical and demographical characteristics of the patients

Mean age in years	58.57 ± 17.29
Months on dialysis	76 ± 58
Body mass index	21.8 ± 3.2
Mean blood pressure (mm Hg)	144 ± 6
Erythropoietin/Darbepoetin (n)	6/1
Erythropoietin (U/week)	11,000 ± 6,000
Darbepoetin (µg/15 days)	30
Vitamin D (n)	5

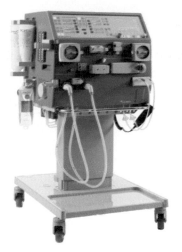 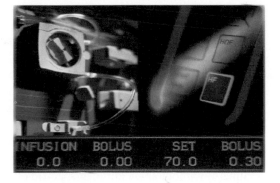

Fig 1. AK200 Ultra system.

Figure 1 depicts the devices used in our study.

In all treatments the same machine type (AK200 Ultra) was used, preparing ultrapure dialysis fluid by stepwise ultrafiltration of water and inlet dialysate using polyamide ultrafilters. The substitution fluid, either in HF or HDF, was prepared on-line by an additive ultrafiltration step just before reinfusion into bloodlines.

The same fluid electrolyte concentration (mmol/l) was used for each patient throughout the study period: sodium 138–140, potassium 1–2, chloride 108–109.5, calcium 1.50–1.75, magnesium 0.5, bicarbonate 30–40, acetate 3 and glucose 5.55. The blood flow was initially set at 350 ml/min and during the HF period (90 min) the infusion flow was set to correspond to the blood flow (predilution). In the second part of treatment (HDF), lasting 120 min, the postdilution volume infusion was scheduled as 50% of that infused in HF.

Polyflux 210 dialyzer (surface 2.1 m^2) equipped with a high-permeability hydrophilic-hydrophobic membrane with an asymmetric 3-layer microstructure was used.

Quantitation of Treatment Efficacy

Treatment efficacy was compared with the previous period of high-flux hemodialysis looking at classical measures such as single pool Kt/V for small molecules, and for plasma concentration of larger low-molecular-weight substances, β_2-microglobulin and cystatin, measured at the beginning of the first dialysis of the week.

Sample for analysis of C-reactive protein, calcium, phosphorus, hemoglobin and hematocrit were, also taken every 3 months. Descriptive analysis of the results was performed by calculating mean values ± standard deviation.

Results

All patients completed the 6 months of SHF-HDF treatment period and followed it by a switch to SHDF-HF for the same period without suffering from major adverse symptoms, such as headache or hypotension. In particular the incidence of symptoms was reduced although there were insufficient data to make the analysis statistically significant.

The opinion of nurses about the feasibility of such a technique (HF-HDF or vice versa) was excellent, due to the absolutely simple way of running the treatment, without any additional specific operation (fig. 1).

The time of treatment was unchanged during both techniques (210 min). All sessions left the patients in a state of well-being making fulltime work and rehabilitation possible.

Table 2 shows the descriptive analysis of the lab results performed by calculating mean values ± standard deviation. No difference was observed between the various periods of treatment, but a reduction in predialysis levels of calcium-phosphorus product, C-reactive protein and β_2-microglobulin, at the end of the first 6-month period was observed. This effect was maintained in the second semester with stable values at 12 months of treatment.

Discussion

HF is a renal replacement therapy specially designed to maximize convective transport. Due to the iso-osmotic fluid removal, a peculiar hemodynamic stability has been advocated for this technique. In fact, this technique has also been used in specific areas such as sepsis, decompensated heart failure, etc [4, 7]. However its use in chronic dialysis was limited in the past by the inadequate

Table 2. Laboratory results of the main parameters during the therapy modes

	Hemodialysis	3-month SHDF-HF	6-month SHDF-HF
Urea (mg/dl)	146 ± 32	153 ± 31	156 ± 39
Creatinine (mg/dl)	11.3 ± 1.8	11.6 ± 2.3	11.8 ± 2.8
Albumin (mg/dl)	4.2 ± 0.2		4.1 ± 0.4
Hemoglobin (g/dl)	11 ± 1.6	11 ± 1.5	11.1 ± 1.3
Calcium-phosphorus	36.8 ± 3.7	34.58 ± 5.2	36.7 ± 5.5
C-reactive protein (IU/ dl)	1.24 ± 0.40	2.19 ± 2.06	0.78 ± 0.35
β_2-microglobulin (mg/dl)	2.9 ± 1.5	2.6 ± 0.2	2.6 ± 0.4
Kt/V	1.6 ± 0.40	1.4 ± 0.1	1.3 ± 0.2

efficiency for clearance of low-molecular-weight substances and the prolongation of treatment times which became necessary. With the advent of on-line fluid preparation, these limitations no longer exist since predilution with high-volume exchange can be performed.

Postdilution HDF with on-line fluid preparation is regarded as the most efficient extracorporeal technique available for clinical routine. HDF on-line with appropriate filters provides clearances of both small and large solutes and improves hemodynamic instability, offering a very comprehensive and well-tolerated renal replacement therapy [5]. Somehow, HDF makes a compromise by high-efficiency dialysis.

The SHDF-HF (or vice versa) procedure is a very promising technique that utilizes the benefits of both methods, so as to achieve both an adequate dialysis of uremic toxins of all sizes and hemodynamic stability. Nowadays, technology gives us the opportunity to apply this process in a simple, safe and clever way.

In our experience, which lasted a year, using a dedicated machine and a hydrophobic-hydrophilic membrane such as polyamide [8], we were able to run SHF-HDF therapy (and vice versa) without any increased burden in terms of cost of devices and/or nurses. We did not notice any significant difference in the Kt/V value, requirements of erythropoietin, and nutritional needs. On the other hand there was an improvement in the C-reactive protein values and calcium-phosphorus product. A reduction was also observed in the hypotensive events during dialysis, although more observations are required to make the data significant.

A clinical multicenter prospective study comparing the outcomes of such a technique on quality-of-life measures, nutritional parameters, better removal of uremic retention solutes of medium size and hemodynamic conditions is proposed to better explore the potential of such a therapeutic approach.

References

1   Wehle B, Asaba H, Castenfors J, Furst P, Gunnarsson B, Shaldon S, Bergstrom J: Hemodynamic changes during sequential ultrafiltration and dialysis. Kidney Int 1979;15:411–418.
2   Rouby JJ, Rottembourg J, Durande JP, Basset JY, Degoulet P, Glaser P, Legrain M: Hemodynamic changes induced by regular hemodialysis and sequential ultrafiltration hemodialysis: A comparative study. Kidney Int 1980;17:801–810.
3   Di Maggio A, Basile C, Scatizzi A: Plasma volume changes induced by sequential ultrafiltration-hemodialysis and sequential hemodialysis-ultrafiltration. Int J Artif Organs 1987;10:291–294.
4   Locatelli F, Di Filippo S, Manzoni C: Removal of small and middle molecules by convective techniques. Nephrol Dial Transplant 2000;15(suppl 2):37–44.
5   Passlick-Deetjen J, Pohlmeier R: On-line hemodiafiltration. Gold standard or top therapy? Contrib Nephrol 2002;(137):201–211.
6   Marinez de Francisco AL, Ghezzi PM, Brendolan A, Fiorini F, La Greca G, Ronco C, Arias M, Gervasio R, Tetta C: Hemodiafiltration with online regeneration of the ultrafiltrate. Kidney Int Suppl 2000;76:S66–S71.
7   Tetta C, Bellomo R, Kellum J, Ricci Z, Pohlmeiere R, Passlick-Deetjen J, Ronco C: High volume hemofiltration in critically ill patients: Why, when and how? Contrib Nephrol 2004;144:362–375.
8   Ronco C, Crepaldi C, Brendolan A, Bragantini L, d'Intini V, Inguaggiato P, Bonello M, Krause B, Deppisch R, Goehl H, Scabardi A: Evolution of synthetic membranes for blood purification: The case of the Polyflux family. Nephrol Dial Transplant 2003;18(suppl 7):vii10–20; discussion vii55.

Prof. Marcello Amato
Ospedale Civile di Prato, Unità Operativa di Prato
Piazza Ospedale 1
IT–50047 Prato (Italy)
Tel. +39 0574 43 44 13, Fax +39 0574 4344 13, E-Mail mamato@usI4.toscana.it

Ronco C, Brendolan A, Levin NW (eds): Cardiovascular Disorders in Hemodialysis.
Contrib Nephrol. Basel, Karger, 2005, vol 149, pp 121–130

· ·

Should Nephrologists Be in Charge?

Klaus Konner

Department of Internal Medicine I, Cologne General Hospital University of Cologne,
Cologne, Germany

Abstract

Once, vascular access (VA) for hemodialysis treatment was initiated by nephrologists:
Scribner introduced the arteriovenous shunt, Shaldon the central-venous catheters and
Brescia-Cimino the arteriovenous fistula. Later on, creating VA became a domain of surgery.
Many nephrologists felt out of responsibility. Interventional procedures, angioplasty and
stent insertion are mostly performed by radiologists. In 2005, the role of the nephrologist in
comprehensive VA care must be newly identified. We know about the value of early referral
to nephrologist and access surgeon to preserve venous vasculature. The nephrologist cares
for clinical examination of vessels, for an ultrasound Doppler evaluation before the creation
of primary VA with clear preference to native arteriovenous fistulae to aim at an early failure
rate. Surveillance and monitoring require the exclusive responsibility of the nephrologist and
his team. Early diagnosis of VA dysfunction allows elective revision before the onset of
thrombosis. There should be an agreement on strategies between nephrology, surgery and
radiology. Surgical techniques and skills are to be assessed from time to time. Worldwide,
new organizational structures in creation, control and documentation of VA are needed.
Flexibility between the disciplines involved as well as educational programs for nephrologists, surgeons and radiologists are future challenges.

Copyright © 2005 S. Karger AG, Basel

Introduction

Nephrologists traditionally are educated in kidney physiology and pathophysiology as well as the management of kidney diseases. They are familiar
with renal anemia, osteodystrophy and many other disorders, they initiate and
supervise dialysis and care for patients before and after kidney transplantation.
All nephrologists recognize the pivotal role of vascular access (VA) for their
hemodialysis patients, but VA is still seen as the Achilles heel of hemodialysis
therapy.

In the early years most nephrologists created Scribner shunts and arterio-venous fistulae (AVFs). With the introduction of arteriovenous grafts (AVGs) and more complex revisions, access care soon was identified almost exclusively with surgeons, mostly vascular or transplant surgeons.

Now, time is changing this. The start of the 21st century presents different challenges: the predominance of older and diabetic patients requires a new type of early preoperative vascular evaluation, the imbalance in the use of native AVFs versus AVGs of various types, the increasing use of temporary and atrial catheters, the introduction of VA monitoring and surveillance techniques as well as development of a broad spectrum of radiologic interventional proce-dures. Thus, it seems the time to look for a new definition of the responsibility of the nephrologist with regard to VA.

The Predialysis Period

Early Referral

The timing and duration of pre-ESRD care by the nephrologist has an impact on the quality of VA. Early referral to the nephrologist should result in early referral to the VA surgeon and this, with a well-defined preservation of the venous vasculature, are important contributors for better outcome in placement of the initial VA. Whenever possible this should be a native AVF.

Routine use of ultrasound (US) Doppler evaluations of arterial and venous vessels provides essential information for the operating surgeon [1]. In our institution, a step-by-step procedure has proven helpful: a patient with chronic kidney disease undergoes a careful clinical examination and an US Doppler evaluation of the vessels if questions remain after the clinical exami-nation. This is used to provide a description of the arterial vasculature, diam-eter and flow rate of the brachial arteries, flow direction in the forearm arteries and measurement of the resistance index as has been described by Malovrh [2].

In most patients, these informations will be sufficient so that further angio-(veno-)graphic evaluation is unnecessary, except in rare, unusual cases where central venous stenoses are suggested.

Referral to Access Surgeon

At this point, the patient is referred by the nephrologist to the VA surgeon who will identify the type and site of the first VA to give the best preservation of venous vasculature. The patient is informed in detail, as he or she is the best partner in venous preservation. In principle, any venous cannulation or blood drawing should be performed as peripherally as possible – in both arms.

The First VA Operation

The nephrologist determines when to place the first VA. In patients with good vasculature this should be about 2 months before the expected start of hemodialysis; in patients with poor vessels this time interval should be about 4 months to allow a longer maturation period.

Absolute preference must be given to an anastomosis of native vessels and, despite any guidelines, initially all patients should be regarded as candidates for a native AVF. It is the clear responsibility of the nephrologist to establish a close cooperation with his surgeon with the aim of timely creation of a native fistula [3].

Nowadays, we know that this strategy avoids or reduces the use of potentially risky temporary central venous catheters at the start of hemodialysis treatment, decreases morbidity and mortality and decreases costs and hospitalizations [4].

Furthermore, a high quality of the first VA, preferably a native AVF, can reduce early failure rate dramatically so that fewer revisions are needed [5]. All these benefits of timely placement of an AVF are seen in patients with a low early failure rate [6]. The type, site, side and quality of the first VA have a substantial impact on patient's morbidity and mortality. Consequently, optimization of collaboration between the nephrologist and access surgeon is one of the best contributions to patient care.

Surveillance and Monitoring of VA

Arteriovenous VA of any type is a dynamic high-flow, low-pressure, highly unphysiologic structure that is regularly and repeatedly violated by large-bore cannulae. Most complications of native AVFs, such as aneurysms and stenoses, except perianastomotic narrowings, are caused by poor cannulation technique. Graft-vein stenoses due to intimal hyperplasia reduce dialysis quality. Ultimately, access dysfunction results in thrombosis and this endpoint of access function should be considered an avoidable event.

Surveillance and monitoring has developed as an important field during the last 2 decades. Differences exist between native AVFs and AVGs with regard to type, site and clinical phenomena that may develop. Monitoring includes physical examination and observation of trends in dialysis-related parameters like venous pressure and/or bleeding time after removal of the cannulae.

Surveillance techniques include measurement of blood flow and static or dynamic venous pressure using various techniques. If questions remain, preferably US Doppler imaging may be used and, if necessary, angiography. Today, angiography is used when US results are not sufficient. US and angiography are regarded as complementary techniques. Unfortunately, US is not reimbursed adequately in all countries.

Access monitoring is an integral part of VA care and should be the exclusive responsibility of the nephrologist and his team. Most useful are 'simple' clinical parameters that can be followed routinely as bedside procedures without any technical assistance – an underestimated opportunity that is offered to the nephrologist and his staff three times a week. To get the best results, continuous education and training is essential to develop and maintain the necessary skills. Experience at our institution has shown that the majority of problems with VA dysfunction can be solved in this way. Nurses must know the characteristics of the individual VAs and if there is any suspicion of a trend towards dysfunction, the nephrologist must be contacted to determine need for further investigation based on the clinical findings.

Surveillance also depends on the availability of technical equipment. Static and dynamic venous pressures were used widely and access flow measurements are the best predictors of access thrombosis in AVF and AVG [7]. It is the decision of the dialysis center which surveillance test best meets their individual conditions and needs. For many dialysis institutions, particularly in underdeveloped countries, the question arises whether the routine use of high-tech surveillance tests is indispensable or can be replaced by careful and regular monitoring and simple observation of arterial and venous pressure. When available, US techniques can provide anatomic and functional data. The use of diagnostic angiography is now mainly used to diagnose central venous stenoses.

The goal of all these nephrology-initiated monitoring and surveillance activities is early diagnosis of access dysfunction aimed at early elective repair by either surgical or interventional techniques. The time should be past when access management was exclusively a crisis management.

There are no rules or guidelines about which type of repair – surgical or interventional – is to be preferred. Advocates of both techniques reclaim priority, success, patient convenience and cost-effectiveness. This issue will remain an open question since randomized and controlled trials are difficult to undertake. The determining factor will be local expertise and dedication, regardless of the discipline involved [8]. However, interventional procedures are high-tech based and not available in any institution.

Consequently, monitoring and surveillance of VA remains a central and important issue for nephrologists and his staff now and in the future.

Nephrology and the Interdisciplinary Approach

In 2005, we know about the benefit of:

- *early referral* of the patient to a nephrologist and an access surgeon to establish early and detailed preservation of venous vasculature;
- *timely creation* of a primary native arteriovenous fistula whenever possible; any patient is, in principle, a candidate for a native arteriovenous fistula;
- routinely performed *monitoring and surveillance* of VA, using clinical examination and US evaluation; and
- *flow measurements* as the best predictive parameter for VA dysfunction.

The interdisciplinary character of VA creation and care has only one common factor: the responsibility of nephrologists to manage VA problems for their hemodialysis patients.

Even so, there is no agreement worldwide about a number of remaining problems and open questions:

(1) Who does US – nephrologist, access surgeon, interventionalist?
(2) Who performs access surgery – surgeons, nephrologists?
(3) Who does the repair or revision in the event of complications?
(4) How should the management of VA care best be organized?
(5) Is there agreement on documentation and quality measurements?
(6) Is VA accepted as part of education and a qualification for a nephrologist, an access surgeon or a radiologist?

Solutions are difficult to realize: human beings are involved with all their personal traditions and changing moods, there are structural and organizational differences in different disciplines and departments, financial jealousies and so on. Furthermore, the systems that exist seem to be fixed for eternity and flexibility is often missed. In many countries financial resources are limited and reimbursement patterns differ widely. Guidelines for VA care have been welcomed but say nothing about these issues and articles on these are few.

Now is the time to consider change. The following statements are based on the author's experience in comprehensive care in VA over 3 decades and involvement in the three disciplines as well as numerous contacts and correspondence over many years with many dedicated nephrologists, surgeons and radiologists. One aspect should be mentioned: the non-nephrology specialist, while often a part of a great team, meets the patient only during the procedure; a nephrologist busy in performing access care meets the patient three times a week. This creates a different type of responsibility. In principle, as with many issues in life, things should be done by the most experienced expert who guarantees success and the best result for the benefit of the hemodialysis patient.

Who Performs US Evaluation – Nephrologist, Access Surgeon or Interventionalist?

The nephrologist, as US investigator, knows the history of access dysfunction and is familiar with the individual problems of the patient. The access surgeon, as US investigator, must be informed in detail about the type of access dysfunction; on the other hand, he interprets the US findings in the context of the surgical options. The radiologist, as US investigator, should be familiar with, for example, the process of cannulation of the access and other dialysis-related issues; this would be beneficial for the access surgeon, too.

It is most important to integrate the result of US evaluation with the patient's record by a simple sketch, which will be helpful for further procedures.

Who Performs Access Surgery – Surgeons, Nephrologists?

Historically seen, at the launch of hemodialysis therapy, nephrologists used to create VAs and surgeons had not integrated these techniques into their repertoire at that time; to read the 'Preface' by Dr. Shaldon should be a must [9].

Different traditions exist in different countries and in part relate to reimbursement. In the USA, most of the access operations are performed by transplant surgeons, followed by vascular and general surgeons. Most European countries rely on vascular surgeons although the leading access surgeon in France originally was a urologist. Germany has a mix from several surgical subspecialists active in access surgery, mainly vascular, but also transplant and general surgeons and, a tradition in a few university hospitals, urologists; in addition, Germany has a small group of highly experienced nephrologists who do VA surgery. Italy is different, nephrologists care for more than 80% of primary accesses and exclusively create native AVFs; only more complex cases are transferred to surgeons [10]. This tradition is transferred from the senior to the junior nephrologist, individualized per institution; a structured education does not exist at this time [11]. It is the author's opinion that these European traditions are not an optimal basis for the future; it is likely that the (nonexisting) European 'system' is usually overestimated. We seem to live in the past and are not yet prepared to overcome the challenges in VA in the future.

Worldwide, most nephrologists are not very happy with 'their' surgeons. They feel that they have no special partner in the group of surgeons, access surgery is often performed by younger surgeons in training, continuity is missed, different surgeons use different techniques, schedules are overcrowded and access operations are often cancelled as 'semi-urgent' cases. Surgeons consider an access surgery as only a small and less spectacular part of their work and many vascular surgeons want to remain just that and not mutate into an access surgeon. Access surgery is not rewarding to reputation and is time-consuming, and therefore, unpopular. Surgeons identify themselves much more with their own specialty than with access surgery. A prominent opposite opinion comes from Prof. Sollinger, one of the world's most experienced kidney and pancreas transplant specialist: 'Access surgery is more challenging and varying than transplant surgery' [pers. commun.]. Fortunately, there are a number of dedicated access surgeons around the world who achieve an admirable work. Nevertheless, the general impression is not so optimistic; the skills required to create native AVFs need greater emphasis in surgical training programs, and many surgeons need more experience in these techniques [12].

Would it be a realistic advance if nephrologists started to perform access surgery? YES and NO. There is no reason a nephrologist should not be able to learn access surgery. Nephrologists with a talent for manual skill and patience should be ready to face the challenge. Mandatory education by an experienced access surgeon should cover creation of primary native AVFs and standard revisions in about 100 cases – in Germany, vascular surgeons become specialists having done only about 30 VA procedures; not enough, by far. Over time, AVFs in the proximal forearm and the region of the elbow will contribute to rising experience. Technically, wrist AVFs are far more challenging than elbow operations. The 'simple wrist AVF' is a myth.

All other operations should be performed with, and in an atmosphere of confidential friendship, an experienced VA surgeon. Access surgery starts to be truly interesting for the operating beginner with the 300th or 500th operation! At this point, creativity and individualization can be realized for the benefit of the patient. It may even happen that the access surgeon who taught you will be happy to have a substantial reduction in the number of 'simple' VA operations. The operating nephrologist should join his teacher from time to time to perform any other type of VA so as to widen his spectrum. Later on, 100 VA operations per year should be the minimum.

When should a nephrologist begin placing AVF? This remains an open question, but it must be emphasized that this should be based on an exemplary partnership with a VA surgeon, a scholar-teacher relationship without any animosity. Nephrologists should be aware of the risk of producing access failures; winning and losing are close neighbors; self-criticism and modesty is required. Good luck!

Who Achieves Repair/Revision in Case of Complication?

The answer touches one unsolved problem in the field of VA: surgery or interventional procedures? As a nephrologist access surgeon, I tend to prefer surgery. In AVFs, most revisions can be done on an outpatient basis under local anesthesia; the result is, with few exceptions, a perfect reconstruction of the venous lumen with a low acute complication rate and a long-term complication-free interval. Following this strategy, interventional procedures are limited to repair of complications along the (cephalad) upper arm and the central veins. Here, angioplasty, may be insertion of a stent, is highly welcomed. However, interventional procedures suffer from the need for frequent reinterventions because of restenoses. Here, methodological limitations should be respected.

In many institutions, the question surgical versus interventional technique remains academic. A highly experienced and dedicated interventionalist may be preferable to a less motivated and dedicated surgeon – and vice versa [8].

In USA, nephrologists have learned to perform interventional access procedures, mainly to overcome the overwhelming number of complications caused by the long-standing preference to insert AVGs instead of creating AVFs. The author repeatedly observed outstanding skill and creativity while visiting such institutions. The numerous salvage procedures in dysfunctional or nonmaturing AVFs should initiate efforts to look for factors before and when placing initial AVF – choice of artery and vein, surgical technique and others.

With reduction of graft implantation rates, efforts should be directed to starting access surgery by nephrologists aiming at high AVF rates, respecting the above-mentioned conditions.

How Can the Management of VA Care Be Organized?

It is almost impossible to answer this question. There are no guidelines, no comparable experiences in this field. A few models were mentioned in the literature: establishment of VA work groups or access coordinators and with weekly conferences; experienced staff members may become access coordinators – a welcomed solution bringing together different disciplines with different timing schedules. In most institutions, a by-chance handling of these problems is still practised.

A major problem reported from many countries arises from a shortage of funding and nephrologists available per institution (ideally, 1 or 2 members of the nephrology team should be dedicated to VA activities). Surgical departments may also suffer from shortage of personnel and availability of operation room times, etc. In any case, increased dedication to VA should not be punished by political decisions, as patients would be the victims. Sometimes private institutions, especially in USA, seemed to initiate new structures so that interventional and other nephrologists would start access creation.

Is There Agreement on Documentation and Quality Measurements?

As care for VA is an interdisciplinary issue, documentation and quality measurement are mandatory. Documentation means the delivery of a sketch after the clinical examination as a hint for the US Doppler evaluation and the access surgeon. Any operation must be documented by this type of drawing that can be transferred via facsimile or electronically from surgery to nephrology.

Sketchy drawings are rapidly done and can give detailed information, far better than a photograph; in our institution, this drawing is sent the same day. Thus, dialysis staff is informed immediately and does not have to wait for weeks for a written report that in most cases cannot be understood by nonsurgeons. In addition, computer-based programs can provide continuous documentation of various access-related parameters. The documentation is essential as information if evaluated regularly and discussed during sessions of VA work groups or similar structures.

The next step is data collection to analyze the type of primary VA and the rates and types of all complications. Early failure rates, use of temporary and atrial cuffed catheters and survival rates of the different types of accesses can be monitored easily, may be in collaboration with other dialysis centers locally or regionally. These data should be accessible to all disciplines involved for analysis of successes and failures. This strategy represents a continuous learning curve – are we as good as we believe to be?

Is VA Accepted as Part of Education/Qualification for Becoming a Nephrologist, an Access Surgeon or a Radiologist?

This seems to be a dark chapter worldwide. There are neither rules nor governmental regulations; even the K/DOQI Vascular Access guidelines as well as the European guidelines fail to recommend a detailed education in the field of VA.

The author understands that first attempts to introduce educational courses are occurring in USA, Italy and Germany; certification is the goal. In addition, VA must be a topic in the final nephrology examination. Similar conditions should be established for surgeons and radiologists. Any further initiative would be welcomed.

Education is practised today in many institutions along the learning-by-doing method from the senior to the junior nephrologist, surgeon or radiologist. In the end, this type of education is very similar to uncontrolled improvization and far from quality control and national standards; a systematic approach is missed.

A lot of work is waiting to be defined and achieved in detail.

Conclusions

Nephrologists have to care for their hemodialysis patients. Beyond the care for complications, VA remains the Cinderella of hemodialysis therapy despite the strong, repeatedly emphasized opinion of renowned nephrologists about the great importance of VA. Unfortunately, the number of randomized, controlled trials in nephrology and particularly in the field of VA is very low, as recently reported; strong efforts will be needed immediately [13].

Nephrologists bear the responsibility for the key-decisions in patient care; they are responsible for the VA. This starts with the first referral of the patient to the nephrologist and ends when the patient dies. A lot of activities are required and continuous cooperation with access surgeons and radiologists is essential. Flexibility between the disciplines involved is needed. Documentation, data collection and analysis of failures are indispensable components. New goals and new educational structures have to be defined. It seems likely that we are just at the beginning of a new era.

References

1   Allon M, Lockhart ME, Lilly RZ, Gallichio MH, Young CJ, Barker J, Deierhoi MH, Robbin ML: Effect of preoperative sonographic mapping on vascular access outcomes in hemodialysis patients. Kidney Int 2001;60:2013–2020.
2   Malovrh M: Native arteriovenous fistula: Preoperative evaluation. Am J Kidney Dis 2002;39: 1218–1225.
3   Besarab A, Brouwer D: Aligning hemodialysis treatment practices with the National Kidney Foundation's K/DOQI Vascular Access Guidelines. Dial Transplant 2004;33/11:694–711.
4   Saran R, Pisoni RL, Weitzel WF: Epidemiology of vascular access for hemodialysis and related practice patterns. Contrib Nephrol. Basel, Karger, 2004, vol 142, pp 14–28.
5   Allon M, Robbin ML: Increasing arteriovenous fistulas in hemodialysis patients: Problems and solutions. Kidney Int 2002;62:1109–1124.
6   Konner K, Hulbert-Shearon TE, Roys EC, Port FK: Tailoring the initial vascular access for dialysis patients. Kidney Int 2002;62:329–338.
7   Spergel LM, Holland JE, Fadem SZ, McAllister CJ, Peacock EJ: Static intra-pressure ratio does not correlate with access blood flow. Kidney Int 2004;66:1512–1516.
8   Konner K: Interventional strategies for haemodialysis fistulae and grafts: Interventional radiology or surgery? Nephrol Dial Transplant 2000;15:1922–1923.
9   Shaldon S: Preface in 'Hemodialysis Vascular Access and Peritoneal Dialysis Access'. Contrib Nephrol. Basel, Karger, 2004, vol 142, pp X–XII.
10 Bonucchi D, D'Amelio A, Capelli G, Albertazzi A: Management of vascular access for dialysis: An Italian survey. Nephrol Dial Transplant 1999;14:2116–2118.
11 Ravani P, Marcelli D, Malberti F: Vascular access managed by renal physicians: The choice of native arteriovenous fistulas for hemodialysis. Am J Kidney Dis 2002;40:1264–1276.
12 Rayner HC, Besarab A, Brown WW, Disney A, Saito A, Pisoni RL: Vascular access results from the Dialysis Outcomes and Practice Patterns Study (DOPPS): Performance against Kidney Disease Outcomes Quality Initiative (K/DOQI) Clinical Practice Guidelines. Am J Kidney Dis 2004;44(suppl 3):S22–S26.
13 Strippoli GFM, Craig JC, Schena FP: The number, quality, and coverage of randomized controlled trials in nephrology. J Am Soc Nephrol 2004;15:411–419.

Dr. Klaus Konner
Schau ins Land 24
DE–51429 Bergisch Gladbach (Germany)
Tel. +49 2204 566 25, Fax +49 2204 609 691, E-Mail klaus.konner@uni-koeln.de

Ronco C, Brendolan A, Levin NW (eds): Cardiovascular Disorders in Hemodialysis.
Contrib Nephrol. Basel, Karger, 2005, vol 149, pp 131–137

....

Challenges in Interventional Nephrology

Miguel C. Riella

Renal Division, Catholic University of Parana, Brazil,
and Chairman of the Interventional Nephrology Committee
of the International Society of Nephrology

Abstract

Lately, there has been a progressive decrease in the interest of nephrology as a medical sub-specialty reflected primarily in the decreasing number of renal fellows. Rising costs in establishing and running dialysis clinics and the 'lost' of nephrologic procedures previously performed by nephrologists are among the many reasons for this disillusionment with the specialty. The care of chronic kidney patients frequently involve many diagnostic and interventional radiological procedures such as: diagnostic renal ultrasonography, ultrasound-guided kidney biopsies, placement of tunneled hemodialysis catheters or peritoneal catheters, sonographic and radiological investigation of vascular access dysfunction, etc. Most of these procedures are nowadays performed by radiologists, vascular surgeons and surgeons in general. This fragmentation does not optimize medical care and it is inconvenient to the patient. This has led many nephrologists to introduce a new paradigm in kidney patients management, often referred as 'interventional nephrology' (IN). This new breed of nephrologists have acquired diagnostic and interventional skills for procedures usually done by others with an added clinical perspective. To train nephrologists in these procedures and avoid the fragmented care of renal patients, the American Society for Diagnostic and Interventional Nephrology was established in 2000 and the International Society of Nephrology in 2004 introduced a new committee to address the issues of IN. It is hoped that concerted efforts will help to rescue these activities for the nephrologists and improve quality of patient care.

Introduction

Over the years, we have witnessed a progressive decrease in the interest of nephrology as a medical sub-specialty reflected primarily in the decreasing number of renal fellows in training programs [1–3].

Among many reasons for the unattractiveness of nephrology, one may single out barriers in dialysis. Stricter dialysis regulations and cost of setting up and running a dialysis clinic, have turned free-standing dialysis facilities a luxury of dialysis chains of large corporations.

Another aspect is the 'lost' of nephrologic procedures previously performed by nephrologists. Medical sub-specialties with many procedures (gastroenterology, cardiology, etc.) continue to attract young physicians, since these procedures represent a significant part of their total income.

The care of chronic kidney patients frequently involve many diagnostic and interventional radiological procedures such as: diagnostic renal ultrasonography, ultrasound-guided kidney biopsies, placement of tunneled hemodialysis (HD) catheters or peritoneal catheters, sonographic and radiological investigation of vascular access dysfunction, etc.

Most of these procedures are nowadays performed by radiologists, vascular surgeons and surgeons in general and from their perspective, most procedures are considered elective and therefore of relatively low priority [4].

This approach does not optimize medical care and it is inconvenient to the patient. While delay and inconvenience are important driving forces for a change, the ultimate concern should be quality of care [4].

This has led many nephrologists to introduce a new paradigm in kidney patients management, often referred as 'interventional nephrology'(IN) [5]. This new breed of nephrologists have acquired diagnostic and interventional skills for procedures usually done by others with an added clinical perspective.

Charles O'Neill [6] has pioneered the efforts in nephrology-based diagnostic ultrasonography and has challenged nephrologists to take a more proactive role in the total provision of nephrologic care of their patients. Beathard and colleagues [7–9], Work [10–12], Rasmussen [13, 14], Asif and colleagues [15–18], Ash [19, 20] and others have made significant contributions in this area.

Why the Need for a 'Procedural Nephrologist'?

The procedures mentioned above are integral part to the care of our patients and no one is better suited to perform and interpret them than nephrologists [21].

We were the leaders in developing these procedures: the first HD access [22], the arteriovenous fistula [23], dual lumen HD catheters [24], and the cuffed peritoneal dialysis (PD) catheter [25] and the peritoneoscopic insertion technique [26] were all developed by nephrologists. Nephrologists pioneered the percutaneous renal biopsy [27] and the subsequent refinements [28] and Holmes [29], another nephrologist, is considered by many to be the father of diagnostic ultrasonography.

Nephrologists have shown benefits and their competence with these procedures [6, 30, 31]. Performance of nephrology-related procedures by an interventional nephrologist has been clearly shown to avoid unnecessary delays [17].

Establishing an IN Program

IN should be an integral part of our fellowship programs. It is generally felt that there is no need for a training program to start offering all procedures. Formal instruction and practical experience with real-time ultrasound guidance for native and allograft kidney biopsies and ultrasound-guided central line placement are obvious starting points [4]. Tunneled catheters and vascular access management through imaging, angioplasty, thrombectomy are logical next steps [4].

Some of the procedures currently performed by the 'procedural nephrologist' are listed.

Procedures

Renal Ultrasonography-Diagnostic and Ultrasound-Guided Renal Biopsy

The concept of 'renal ultrasound by the nephrologist' was introduced by O'Neill [6] in the 90s when he demonstrated that ultrasonography performed by the nephrologist yields prompt diagnostic and fast therapy formulation.

Both procedures can be performed at the bed-side, thus avoiding scheduling, transportation and radiology suite restrictions. It will also reduce cost by avoiding the use of a radiology suite and reducing delays.

Recent data have shown that the time required to obtain a renal ultrasound on an outpatient basis was markedly reduced when performed by the IN service, from 46.5 ± 2.4 days (mean $\pm$ SE) using radiology to 4.7 ± 0.7 days using IN [17].

Ultrasound is a safe and simple tool that can be employed effectively by nephrologists as a critical element in the diagnostic work-up of a patient with renal disease.

Guidelines for training, certification and accreditation in renal ultrasonography have been published [32].

PD Access Procedures

Among the many reasons for the low prevalence of PD in the dialysis population, particularly in the United States, one will find timely placement of the PD catheter as a contributor factor [33]. Asif et al. [33] have shown that at their institution, the average time interval between first contact and actual placement of the PD catheter by the nephrologist was 7 days, while prior to the establishment of an IN program, referral to a surgeon required about 1–2 weeks. Up to 4 weeks were often necessary for chronic ambulatory peritoneal dialysis catheter placement.

The procedure can safely be performed in a procedure room or interventional laboratory under local anesthesia. The three commonly used techniques for PD catheter insertion are surgical, blind trocar and laparoscopic. The latter technique allows direct visualization and recently a randomized trial comparing peritoneoscopic versus surgical technique clearly demonstrated prolonged catheter survival with peritoneoscopic placement [34].

Guidelines for training, certification and accreditation in placement of permanent tunneled and cuffed PD catheters have been published [35].

Vascular Access: Endovascular Procedures and Tunneled HD Catheter

Vascular access is the *life line* of dialysis patients. Therefore, it is important to maintain vascular access in good functional shape at all times. There is no one better than the nephrologist to understand the magnitude and seriousness of this problem and to tackle it effectively and promptly.

An interventional nephrologist can perform percutaneous angiography and balloon angioplasty for failing accesses, declot failed AV grafts using mechanical thrombolysis and place tunneled HD catheters [36].

The establishment of a vascular access program is a little more complicated due to the difficulty of finding a suitable location to initiate the program. Interventional radiology suites or cardiac catheterization laboratories are usually busy. Lack of space to set up a room with radiation protection and the cost of equipment (fluoroscope, C-arm) are additional problems.

Nonetheless, vascular access procedures performed by the nephrologist ensures prompt delivery of care and improves patient outcome.

Guidelines for training, certification and accreditation for HD vascular access and endovascular procedures have been published [37].

Percutaneous Bone Biopsy

Renal osteodistrophy encompasses a wide spectrum of bone disorders and is often classified on the basis of the predominant histopathologic patterns [38]. The nephrologist will have to decide when it is appropriate to perform a bone biopsy which should be based on the presence of symptomatic manifestations and whether the possibility of aluminum accumulation exists.

The performance of bone histomorphometry requires obtaining undecalcified sections of bone of excellent quality. A transcortical iliac crest biopsy is the procedure of choice and performed under local anesthesia. This procedure continues to be used for diagnosis and research of metabolic bone diseases. The technical perfomance is not specially difficult and could easily be learned by nephrologists [38].

At this moment, there are no guidelines for training, certification and accreditation.

The American Society of Diagnostic and Interventional Nephrology

The American Society of Diagnostic and Interventional Nephrology (ASDIN) was founded in the year 2000 to promote the procedural aspects of nephrology through education, training, research, certification and accreditation. Several guidelines have already been published in the area of renal ultrasonography, vascular access and endovascular procedures and PD access [32, 35, 37].

More recently, Seminars in Dialysis became the official Journal of ASDIN and the appropriate forum for publications in this area.

The IN Committee of the International Society of Nephrology

The International Society of Nephrology (ISN) has been following for quite some time the discussion on the role of diagnostic and IN and felt the need to intervene on a worldwide basis. Therefore, in June, 2004 the ISN decided through the executive committee to propose the establishment of a new committee, IN committee, to address the issues of education, training, certification and accreditation of this new field.

Together with ASDIN, the committee hopes to contribute to the establishment of training centers around the world, to promote symposia and training courses, stimulate research and contribute for the improvement of guidelines for training, certification and accreditation in the areas mentioned.

Future Challenges

According to O'Neill, 'our patient's needs are simply not being met and to remedy this, nephrologists must take charge, embracing new techniques and knowledge. The new nephrologist will need to be proficient in diagnostic ultrasonography, renal biopsy and insertion and removal of cuffed dialysis catheters.

The new nephrologist must be knowledgeable in the vascular anatomy of the upper extremity and the care and evaluation of vascular accesses. In addition, every nephrology practice should have available expertise in angiography, angioplasty and the insertion of PD catheters.

Our challenge is to identify existing training centers, stimulate the development of new training centers and to establish guidelines, certification and criteria for credentialing.

References

1   Council of American Kidney Societies predicts 'critical shortage' of nephrologists by 2010. Nephrol News Issues 1997;11:12–14.
2   Levison SP: The changing face of nephrology. Semin Nephrol 1999;19:98–104.
3   Kletke PR: The changing supply of renal physicians. Am J Kidney Dis 1997;29:781–792.
4   Allon M, Warnock DG: Interventional nephrology: Work in progress. Am J Kidney Dis 2003;42: 388–391.
5   O'Neill WC: Seminars in nephrology. Introduction. Semin Nephrol 2002;22:181–182.
6   O'Neill WC: Renal ultrasonography: A procedure for nephrologists. Am J Kidney Dis 1997;30: 579–585.
7   Beathard GA: The treatment of vascular access graft dysfunction. A nephrologist's view and experience. Adv Ren Replace Ther 1994;1:131–147.
8   Beathard GA, Welch BR, Maidment HJ: Mechanical thrombolysis for the treatment of thrombosed hemodialysis access grafts. Radiology 1996;200:711–716.
9   Beathard GA: Angioplasty for arteriovenous grafts and fistulae. Semin Nephrol 2002;22:202–210.
10   Work J: Hemodialysis catheters and ports. Semin Nephrol 2002;22:211–220.
11   Work J: Does vascular access monitoring work? Adv Ren Replace Ther 2002;9:85–90.
12   Work J: Chronic catheter placement. Semin Dial 2001;14:436–440.
13   Rasmussen RL: Establishing a dialysis access center. Nephrol News Issues 1998;12:61–63.
14   Rasmussen RL: Establishing an interventional nephrology suite. Semin Nephrol 2002;22:237–241.
15   Asif A, Byers P, Vieira CF, Preston RA, Roth D: Diagnostic and interventional nephrology. Am J Ther 2002;9:530–536.
16   Asif A, Byers P, Vieira CF, Merrill D, Gadalean F, Bourgoignie JJ, Leclercq B, Roth D, Gadallah MF: Peritoneoscopic placement of peritoneal dialysis catheter and bowel perforation. Experience of an interventional nephrology program. Am J Kidney Dis 2003;42:1270–1274.
17   Asif A, Byers P, Vieira CF, Roth D: Developing a comprehensive diagnostic and interventional nephrology program at an academic center. Am J Kidney Dis 2003;42:229–233.
18   Asif A: Interventional nephrology: A call to action. Int J Artif Organs 2003;26:447–451.
19   Ash SR: The evolution and function of central venous catheters for dialysis. Semin Dial 2001;14: 416–424.
20   Ash SR: Chronic peritoneal dialysis catheters: Procedures for placement, maintenance, and removal. Semin Nephrol 2002;22:221–236.
21   O'Neill WC: The new nephrologist. Am J Kidney Dis 2000;35:978–979.
22   Quinton WE, Dillard DH, Cole JJ, Scribner BH: Eight months' experience with silastic-teflon bypass cannulas. Trans Am Soc Artif Intern Organs 1962;8:236–245.
23   Brescia MJ, Cimino JE, Appel K, Hurwich BJ: Chronic hemodialysis using venipuncture and a surgically created arteriovenous fistula. N Engl J Med 1966;275:1089–1092.
24   Uldall PR, Woods F, Bird M, Dyck R: Subclavian cannula for temporary hemodialysis. Proc Clin Dial Transplant Forum 1979;9:268–272.
25   Tenckhoff H, Schechter H: A bacteriologically safe peritoneal access device. Trans Am Soc Artif Intern Organs 1968;14:181–187.

26 Ash S: Bedside peritoneoscopic peritoneal catheter placement of Tenckhoff and newer peritoneal catheters. Adv Peritoneal Dial 1998;14:75–79.

27 Iversen P, Brun C: Aspiration biopsy of the kidney. Am J Med 1951;11:324–330.

28 Kark R: The development of percutaneous renal biopsy in man. Am J Kidney Dis 1990;16: 585–589.

29 Holmes J: Early diagnostic ultrasonography. J Ultrasound Med 1983;2:33–43.

30 Beathard GA, Marston WA: Endovascular management of thrombosed dialysis access grafts. Am J Kidney Dis 1998;32:172–175.

31 Nass K, O'Neill WC: Bedside renal biopsy: Ultrasound guidance by the nephrologist. Am J Kidney Dis 1999;34:955–959.

32 Guidelines for training, certification, and accreditation in renal sonography. Semin Dial 2002;15: 442–444.

33 Asif A, Byers P, Gadalean F, Roth D: Peritoneal dialysis underutilization: The impact of an interventional nephrology peritoneal dialysis access program. Semin Dial 2003;16:266–271.

34 Gadallah MF, Pervez A, el-Shahawy MA, Sorrells D, Zibari G, McDonald J, Work J: Peritoneoscopic versus surgical placement of peritoneal dialysis catheters: A prospective randomized study on outcome. Am J Kidney Dis 1999;33:118–122.

35 Guidelines for training, certification, and accreditation in placement of permanent tunneled and cuffed peritoneal dialysis catheters. Semin Dial 2002;15:440–442.

36 Asif A, Merrill D, Brouwer D, Roth D, Ash SR: Procedural nephrology: Changing the face of renal disease care. Dial Transplant 2004;33:258–260.

37 Guidelines for training, certification, and accreditation for hemodialysis vascular access and endovascular procedures. American Society of Diagnostic and Interventional Nephrology. Semin Dial 2003;16:173–176.

38 Ho LT, Sprague SM: Percutaneous bone biopsy in the diagnosis of renal osteodistrophy. Semin Nephrol 2002;22:268–275.

Prof. Miguel C. Riella, MD, PhD
Iguassu Ave 2689
Curitiba, 80240–030, Parana (Brazil)
Tel. +55 41 342 5849, Fax +55 41 244 5539, E-Mail meriella@pro-renal.org.br

Ronco C, Brendolan A, Levin NW (eds): Cardiovascular Disorders in Hemodialysis.
Contrib Nephrol. Basel, Karger, 2005, vol 149, pp 138–149

....

Vascular Access Education, Planning and Percutaneous Interventions by Nephrologists

Arif Asif, Donna Merrill, Phillip Pennell

Department of Medicine Division of Nephrology,
University of Miami Miller School of Medicine, Miami, Fla., USA

Abstract

To optimize vascular access care of patients with end stage renal disease, nephrologists themselves are taking a keen interest in the management of vascular access-related issues. Because of their unique clinical perspective on dialysis access and better understanding of the intricacies of renal replacement therapy, nephrologists are ideally suited for this activity. Two areas are the main focus of attention by these specialists: vascular access education and access-related percutaneous interventions. Vascular access-related procedures commonly performed by nephrologists include percutaneous balloon angioplasty for vascular access stenosis, thrombectomy procedure for a thrombosed arteriovenous access, tunneled hemodialysis catheter-related procedures and vascular mapping to determine the patient's optimal vascular access. While the performance of these procedures by nephrologists offers many advantages, appropriate training in order to develop the necessary procedural skills is critical. Recent data have emphasized that a nephrologist can be successfully trained to become a competent interventionalist. In addition to documenting excellent outcome data, multiple reports have demonstrated safety and success when these procedures are performed by nephrologists. This chapter focuses on vascular access education and hemodialysis access-related procedures performed by nephrologist and calls for a proactive approach in optimizing this aspect of patients care.

Introduction

Over the past decade, significant advances have been made by nephrologists in the performance of hemodialysis access-related procedures [1–16].

Because of their training and experience, nephrologists have a unique clinical perspective on vascular access-related issues, renal disease, renal replacement therapy and dialysis outcomes. This perspective makes them ideally suited to perform vascular access-related procedures. Recent data have emphasized that nephrologists can assume a greater role in the procedural aspects of vascular access and do so with effectiveness, efficiency and safety [2–18].

The purpose of this chapter is to review vascular access education and planning as well as a variety of vascular access-related procedures performed by nephrologists, including angioplasty for venous stenosis, treatment of thrombosed vascular access, salvage of undeveloped arteriovenous fistulae (AVFs), management of tunneled dialysis catheters (TDCs) and other related procedures. In addition, procedure-related complications when performed by nephrologists are also highlighted.

Vascular Access Education and Planning

Vascular access always has been the Achilles heel of extracorporeal dialysis. Historically, nephrologists took the lead in the development and clinical application of innovations in vascular access. However, surgeons and radiologists have become the clinicians who perform most vascular access-related procedures and, perhaps by default, have become the clinicians who have made most of the decisions regarding access planning and placement. As a result, the vascular access options available to nephrologists caring for hemodialysis patients typically have been limited by the diagnostic and percutaneous technologies offered by their radiology colleagues and by the repertoire of their surgical colleagues. The terrain is changing in those medical communities where interventional nephrology has emerged and widened the horizons of both diagnostic and procedural options available for nephrologists to offer their hemodialysis patients. A consequence is the resurgence of nephrologists again assuming the leadership role in clinical decision making regarding vascular access planning, placement, and maintenance.

Vascular Access Education

In our university medical center, interventional nephrologists have spurred the dialogue regarding vascular access. There is a flurry of enthusiasm for learning about vascular access that extends from the old timers through the junior faculty to the fellows and beyond to embrace the house staff and medical students. The clinical staffs of our dialysis units have a new enthusiasm and commitment to access surveillance, management and maintenance.

These changes have resulted from the direct involvement of interventional nephrologists in patient care, working together with physicians and other members of the health care team in clinical settings. New dialysis patients are examined for purposes of access planning. Clinical problems in prevalent hemodialysis patients such as reduced access flow rates, swollen arms, and steal syndromes are assessed in the dialysis clinics. Physical examination techniques are demonstrated and often correlated with vascular radiographic images. Results of diagnostic and therapeutic interventions are shared with the patient's care team.

Interventional nephrologists are having a very positive impact on the educational environment of our university medical center and importantly are raising the level of concern about vascular-access health care and positively influencing access outcomes at all levels. One of the most important effects has been to stimulate early referral of patients for vascular access evaluation, planning and placement. Perhaps just as important has been the positive impact on patient education and awareness of vascular access issues (vide infra).

Vascular Access Planning

The National Kidney Foundation Kidney Disease Outcomes Quality Initiative (K/DOQI) recommends that prospective end-stage renal disease patients as well as prevalent dialysis patients should be clinically evaluated by history and physical examination to determine the most suitable type and location of vascular access [1]. When clinical evaluation by history and physical examination reveals potential problems known to be associated with vascular impairment, diagnostic evaluation by venography and/or vascular ultrasound is recommended to detect underlying defects in vascular structure and flow. Such problems include edema, collateral veins, prior subclavian vein cannulation, a transvenous pacemaker, prior trauma or surgery in the area of venous drainage, multiple previous accesses, and diminished arterial pulses. Venography is cited as most useful for vein assessment, especially central veins, while ultrasonography may be particularly useful for vascular evaluation when contrast agents are to be avoided [1]. A major goal of our interventional program is to foster the placement of AVFs while minimizing the placement of arteriovenous grafts (AVGs) and limiting TDCs to short-term use as bridge accesses while the preferred arteriovenous accesses (AVFs) mature. In our view, vascular mapping prior to any vascular surgery procedure has proven to be essential for achieving this goal.

Vascular Mapping

One of the most important impacts of the interventional nephrology team has been to demonstrate the importance of vascular mapping prior to any surgical procedures. This experience has confirmed that physical examination is not adequate to demonstrate either the presence or absence of suitable veins for creation of an arteriovenous fistula. The final outcome has been the identification of vessels suitable for an AVF in patients previously thought to be candidates only for grafts or for consignment to permanent use of TDCs.

It has become routine at our medical center for all patients needing vascular access surgery to be referred to interventional nephrology for preoperative venography [3, 18]. The procedure involves cannulation of a peripheral vein on the dorsum of the hand for injection of a small quantity of low osmolarity contrast medium (Isovue 370; Bracco, Minneapolis, Minn., USA) (10–20 cc) diluted with 10–20 cc of normal saline. Using this protocol, we have not observed impairment of renal function [18]. Images from the wrist to the right atrium are obtained by fluoroscopy (GE 9800, GE Medical Systems) using the pulse and road map feature (15 frames per second) with calibration for measurement using a radiopaque ruler. During venography, a tourniquet to distend veins by occluding venous outflow is not applied in order to avoid overestimating vein size. Based on a prior benchmark study [19], the criteria for suitability of veins include size of at least 2.5 mm, absence of stenosis and continuity with downstream patent veins. Vein mapping is performed on both upper extremities with evaluation of peripheral as well as central veins in all patients. A stenosis equal to or exceeding 50% compared to normal adjacent vessel is considered to be significant, based on K-DOQI guidelines [1]. According to the recommendations delineated by the NKF-DOQI vascular access guidelines [1], the arterial system is evaluated by means of a detailed physical examination of blood pressure, arterial pulses, capillary refill and the Allen test in both extremities [1]. Specifically required are a blood pressure differential of <20 mm Hg and a negative Allen test (patent palmar arch).

Our interventional nephrology team has elected to evaluate vessels for access placement by using venography rather than sonography [3, 18]. There have not been randomized studies to determine whether venography is superior to sonography for evaluation of the vasculature prior to access surgery. Venography offers the clear advantage of directly imaging central veins instead of indirect assessment provided by Doppler evaluation. Although Doppler ultrasound offers the advantage of noninvasive arterial evaluation, we have found that physical examination of the arterial system provides ample information for AVF creation. Indeed, K/DOQI guideline #2 offers the opinion-based recommendation that Doppler examination or arteriography should be performed only in the presence of markedly diminished arterial pulses on physical examination [1]. Our approach is based on this recommendation.

Upon conclusion of vascular mapping, in lay terminology patients are educated about possible site(s) for fistula creation. We have found that this is an opportune time to educate patients regarding vascular access types and their associated complications, including the risks of morbidity and mortality. The AVF is highlighted as the best available access with the lowest incidence of complications and best associated patient survival. Patients are encouraged to request the creation of an AVF at their appointment with the surgeon.

Preoperative vascular assessment utilizing ultrasonography or venography has been demonstrated to be superior to physical examination (inspection of veins by the naked eye using the tourniquet placed on the upper arm) in evaluating vessels suitable for arteriovenous fistula creation [19–22]. Relying on physical examination to assess the vascular system may exclude patients in whom venography would demonstrate adequate veins for AVF creation. One study demonstrated that a dramatic increase in arteriovenous fistula creation resulted from using preoperative vascular mapping by sonography rather than the traditional physical examination (preoperative physical examination = 34%, preoperative sonographic vascular mapping = 64%; $p < 0.001$) with a doubling of the patients dialyzing successfully with a fistula (preoperative physical examination = 16%, preoperative sonographic vascular mapping = 34%; $p < 0.001$) [20]. In another study, there was significant improvement in AVF creation (from 14 to 63%), reduction in AVG placement (from 62 to 30%) and reduction in TDC insertion (from 24 to 7%) when preoperative mapping of the arteries and veins was performed by Doppler duplex ultrasonography [19].

The value of vascular mapping in fostering creation of AVFs is illustrated in two case series we recently have reported [3, 23]. In the first instance [3], among 86 patients consigned to TDCs, vascular mapping demonstrated that 94% of patients with no prior arteriovenous accesses (64 of 66) had suitable veins for arteriovenous access placement and ninety percent of patients with previously failed arteriovenous accesses (18 of 20) had suitable veins, all basilic veins in the latter cases. Fistulae were created in 94% of those patients who agreed to proceed to vascular access surgery (68 of 72 patients), the other 4 patients having required AVG placement. These findings clearly demonstrate the value of vascular mapping in detecting vessels suitable for arteriovenous access placement, even in patients previously consigned to percutaneous catheters because of prior vascular access failures. The most striking outcome of this experience was that AVFs comprised 94% of the arteriovenous accesses created. Indeed, these results virtually mandate the search for patent veins suitable for arteriovenous access placement in every patient dialyzing with a percutaneous catheter.

The second series [23] of patients was comprised of 10 patients referred to interventional nephrology with problematic arteriovenous accesses: 5 patients with AVGs that were either thrombosed or had poor flow due to venous stenosis, 3 patients who had mega fistulae with thin-walled aneurysms and outflow stenosis, one patient with inadequate arterial flow to an AVF, and a final patient with multiple thromboses of an AVF related to prior stent placement. Management options for thrombosed AVGs and stenotic lesions traditionally have included thrombectomy, angioplasty, stent placement and, when these fail, placement of a 'jump' AVG. Similarly, the treatment of 'serpentine' mega AVFs, after resection, generally has been replacement by an AVG. The end result regretfully has been an AVG. During the interventional procedures for access problems in these 10 patients, vascular anatomy suitable for creation of AVFs was identified, and all 10 patients successfully had creation of secondary AVFs. Our experience with these patients demonstrates that secondary AVFs, rather than 'jump' AVGs, often can be placed successfully in patients with problematic arteriovenous accesses. As recommended by K/DOQI guideline # 29 [1], we recommend that during percutaneous interventions, patients routinely should have identification of vessels suitable for creation of secondary AVFs.

These two clinical experiences illustrate how a dedicated interventional nephrology team can be effective in optimizing the vascular access health status of dialysis patients by minimizing AVG use and maximizing creation of AVFs.

Vascular Access-Related Procedures

Percutaneous Balloon Angioplasty

Both AVGs and AVFs may develop vascular stenosis. The pathophysiologic mechanisms of stenosis are complex; however, neointimal hyperplasia appears to play a pivotal role [24, 25]. Traditionally, it has been highlighted that stenosis occurs most frequently at the venous anastomosis (60%). Recent information from our center, however, has emphasized that lesions may occur anywhere within the access system and can coexist as single or multiple [7]. Regardless of the location of the stenosis, percutaneous balloon angioplasty has become a standard treatment for the management of arteriovenous dialysis access (grafts, fistulae) stenosis [4–8].

Access stenosis should be treated if the stenosis is 50% or greater and is associated with clinical or physiological abnormalities [1]. The abnormal clinical parameters used to suspect the presence of stenosis should return to within acceptable limits following intervention. In a series of 1,120 cases of venous stenosis treated by angioplasty, Beathard [26] reported the initial success rate to be 94%. Primary (unassisted) patency determined by life table analysis was as follows: 1 month – 87.4%, 2 months – 84.8%, 3 months – 77.2%, 6 months – 66.4%, one year – 44.5%. These results are comparable to those reported by interventional radiology [27, 28].

A significant number (10–25%) of AVFs do not adequately develop and fail to sustain dialysis therapy. Recent data have classified fistulae failure into early and late failure [8]. Early failure refers to the fistulae that never develop to the point where they can be used or that fail within the 3 months of successful usage. In contrast, late failure denotes failure after 3 months of successful usage. Traditionally, these undeveloped fistulae were abandoned. Recently, Beathard et al. [8] provided invaluable information regarding how to improve the function of an AVF that is not developing properly. In this prospective observational study, 100 patients with early failure underwent evaluation and treatment at six free-standing outpatient vascular access centers. Vascular stenosis and the presence of a significant accessory vein (an accessory vein is described as a branch coming off the main venous channel that comprised the fistula) alone or in combination were found to be the culprits. Venous stenosis was present in 78% of the cases. A majority of these lesions (48%) were found to be close to the anastomosis (juxta-anastomotic lesion). A significant accessory vein was present in 46% of the cases. Percutaneous balloon angioplasty and accessory vein obliteration using any of the three techniques (percutaneous ligation using 3/0 nylon, venous cutdown, coil insertion) were used to salvage the failed fistulae. Angioplasty was performed with 98% and vein obliteration with 100% success rate. Postintervention, it was possible to initiate dialysis using the fistula in 92% of the cases. Actuarial life table analysis showed that 84% were functional at 3 months, 72% at 6 months and 68% at 12 months.

A nephrologist can effectively establish a surveillance program to monitor vascular access function, identify the failing access and perform percutaneous intervention to correct stenosis and maintain a healthy vascular access and avoid access failure. By use of an aggressive approach and employment of two basic techniques, balloon angioplasty and vein obliteration, nephrologists can successfully salvage and subsequently utilize an otherwise failed fistula.

Thrombectomy Procedure for Access Thrombosis

The most frequent complications associated with AVGs are stenosis and thrombosis. Since the most common cause of access thrombosis is stenosis these should not be considered separate problems. Venous stenosis and access thrombosis share the relationship of cause and effect. Indeed, the frequency of stenosis identified at thrombolysis has been found to exceed 90% [10].

Nephrologists are routinely performing thrombectomy procedures for a thrombosed AVG and AVF [5, 6, 9, 10]. Both mechanical and pharmacomechanical thrombolysis for the treatment of thrombosed dialysis access can be successfully performed by nephrologists [6, 9, 10]. Although the initial success rate (95%) is similar to angioplasty procedure, primary patency rates after thrombectomy of a clotted vascular access are markedly reduced when compared to the primary patency following angioplasty of access stenosis. In a prospective analysis of 1,176 cases, 3-month, 6-month and 12-month primary patency following mechanical thrombolysis of thrombosed grafts was found to be 52, 39 and 17%, respectively [10]. These results are comparable to those reported by interventional radiology [27, 28].

Thrombosis leading to access failure is a major issue. It often causes unnecessary hospitalization, missed dialysis, frustration on the part of the dialysis staff and patients and exposes the patient to temporary catheters. Ideally, this complication should be managed rapidly, under local anesthesia and on an outpatient basis. Recent data have clearly shown that such care is being delivered successfully by nephrologists at many centers on an outpatient basis [4–6, 10]. Nevertheless, aggressive detection and early correction of hemodialysis access stenosis is of utmost importance to decrease graft thrombosis and improve access survival.

Tunneled Hemodialysis Catheter Procedures

Hemodialysis cuffed tunneled catheters are commonly used as a bridge access to allow time for placement or maturation of a permanent vascular access (AVF, AVG). In addition, they may be used as a temporary access for hemodialysis for patients with acute renal failure. Finally, tunneled catheters may also be used as a permanent vascular access for patients who have exhausted all other options to receive hemodialysis. Traditionally, these catheters were placed by surgeons followed by interventional radiologists [29]. However, nephrologists have recently begun performing this procedure routinely both on an inpatient as well as outpatient basis [11, 13–15]. In addition to the catheter insertion, catheter exchange and removal procedures, nephrologists are also engaged in developing optimal catheter design to achieve adequate blood flow to sustain dialysis treatment [15].

Tunneled catheters play a major role in the delivery of hemodialysis therapy to a large portion of the dialysis population. However, their use is associated with many complications [1, 16]. In addition to catheter related-infection, fibroepithelial sheath formation is associated with these catheters and leads to catheter malfunction and occlusion [16, 30]. A recent study [30] evaluating 947 cases of catheter dysfunction (inability to sustain a blood flow greater than 300 ml/min) documented that fibrin was detected in 368 cases (38.8%). The presence of a fibrin sheath was determined at the time of exchange using radio-contrast material administered through the venous port of the old catheter. In this study, an angioplasty balloon catheter was inserted over a guide wire through the catheter tunnel, and therefore, through the lumen of the fibrin sheath and was then inflated to disrupt the fibro-epithelial sheath. The investigators used an 8-mm diameter balloon with 100% success rate. Removal of the sheath was confirmed by a repeat radiocontrast injection at the time of insertion of the new catheter over the guide wire. Catheter blood flow rates sufficient for dialysis were achieved in 99%. The presence of a fibrin sheath is a relatively common cause of tunneled catheter dysfunction [16, 40]. Nephrologists can successfully manage this complication on an outpatient basis by using percutaneous balloon angioplasty and over the wire catheter exchange techniques.

Complications of Percutaneous Interventions by Nephrologists

Recently, in the largest prospective series published to date (n = 14,047), Beathard reported the complications of endovascular procedures performed by interventional nephrologists [17]. In this report, data on basic hemodialysis procedures [tunneled hemodialysis catheter (THC) insertion and exchange, percutaneous transluminal angioplasty (PTA) of grafts and fistulae, and thrombectomy of both grafts and fistulae] were analyzed for safety and effectiveness.

In 5,121 PTA procedures (fistulae n = 1,561; grafts n = 3,560), the complication rate in cases of fistulae and grafts included 3.35 and 0.76% grade 1 hematoma (stable, does not affect flow), 0.4 and 0.11% grade 2 hematoma (stable, slows or stops flow) and 0.19 and 0.05% grade 3 hematoma (represents a complete vascular rupture, expands rapidly and leads to access loss), respectively [17]. These results are far superior to those reported previously (1.7–6.6%) [31–34].

Amongst 4,899 thrombectomy cases (fistulae, n = 228; grafts, n = 4,671), the complication rate in cases of fistulae and grafts included 5.7 and 3.32% grade 1 hematoma (stable, does not affect flow), 0.88 and 0.83% grade 2 hematoma (stable, slows or stops flow) and 0.43 and 0.41% grade 3 hematoma (represents a complete vascular rupture, expands rapidly, leads to access loss), respectively [17]. Peripheral artery embolism occurred in 0.38% of cases. These complication rates are lower than those reported previously (10–16%) [28, 35, 36].

A total of 4,027 THC-related procedures were evaluated for complications [17]. In the THC insertion procedure group (n = 4,027), complications included minor oozing at the cannulation and exit site (0.36%) and major adverse events like pneumothorax (0.06%). In contrast, both the surgical and radiological literature have documented a much higher incidence rate of complications including pneumothorax (2.5%) [29], hemothorax (0–0.6%), bleeding requiring exploration and/or transfusion (0–4.7%) and recurrent laryngeal nerve palsy (0–1.6%) [37–39]. When cases with THC exchange were analyzed (n = 2,262), only 1.41% had minor complications and there were no major complications identified [17].

Conclusion

As nephrologists, we bear the ultimate responsibility for the outcomes of our patients. Data clearly indicate safety, success, quality and excellent outcomes when vascular access-related procedures are performed by nephrologists. In addition, the all-too-frequent delays are minimized and procedural care more efficiently delivered by a nephrologist trained in hemodialysis vascular access procedures. We suggest that nephrologists should play a more proactive role in the planning and procedural management of hemodialysis access.

References

1   National Kidney Foundation: K-DOQI Clinical Practice Guidelines For Vascular Access: Update 2000. Guideline 10: Monitoring, surveillance, and diagnostic testing. Am J Kidney Dis 2000;37 (suppl 1):S150–S164.
2   Beathard GA: Interventionalist's role in identifying candidates for secondary fistulas. Semin Dial 2004;17:233–236.
3   Asif A, Cherla G, Merrill D, Cipleu D, Briones P, Pennell P: Conversion of tunneled hemodialysis consigned patients to arteriovenous fistula. Kidney Int, in press.
4   Beathard GA: Percutaneous transvenous angioplasty in the treatment of access vascular stenosis. Kidney Int 1992;42:1390–1397.
5   Jackson JW, Lewis JL, Brouillette JR, Brantley RR: Initial experience of a nephrologist-operated vascular accesscenter. Semin Dial 2000;13:354–358.
6   Schon D, Mishler R: Pharmacomechanical thrombolysis of natural vein fistulas: Reduced dose of TPA and long-term follow-up. Semin Dial 2003;16:272–275.
7   Asif A, Gadalean F, Merrill D, Cherla G, Cipleu C, Epstein DL, Roth D: Inflow stenosis in arteriovenous fistulae and grafts: A prospective, multi-center study. Kidney Int 2005;67:1–8.
8   Beathard GA, Arnold P, Jackson J, Litchfield T: Aggressive treatment of early fistula failure. Physician Operators Forum of RMS Lifeline. Kidney Int 2003;64:1487–1494.
9   Schon D, Mishler R: Salvage of occluded autologous arteriovenous fistulae. Am J Kidney Dis 2000;36:804–810.
10   Beathard GA, Welch BR, Maidment HJ: Mechanical thrombolysis for the treatment of the thrombosed hemodialysis grafts. Radiology 1996;200:711–716.
11   Asif A, Byers P, Vieira CF, Roth D: Developing a comprehensive diagnostic and interventional nephrology program at an academic center. Am J Kidney Dis 2003;42:229–233.
12   American Society of Diagnostic and Interventional Nephrology: Available at www.asdin.org. Accessed: 2001.

13 Work J: Hemodialysis catheters and ports. Semin Nephrol 2002;22:211–219.

14 Mankus RA, Ash SR, Sutton JM: Comparison of blood flow rates and hydraulic resistance between the Mahurkar catheter, the Tesio twin catheter, and the Ash Split Cath. ASAIO J 1998;44: M532–M534.

15 Ash SR: The evolution and function of central venous catheters for dialysis. Semin Dial 2001;14: 416–424.

16 Schon D, Whittman D: Managing the complications of long-term tunneled dialysis catheters. Semin Dial 2003;16:314–322.

17 Beathard GA, Litchfield T: Effectiveness and safety of dialysis vascular access procedures performed by interventional nephrologists. Kidney Int 2004;66:1622–1632.

18 Asif A, Cherla G, Merrill D, Cipleu CD, Lenz O: Venous mapping using venography and the risk of radiocontrast-induced nephropathy. Semin Dial, in press.

19 Silva MB Jr, Hobson RW 2nd, Pappas PJ, Jamil Z, Araki CT, Goldberg MC, Gwertzman G, Padberg FT Jr: A strategy for increasing use of autogenous hemodialysis access procedures: Impact of preoperative noninvasive evaluation. J Vasc Surg 1998;27:302–307.

20 Allon M, Lockhart ME, Lilly RZ, Gallichio MH, Young CJ, Barker J, Deierhoi MH, Robbin ML: Effect of preoperative sonographic mapping on vascular access outcomes in hemodialysis patients. Kidney Int 2001;60:2013–2020.

21 Huber TS, Ozaki CK, Flynn TC, Lee WA, Berceli SA, Hirneise CM, Carlton LM, Carter JW, Ross EA, Seeger JM: Prospective validation of an algorithm to maximize native arteriovenous fistulae for chronic hemodialysis access. J Vasc Surg 2002;36:452–459.

22 Miller A, Holzenbein TJ, Gottlieb MN, Sacks BA, Lavin PT, Goodman WS, Gupta SK: Strategies to increase the use of autogenous arteriovenous fistula in end-stage renal disease. Ann Vasc Surg 1997;11:397–405.

23 Asif A, Cherla G, Merrill D, Cipleu CD, Briones P, Pennell P: Maximizing fistula creation in prevalent hemodialysis population (abstract). American Society of Diagnostic and Interventional Meeting. Phoenix, 2005.

24 Roy-Chaudhury P, Kelly BS, Miller MA, Reaves A, Armstrong J, Nanayakkara N, Heffelfinger SC: Venous neointimal hyperplasia in polytetrafluoroethylene dialysis grafts. Kidney Int 2001; 59:2325–2334.

25 Sukhatme VP: Vascular access stenosis: Prospects for prevention and therapy. Kidney Int 1996;49:1161–1174.

26 Beathard GA: Percutaneous angioplasty for the treatment of venous stenosis: A nephrologist's view. Semin Dial 1995;8:166–170.

27 Gray RJ, Sacks D, Martin LG, Trerotola SO: Reporting standards for percutaneous interventions in dialysis access. Technology assessment committee. J Vasc Interv Radiol 1999;10:1405–1415.

28 Aruny JE, Lewis CA, Cardella JF, et al: Quality improvement guidelines for percutaneous management of the thrombosed or dysfunctional dialysis access. J Vasc Interv Radiol 1999;10: 491–498.

29 Lund GB, Trerotola SO, Scheel PF Jr, et al: Outcome of tunneled hemodialysis catheters placed by radiologists. Radiology 1996;198:467–472.

30 Beathard GA, Arnold P, Litchfield T: Management of fibrin sheath associated with tunneled hemodialysis catheters (abstract). J Am Soc Nephrol 2003;14:A241.

31 Longwitz D, Pham TH, Heckemann RG, Hecking E: [Angioplasty in the stenosed hemodialysis shunt: experiences with 100 patients and 166 interventions]. Rofo Fortschr Geb Rontgenstr Neuen Bildgeb Verfahr 1998;169:68–76.

32 Rundback JH, Leonardo RF, Poplausky MR, Rozenblit G: Venous rupture complicating hemodialysis access angioplasty: Percutaneous treatment and outcomes in seven patients. Am J Roentgenol 1998;171:1081–1084.

33 Raynaud AC, Angel CY, Sapoval MR, Beyssen B, Pagny JY, Auguste M: Treatment of hemodialysis access rupture during PTA with Wallstent implantation. J Vasc Interv Radiol 1998;9: 437–442.

34 Turmel-Rodrigues L, Pengloan J, Baudin S, Testou D, Abaza M, Dahdah G, Mouton A, Blanchard D: Treatment of stenosis and thrombosis in haemodialysis fistulas and grafts by interventional radiology. Nephrol Dial Transplant 2000;15:2029–2036.

35 Haage P, Vorwerk D, Wildberger JE, Piroth W, Schurmann K, Gunther RW: Percutaneous treatment of thrombosed primary arteriovenous hemodialysis access fistulae. Kidney Int 2000;57: 1169–1175.

36 McCutcheon B, Weatherford D, Maxwell G, Hamann MS, Stiles A: A preliminary investigation of balloon angioplasty versus surgical treatment of thrombosed dialysis access grafts. Am Surg 2003;69:663–667.

37 Bour ES, Weaver AS, Yang HC, Gifford RR: Experience with the double lumen Silastic catheter for hemoaccess. Surg Gynecol Obstet 1990;171:33–39.

38 McDowell DE, Moss AH, Vasilakis C, Bell R, Pillai L: Percutaneously placed dual-lumen silicone catheters for long-term hemodialysis. Am Surg 1993;59:569–573.

39 Schwab SJ, Beathard G: The hemodialysis catheter conundrum: Hate living with them, but can't live without them. Kidney Int 1999;56:1–17.

40 Beathard GA, Arnold P, Litchfield T: Management of fibrin sheath associated with funneled hemodialysis catheters (abstract). J Am Soc Nephrol 2003;14:241A.

Arif Asif, MD
Director, Interventional Nephrology
Associate Professor of Medicine, Department of Medicine Division of Nephrology
University of Miami Miller School of Medicine, 1600 NW 10th Ave (R 7168)
Miami, FL 33136 (USA)
Tel. +1 305 243 3583, Fax +1 305 243 3506, E-Mail Aasif@med.miami.edu

Ronco C, Brendolan A, Levin NW (eds): Cardiovascular Disorders in Hemodialysis.
Contrib Nephrol. Basel, Karger, 2005, vol 149, pp 150–161

....

Whole Body – Single Frequency Bioimpedance

Antonio Piccoli

Department of Medical and Surgical Sciences,
Nephrology Clinic, University of Padova, Padova, Italy

Abstract

Background: The postdialysis target weight is determined as the lowest weight a patient can tolerate without intradialytic symptoms or hypotension. Patterns of electrical properties of tissues allow a direct monitoring of fluid status without the need of the body weight. **Methods:** Whole body impedance is measured from skin electrodes on hand and foot. Impedance (Z vector) is a combination of resistance, R, i.e. the opposition to the flow of an injected alternating current, at any current frequency, through intra- and extracellular ionic solutions, and reactance, Xc, i.e. the dielectric component of cell membranes and organelles, and tissue interfaces. Measurements at 50 kHz current frequency are done with the best signal to noise ratio. **Results:** Cyclical tissue hydration changes in hemodialysis patients are detectable as changes in the whole body impedance, which can be utilized with patterns of impedance vector analysis in monitoring the prescription of optimal hydration independent of the body weight. Wet-dry weight prescription based on impedance vector analysis should bring abnormal vectors back into the reference, 75% tolerance ellipse, where tissue electrical conductivity is restored. **Conclusions:** Identification and ranking of normal versus abnormal tissue hydration can be obtained from impedance vector patterns without the need of equations.

Targeting the Optimal Weight before Symptoms

Hemodialysis (HD) often results in bringing the patient either to dehydration, that can be symptomatic (decompensated) or asymptomatic (compensated), or to fluid overload with pitting edema and worsening of hypertension. Edema is not usually detectable until the interstitial fluid volume has risen to about 30% above normal (4–5 kg of body weight), while severe dehydration can develop before clinical signs. The optimal postdialysis target weight is determined clinically, generally as the lowest weight a patient can tolerate without intradialytic symptoms or hypotension (the so-called dry weight).

The routine evaluation of hydration status based on body weight and blood pressure changes over time can be misleading, since changes are not uniquely determined by body fluid volume variations.

We present and discuss a simple methodology, based on electrical properties of tissues (bioelectrical impedance), which allows a direct monitoring of fluid status without the need of the body weight.

Electrical Properties of Tissues

Bioelectrical properties of the heart tissue are measured from skin electrodes on limbs and trunk and the direct measurements are utilized in the clinical setting. Both electrocardiography and bioelectrical impedance analysis (BIA) aim to transform electrical properties of tissues into clinical information [1]. Electrocardiography is based on several established patterns relating electrical measurements to heart disorders. *Conventional BIA* is based on electric models supporting quantitative estimates of body compartments through regression equations. *Vector BIA* is based on patterns of the Resistance-Reactance graph (RXc graph) relating body impedance to body hydration [1–3].

Body impedance is generated in tissues as an opposition to the flow of an injected current and is measured from skin electrodes on hand and foot. Impedance is represented with a complex number (a point) in the real-imaginary plane (Z vector), that is a combination of resistance, R, i.e. the opposition to the flow of an injected alternating current, at any current frequency, through intra- and extracellular ionic solutions (representing the real part of Z) and reactance, Xc, i.e. the dielectric or capacitative component of cell membranes and organelles, and tissue interfaces (representing the imaginary part of Z). The arc tangent of Xc/R is called the phase angle. In simple biological conductors without cells (e.g. saline, urine, ascites, and dialysate), no Xc component can be measured. Contribution of bone to impedance is negligible, and lean contributes more than fat soft tissue because adipocyte droplets of triacylglycerols are nonconductors [1–4].

The impedance (Ω) of a cylindrical conductor is proportional to its impedivity and to its length, and is inversely proportional to its cross-sectional area. Hence, whole body impedance is determined by limbs up to 90% and by trunk up to 10% [4]. Vector normalization by the subject's height (Z/H, in Ω/m) controls for the different conductor length.

Methods of Bioimpedance for Body Fluid Volume Assessment

BIA is a noninvasive method of body composition analysis, specific for the assessment of soft tissue hydration, and is divided along four methods of body fluid volume assessment.

The first and the most validated method is prediction of total body water (TBW) with functions of 50-kHz single-frequency, whole body impedance (either series measures or their parallel equivalents, mostly neglecting the Xc component). Hundreds of excellent validation studies have established a solid relation between whole-body impedance at 50 kHz, through the impedance index H^2/R, and body fluid volume, through isotope dilution, despite the fact that isotope dilution reflects the TBW, whereas bioimpedance reflects the fluid volume of soft tissue, particularly of limbs (90%) [2–4].

The second method is prediction of extracellular water and TBW with functions of low (1–5 kHz)- and high (100–500 kHz)-frequency impedance (dual- or multiple-frequency BIA), with the intracellular water calculated by difference [2–4].

The third and more recommended method in literature is use of many impedance measurements (500 measures in the range 1–1,000 kHz) through bioimpedance spectroscopy (BIS) following the Cole's model approach that extrapolates R values at limit frequencies for prediction of TBW, extracellular- and intracellular water [2–4]. Unfortunately, with BIS and multifrequency BIA, which are accurate in suspended cells, it is impossible to estimate the extracellular electric volume of tissues (and of intracellular by difference from the total volume) because an unknown and variable amount of low-frequency current passes through cells (tissue anisotropy), particularly through muscle fibers (parallel direction) [4, 5]. Differences in compartment estimates obtained with BIS as opposed to single-frequency BIA are caused by different, arbitrary constants of electric models [5].

The fourth and more recent method is use of the direct impedance vector measurement (both R and Xc component) at 50 kHz in a probabilistic graph, namely, the *RXc graph* of *Vector BIA*, which is a stand-alone method of body composition analysis, where the continuous, bivariate, random vector of impedance is evaluated through an ordinal scale of the deviation from a reference population (50, 75, and 95% tolerance ellipses, fig. 1). Clinical information on hydration is obtained through patterns of vector distribution with respect

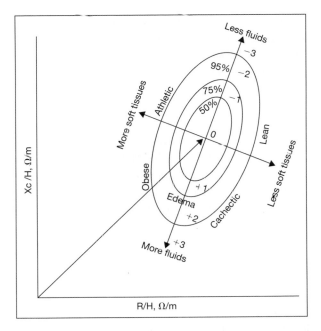

Fig. 1. The intersubject variability of the impedance vector is represented with the bivariate normal distribution, i.e. a graph with elliptical probability regions (50, 75, and 95% tolerance ellipses) on the R-Xc plane normalized by the height (R/H, and Xc/H, in Ω/m). Vector position on the RXc graph is interpreted and ranked following two directions: (1) Vector displacements parallel to the major axis of tolerance ellipses indicate progressive changes in soft tissue hydration; and (2) vectors lying on the left side, above the major axis, or on the right side, below the major axis of tolerance ellipses indicate more or less cell mass, respectively, that is contained in soft tissues. R = Resistance; Xc = reactance; H = height.

to a healthy population of the same race, sex, and class of BMI and age [1–3, 5–12].

Errors in BIA and BIS Equations versus Vector BIA

Prediction Equations of Fluid Volume. The prediction error of BIA and BIS equations is the sum of five errors, namely, the impedance measurement error, the regression error (standard error of the estimate against the reference method), the intrinsic error of the reference method (e.g. dilution methods), the electric-volume model error (anisotropy of tissues and geometry other than cylinder), and the biological variability among subjects (individual composition and geometry) [1, 5]. Impedance measurements are made with a 2–3% precision error. BIA and BIS equations are validated against dilution methods that have their relevant error greater than 3–6% [2–4].

If the hydration of the fat-free mass is not fixed at 73%, as happens in HD patients, an additional bias is introduced in estimates of prediction equations [1–4]. Thus, BIA and BIS prediction equations for fluid volume should not be utilized in HD patients, particularly during ultrafiltration (UF).

Segmental Bioimpedance. Reference, dilution methods cannot be applied ex vivo to segments of the body due to diffusion of the tracer substance across segments through the whole body (a temporary amputation should be necessary for deriving validation equations). Furthermore, the standardization of the type of electrodes used and their placement is still a major concern [2, 4]. As a consequence, the size of prediction error is unknown at any current frequency.

Vector BIA. In contrast with conventional BIA and BIS, *Vector BIA* only needs to take care of the measurement error and of the biological variability of subjects in any clinical condition. The intersubject variability of Z is represented with the bivariate normal distribution, i.e. with elliptical probability regions on the R-Xc plane, which are confidence (95%) and tolerance ellipses (50, 75, 95%) for mean and individual vectors, respectively [1, 5–10].

Clinical Patterns of Vector BIA

The *RXc graph* is a probability chart that classifies an individual vector according to the distance from the mean value of the reference population. The positive correlation coefficient between vector components indicates that factors modifying R values are expected to modify also Xc values, and vice versa because soft tissues are anisotropic media where fluids are allowed to increase and decrease with a definite, 'correlated' adaptation of tissue mass and structure. From clinical validation studies in adults, vectors falling out of the 75% tolerance ellipse indicate abnormal tissue impedance [5–13].

Vector Patterns in the Steady State. Vector position on the *RXc graph* is interpreted and ranked following two directions on the R-Xc plane, as depicted in figure 1: (1) *Vector displacements parallel to the major axis of tolerance ellipses* indicate progressive changes in tissue hydration (dehydration with long vectors, out of the upper poles of the 75 and 95%, and hyperhydration, with apparent edema, with short vectors, out of the lower poles of 75 and 95%); and (2) *peripheral vectors* lying on the left side, above the major axis, or on the right side, below the major axis of tolerance ellipses indicate more or less cell mass, respectively, that is contained in soft tissues (i.e. vectors with a comparable R value and a higher or lower Xc value, respectively).

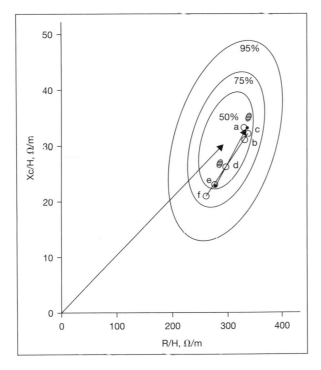

Fig. 2. The impedance vector migration associated with the hemodialysis cycle over 3 days is depicted on the RXc graph [14]. Reference values for an individual vector (thin arrow to the center of ellipses) are depicted as 50, 75 and 95% tolerance ellipses (male, Italian population) [7]. Solid circles represent vectors at the start and the end of the session. Open circles represent vectors after 30 (a), 60 (b), 120 min (c), and in the next days, after 24 (d), 48 (e), and 68 h (f). The vector lengthening during the hemodialysis session is represented by the bold arrow in the direction of the major axis. The trajectory followed by vector shortening after dialysis is represented by segments of a path still parallel to the major axis of tolerance ellipses. Small, hatched ellipses represent the 95% confidence of the mean, pre (lower ellipse)- to post (higher ellipse)-dialysis vector displacement in a large Italian population [11].

The same information given in polar coordinates needs both the phase angle and length of the vector.

Vector Patterns Associated with the HD Cycle. The standard thrice-weekly HD cycle of body fluids is characterized by fluid removal with the UF during the dialysis session (3–5 h) and a progressive fluid repletion over the short (48 h) or long (72 h) interdialysis period. Figure 2 shows vector migration over a complete HD cycle of 3 days [14], where vector lengthening during UF was parallel to the major axis of tolerance ellipses. Then vectors measured within 120 min after the session randomly fluctuated close to the end-dialysis vector.

After 24, 48, and 68 h vectors progressively shortened back along the same direction associated with the UF. Interestingly, the vector at 48 h reached the baseline vector position at the start of the HD session, indicating the same fluid repletion, while the vector measured after 68 h shortened further, indicating a greater fluid overload during the long interdialysis period.

The post-HD random fluctuation for 120 min excludes 'electric rebound' phenomena and supports the validity of immediate post-HD measurement (patient's time saving).

The linear trajectory of vectors can be observed during the HD session at any current frequency (meaning same information at any frequency) [5]. The linear trajectory both in the intra- and interdialysis period supports validity of monitoring with only pre- and postdialysis measurements. In the pre-post HD cycle we previously described in 1,116 HD patients, the wet-dry weight cycling was associated with a pattern of cyclical, backward-forward displacement of the impedance vector [11].

Consistently, vector distribution in continuous ambulatory peritoneal dialysis patients without edema was close to that of pre-HD vectors. Vectors from patients with edema were shorter and less steep, lying below the lower pole of the 75% tolerance ellipse, in the direction of the major axis [12].

Figure 3 shows vector trajectories spanning within (solid circles) or out (open circles) of the 75% tolerance ellipses during 3–4 h of UF in representative HD patients. Trajectories can be classified into *normal* (within 75%) versus *abnormal* (out of 75%) assuming that a normal hydration in HD patients is the normal hydration of the healthy population, as reflected by impedance vector distribution of the reference population (726 subjects, BMI 17–31 kg/m^2 and age 15–85 years) [7].

Normal versus Abnormal Vector Trajectories in HD. The classification of vectors into normal versus abnormal with respect to the reference third quartile (75% tolerance ellipse) is based on electrical properties of soft tissues, independent of clinical condition and body weight of patients. Vector distribution in asymptomatic HD patients was close to that of healthy subjects in the same BMI range [11].

Vectors from 72% of the 1,116 asymptomatic HD patients were within the 75% reference tolerance ellipse before the HD session. At the end of the session, vectors of 55% of the asymptomatic HD patients were within the 75% reference tolerance ellipse. On average 50% of vectors cycled within the 75% tolerance ellipse meaning that a full restoration of electrical properties of uremic tissues was achieved with HD.

In figure 3, open circles indicate the two dynamic patterns of abnormal vector trajectories which are observed in HD patients undergoing 3–4 h of UF. The first and more frequent pattern is a vector displacement parallel to the

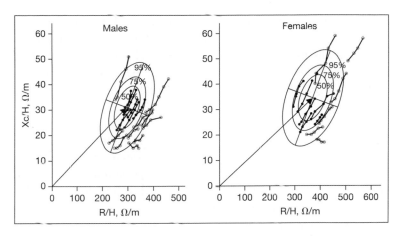

Fig. 3. Solid and open circles indicate the vector trajectories that spanned within (normal) or out (abnormal) of the 75% tolerance ellipses, respectively, during 3–4 h of UF in representative HD patients. Two dynamic patterns of abnormal vector trajectories are characteristic in HD patients undergoing ultrafiltration. The first and more frequent pattern is a vector displacement parallel to the major axis of the tolerance ellipses, leaving or ending out of the 75% tolerance ellipse. Long vectors overshooting the upper poles indicate dehydration (dry vectors), and short vectors migrating across the lower poles indicate fluid overload (wet vectors). Vector trajectories spanning on the left side versus trajectories on the right side of ellipses are from patients with more versus less soft tissue mass, respectively. The second pattern of ultrafiltration is a flat vector migration to the right, due to an increase in R/H without a proportional increase in Xc/H due to loss of cells in soft tissue. This pattern is characteristic of patients with severe malnutrition or cachexia. It is never observed in vectors lying on the left of the ellipses.

major axis of the tolerance ellipses, leaving or ending out of the 75% tolerance ellipse. According to the general patterns defined in figure 1, long vectors migrating across the upper poles indicate dehydration (dry vectors), and short vectors migrating across the lower poles indicate fluid overload (wet vectors). Vector trajectories spanning on the left side versus trajectories on the right side of ellipses are from patients with more versus less soft tissue, respectively. For instance, the lengthening and steepening of vectors in obese subjects (BMI 31–41 kg/m²) undergoing the same UF are comparable to those of lean subjects, in spite of spanning on the left side of ellipses [9].

The second pattern associated with UF is a flat vector migration to the right, due to an increase in R/H without a proportional increase in Xc/H due to severe loss of soft tissue mass. This pattern is characteristic of patients with severe malnutrition or cachexia. It is never observed in vectors lying on the left of the ellipses. Average pre-HD vectors of 251 hypotensive patients were longer and with a smaller phase angle compared to asymptomatic HD patients. The vector displacement induced by a comparable fluid removal was significantly shorter and less steep in hypotensive patients, in whom the concentration of both hemoglobin and albumin was also lower. We speculated that malnutrition was involved through the loss of interstitial gel leading to a reduced volume of interstitial fluid, in both gel and free form, which decreased the effectiveness of the vascular refilling [11].

Adequate versus Inadequate Vector Trajectories in HD. As with any diagnostic test, intervention studies are needed to establish whether dialysis prescription that induces normal versus abnormal trajectories is associated with adequate versus inadequate UF, ultimately leading to different outcomes. In practice, the first target is fixed on the normal range whose assessment for adequacy needs clinical validation and tuning of vector target during feedback with the clinical course of individual patients.

For instance, an abnormal, long vector overshooting the upper 95% pole that is associated with an asymptomatic UF and a good control of blood pressure is an adequate (dry) vector. The same vector in a patient with hypotensive episodes is an inadequate, abnormal, long vector. Yet, a short, abnormal vector in a patient with hypotension may be more adequate than a normal vector associated with more hypotensive episodes.

In a large multicenter, observational study (USA dialysis centers, 3,009 patients), shorter pre-HD vectors, indicating more fluid repletion and probably inadequate UF, were associated with a greater multivariate relative risk of death after one year of follow-up [15].

In a small observational study, longer post-HD vectors out of the upper pole of the 75% tolerance ellipse were associated with a higher frequency (73%) of symptomatic hypotension and cramps. But 57% of (still) asymptomatic patients were classified as dehydrated due to their long vectors. A follow-up of clinical course in the next sessions was not performed. The classification of volume status based on conventional BIA equations was insensitive to the clinical course [13].

Specific Information from Vector BIA in HD

The methodology of *Vector BIA*, applied to whole-body, 50-kHz single frequency impedance measurements obtained from HD patients before and after a dialysis session can contribute to dialysis prescription with three unique indications:

(1) *Classification of tissue hydration into normal or abnormal, independent of body weight.* Patient's impedance is compared with tolerance intervals of a reference population. Abnormality is ranked as a distance from the mean

vector of the reference population. In particular cases, an abnormal hydration can be considered an adequate hydration if it is associated with less morbidity.

(2) *Classification by exclusion of volume-dependent hypertension and hypotension.* A long vector indicating dehydration in a patient with hypertension allows exclusion of volume-dependent hypertension. Further UF is not expected to decrease blood pressure. The same vector in a patient with hypotension is compatible with a volume-dependent hypotension, which can be corrected by reducing UF (feedback with vector shortening). A short vector in a patient with hypertension is compatible with volume-dependent hypertension, which can be corrected by increasing UF (feedback with vector lengthening). The same vector in a patient with hypotension allows exclusion of volume-dependent hypotension. No correction is expected by increasing UF.

(3) *Discrimination between changes in body weight due to fluid versus tissue mass changes.* As shown in figure 1, different vector patterns indicate the nature of body weight change based only on electrical properties of tissues. This contribution is particularly useful in obese patients in whom interpretation of changes in body weight is often impossible to establish with clinical acumen.

Optimal Current Frequency. Although *Vector BIA* can be done on R and Xc components at any current frequency, the optimal performance of the method is obtained with the standard, single frequency, 50-kHz current that allows impedance measurements with the best signal to noise ratio [1, 4, 5].

Correlation of Impedance with UF. The wet-dry weight cycling of HD patients is associated with a cyclical, backward-forward displacement of the impedance vector on the R-Xc plane. Despite this association, the correlation between vector components and the amount of fluid removed by UF is weak in literature, using either single or multiple frequency [5, 11]. This may be accounted for by the time delay between the intravascular fluid removal with UF and the suction of fluid from interstitial gel of tissues in the subsequent vascular refilling process [11, 12].

Other Methods for the Assessment of Body Hydration

Anthropometry. Estimates of TBW with formulas based on anthropometry (Watson, Hume, Chertow, Johanssohn) should not be utilized, as they are insensitive to change in hydration (e.g. pitting edema) [12].

Central Venous Pressure. Measurement of the central venous pressure is invasive and only evaluates the volemic status that is an indirect indicator of soft tissue hydration, which is specifically assessed by *Vector BIA* [10]. Measurement of vena caval diameter is also based on the same principle.

Diluitometry. At the top of TBW measurement, utilization of diluitometry (e.g. tritium and deuterium) is implemented in unique dialysis centers evaluating plasma activity 3–4 h after isotope ingestion. But estimates are obtained with a relevant within-subject measurement error (about 10%) even in the absence of delayed gastric emptying. Furthermore, three available isotopes measure different water spaces (by 3–4%). Therefore, diluitometry cannot be considered an optimal method for monitoring of hydration in the clinical setting, as discussed elsewhere [12].

Conclusions

Cyclical tissue hydration changes in HD patients are detectable as changes in the whole body impedance, which can be utilized with *Vector BIA* patterns in monitoring the prescription of optimal hydration independent of the body weight.

Wet-dry weight prescription based on *Vector BIA* indication would bring abnormal vectors back into the 75% reference ellipse, where tissue electrical conductivity is restored. Longitudinal and interventional studies are required to establish whether patients with vectors cycling within the normal third quartile ellipse have better outcomes than those cycling out of the target interval.

References

1   Piccoli A: Patterns of bioelectrical impedance vector analysis: Learning from electrocardiography and forgetting electric circuit models. Nutrition 2002;18:520–521.
2   Kyle UG, Bosaeus I, De Lorenzo AD, Deurenberg P, Elia M, Gomez JM, Heitmann BL, Kent-Smith L, Melchior JC, Pirlich M, Scharfetter H, Schols AMWJ, Pichard C: Bioelectrical impedance analysis – Part I: Review of principles and methods. Clin Nutr 2004;23:1226–1243.
3   Kyle UG, Bosaeus I, De Lorenzo AD, Deurenberg P, Elia M, Gomez JM, Heitmann BL, Kent-Smith L, Melchior JC, Pirlich M, Scharfetter H, Schols AMWJ, Pichard C: Bioelectrical impedance analysis – Part II: Utilization in clinical practice. Clin Nutr 2004;23:1430–1453.
4   Foster KF, Lukaski HC: Whole-body impedance – What does it measure? Am J Clin Nutr 1996;64 (suppl):S388–S396.
5   Piccoli A, Pastori G, Guizzo M, Rebeschini M, Naso A, Cascone C: Equivalence of information from single versus multiple frequency bioimpedance vector analysis in hemodialysis. Kidney Int 2005;67:301–313.
6   Piccoli A, Rossi B, Pillon L, Bucciante G: A new method for monitoring body fluid variation by bioimpedance analysis: The RXc graph. Kidney Int 1994;46:534–539.
7   Piccoli A, Nigrelli S, Caberlotto A, Bottazzo S, Rossi B, Pillon L, Maggiore Q: Bivariate normal values of the bioelectrical impedance vector in adult and elderly populations. Am J Clin Nutr 1995;61:269–270.

8   Piccoli A, Pillon L, Dumler F: Impedance vector distribution by sex, race, body mass index, and age in the United States: Standard reference intervals as bivariate Z scores. Nutrition 2002;18: 153–167.

9   Piccoli A, Brunani A, Savia G, Pillon L, Favaro E, Berselli ME, Cavagnini F: Discriminating between body fat and fluid changes in the obese adult using bioimpedance vector analysis. Int J Obes 1998;22:97–104.

10 Piccoli A, Pittoni G, Facco E, Favaro E, Pillon L: Relationship between central venous pressure and bioimpedance vector analysis in critically ill patients. Crit Care Med 2000;28:132–137.

11 Piccoli A, for the Italian HD-BIA study group: Identification of operational clues to dry weight prescription in hemodialysis using bioimpedance vector analysis. Kidney Int 1998;53:1036–1043.

12 Piccoli A, for the Italian CAPD-BIA study group: Bioelectrical impedance vector distribution in peritoneal dialysis patients with different hydration status. Kidney Int 2004;65:1050–1063.

13 Guida B, De Nicola L, Trio R, Pecoraio P, Iodice C, Memoli B: Comparison of vector and conventional bioelectrical impedance analysis in the optimal dry weight prescription in hemodialysis. Am J Nephrol 2000;20:311–318.

14 Piccoli A, Codognotto M: Bioimpedance vector migration up to three days after the hemodialysis session. Kidney Int 2004;66:2091–2092.

15 Pillon L, Piccoli A, Lowrie EG, Lazarus JM, Chertow GM: Vector length as a proxy for the adequacy of ultrafiltration in hemodialysis. Kidney Int 2004;66:1266–1271.

Prof. Antonio Piccoli
Department of Medical and Surgical Sciences
Nephrology Clinic, University of Padova
Policlinico IV piano, Via Giustiniani, 2
IT-35128 Padova (Italy)
Tel. +39 335 8256780, Fax +39 049 618157, E-Mail apiccoli@unipd.it

Ronco C, Brendolan A, Levin NW (eds): Cardiovascular Disorders in Hemodialysis.
Contrib Nephrol. Basel, Karger, 2005, vol 149, pp 162–167

....

Use of Segmental Multifrequency Bioimpedance Spectroscopy in Hemodialysis

Nathan W. Levin[a]*, Fansan Zhu*[a]*, Eric Seibert*[a]*,*
Claudio Ronco[b]*, Martin K. Kuhlmann*[a]

[a]Renal Research Institute, New York, N.Y., USA;
[b]Department of Nephrology, St. Bartolo Hospital, Vicenza, Italy

Abstract

Whole body bioimpedance (BI) appears to be accurate and reproducible in the assessment of body composition, but does not appear useful for estimation of dry weight. Segmental BI has been used for the assessment of muscle mass in body segments, such as arms or legs and may be useful for rehabilitation studies. A promising new development is the application of segmental BI for dry weight determination. Changes in extracellular volume of the calf are recorded continuously during HD, thereby allowing the detection of a timepoint at which no further volume is removed from the calf despite ongoing ultrafiltration (UF). Continuation of UF beyond this point is associated with an increased risk of intradialytic hypotension. This new technology may help optimizing the prescription of dry weight and UF rates in hemodialysis patients.

Body composition and hydration status are important issues in treating dialysis patients. So far, these two components of patient care have mainly been assessed by clinical judgment, which is strongly dependent on clinical experience and shows a high interobserver variability. More objective measures for assessment of body composition are therefore needed and bioelectrical impedance (BI) offers the potential as a simple, portable and relatively inexpensive technique for the assessment of body fluid volumes and lean body mass.

Methodological Aspects

BI describes the electrical properties of tissues and is represented as a combination of tissue resistance and reactance towards an alternating current.

The theoretical basis for BI is the observation that different types of tissues differ in their resistance to low-frequency ($<$10 kHz) and high-frequency currents. These differences are due to variations in water content, a good conductor, and cell mass, which act as insulators for low-frequency currents. Low-frequency currents mainly pass through the extracellular (ECV) but not the intracellular fluid compartment, whereas high-frequency currents pass through both, ECF and intracellular fluids. Measured resistance and reactance at different frequencies are thus characteristic for water and cell content of the tissue subjected to the current [1]. A number of different bioimpedance methods have been described and studied in dialysis patients.

'*Bioimpedance analysis (BIA)*' refers to a method where a single 50-kHz frequency is applied to the whole body or a defined body segment. Since the current passes through ECV and intracellular fluid compartment, this method gives one estimate of total body water [2].

Bioimpedance spectroscopy (BIS) uses a whole spectrum of low and high frequencies from 5 to 1,000 kHz and thus yields data for both, ECV and total body water, with intracellular fluid compartment then calculated by the difference [3].

In the basic BI theory the body is viewed as 5 interconnecting cylinders, 2 for the arms, 2 for the legs, and 1 for the trunk. The clinically most popular methods of whole-body BIA or BIS both use wrist-to-ankle protocols with the current passing through the whole body. The algorithms used for calculation of body composition view the whole body as just one cylinder and therefore are prone to errors which depend on body size, shape and regional fluid accumulation. In general it may be assumed that accuracy of any BI measurement increases as the distance between the electrodes decreases and the more the shape of the segment resembles a true cylinder.

Segmental bioimpedance has been developed to overcome these limitations of whole-body impedance. This method applies the BI technology to any segment of the body with electrodes being placed at the far ends of the segment. Each segment is a mathematically treated separate cylinder, and whole-body composition can eventually be calculated as the sum of segments. Segmental BIS has also been developed for very specific applications, such as dry weight analysis in hemodialysis (HD) patients [4].

Applications of Bioimpedance in HD

Assessment of Body Composition
Whole-body BIA and BIS have been used for the assessment of body composition and body fluid volumes in dialysis patients and both methods

have been validated against 'gold standards', such as deuterium for total body water, bromide for ECV and K^{40} for body cell mass. These methods utilize 'linear regression' formulas to predict body masses based on biological and demographic data input into one or more equations. Most studies have been performed in normal populations and validity has been questioned in diseased subjects such as dialysis patients. We have recently validated whole-body BIS against gold standards in 40 HD patients and could demonstrate a good correlation with total body muscle mass estimated from MRI [manuscript submitted].

Segmental BIS may have particular value for body composition estimation in dialysis patients where loss of muscle mass is frequently observed and predictive of outcome. The assessment of segmental muscle mass of arms and legs may offer the opportunity to more closely follow changes in muscle mass during nutritional intervention or physical rehabilitation. The correlation we found for segmental BIS-estimated muscle mass with MRI-measured muscle mass was even stronger than that observed for whole-body BIS.

The algorithms used for calculation of each segment's fluid volumes include a factor for segmental resistivity, which used to be the same for all segments, arms, trunk, and legs. However, since tissue composition in each segment may vary due to hydration, posture and fluid distribution, segment-specific resistivity factors might improve the method. We were recently able to show that the resistivity of each segment indeed differs and that the accuracy of segmental bioimpedance measures can be improved by applying segment-specific resistivity factors [5].

Assessment of Dry Weight

A number of different approaches to use whole-body BIA or BIS for prediction of dry weight have been tried but have failed to gain acceptance due to their limitations in precision. All these methods predict dry weight as an absolute value derived by comparison with a database from healthy individuals of similar age and gender. However, even in a normal population hydration status varies widely between subjects, but is tightly regulated within each individual. The normal range of dry weight from a healthy population thus appears much too wide to be useful in individual dialysis patients. Indeed, in a study by Chamney et al. [6] the dry weight estimated from predialysis whole-body BIS was within ± 1.58 kg of the best clinical estimate. Since dry weight may be a matter of only ± 0.5–2 l from the normal ECV, another approach is necessary to provide precision.

We have recently developed a new method for relative assessment of dry weight in dialysis patients, which uses the patient as his/her own control [6]. Our approach is based on regional BIS of the calf, where changes in calf

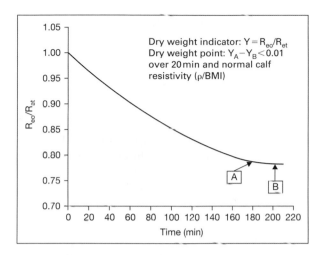

Fig. 1. Principle of segmental calf-BIS for determination of dry weight in HD patients. The change in calf resistivity (R) is recorded continuously during HD with ultrafiltration and plotted as R_{e0}/R_{et}, where R_{e0} is resistance at time 0 and R_{et} the resistance at any time during HD. Dry weight is defined as the weight achieved after flattening of the R_{e0}/R_{et} slope occurred for a period of 20 min despite ongoing UF (A = start of flattening, B = A + 20 min) and with calf resistivity ρ within a normal range.

extracellular resistance (R_e) are recorded continuously during dialysis. We have shown that electrical resistance increases as ECF is removed by ultrafiltration. In order to show that R_e changes analogously to ECF, data are presented as the resistance at time 0 (R_{e0}) to the resistance at any time during the treatment (R_{et}), thus R_{e0}/R_{et} (fig. 1). As ECF is removed from the calf during dialysis, calf ECV declines and so the R_{e0}/R_{et} slope. In the overhydrated individual there is a linear reduction in ECV with a constantly declining R_{e0}/R_{et} slope. At a time during HD when the rate of ECV removal from tissues diminishes, the R_{e0}/R_{et} slope will begin to flatten (fig. 1). This follows closely the classical relationship between decrease of excess ECF and increase in tissue pressure as shown by Guyton [7].

Currently, calf-BIS dry weight is defined as the weight at 20 min after flattening of the R_{e0}/R_{et} slope begins. We are uncertain about what period of observation is necessary before labeling the flat slope as being equivalent to dry weight. By definition, dry weight at constant ultrafiltration should be accompanied by continuing reduction in blood volume, and consequently the development of cramping or a drop in blood pressure. One might think the point of definition is crucial, but from the clinical point of view all that is necessary is to be close to the ideal value. We have recently shown that fluid removal below the BIS-determined dry weight is accompanied by cramps and hypotension [8].

The calf has been chosen for this approach as being the most dependent area of the body with the thought that when it is free of excess ECF there is in fact no excess water elsewhere. We have shown that both calves lose fluid identically (within a range of 0.03–0.13 l during treatment). Flattening of the R_{e0}/R_{et} slope may occur without dry weight being reached, for example when ultrafiltration is stopped during dialysis or removal of ECF from the calf is hindered due to venous thrombosis, lymph edema or similar conditions. In order to differentiate true from artificial flattening of the R_{e0}/R_{et} slope, this new method also includes a backup measure of calf resistivity (ρ). Calf resistivity is defined as calf resistance in relation to calf volume, which can be predicted at any time during dialysis from calf circumference at start of treatment and the R_{e0}/R_{et} slope. Similar to calf resistance, ρ increases when calf ECF is reduced. The absolute values for calf-ρ at any time point during dialysis are then compared to calf-ρ of a healthy population. Only if calf-ρ exceeds the lower threshold of the normal range at the time when the R_{e0}/R_{et} slope flattens, the determination of approximate dry weight is confirmed. It should be pointed out that the individual calf-ρ can be well within the normal range even before flattening of the R_{e0}/R_{et} slope occurs and that resistivity measurement would not, in itself, be a suitable determinant of dry weight since, as in case of any comparison with a normal range, a single figure has a large chance of error. The combination of two criteria for determination of dry weight, i.e. flattening of the R_{e0}/R_{et} slope and calf-ρ in a normal range should diminish the uncertainties of normal variations.

In summary, whole-body BIA or BIS are used for assessment of body fluid volumes and body composition in dialysis patients. Their precision, however, is limited for many reasons, such as regional fluid accumulation and the fact that the equations applied for calculating the results view the body as just one cylinder. Segmental BIS allows a more accurate prediction of tissue masses and fluid volumes, due to shorter distance between the electrodes and a naturally more cylindrical shape of the segments. The new approach of using segmental calf-BIS as a relative measure of dry weight where the patient is his/her own control has apparent advantages. It remains to be seen how practical the method is in routine use.

References

1   Grimnes S, Martinsen OG: Bioimpedance and Bioelectricity Basics. London, Academic press, 2000.
2   Heymsfield SB, Wang ZM, Visser M, Gallagher D, Pierson RN Jr: Techniques used in the measurement of body composition: An overview with emphasis on bioelectrical impedance analysis. Am J Clin Nutr 1996;64(suppl):S478–S484.

3   Hannan WJ, Cowen SJ, Plester CE, Fearon KC, de Beau A: Comparison of bioimpedance spectroscopy and multifrequency bioimpedance analysis for the assessment of extracellular and total body water in surgical patients. Clin Sci 1995;89:651–658.

4   Zhu F, Kuhlmann MK, Sarkar S, Kaitwatcharachai C, Khilnani R, Leonard EF, Greenwood R, Levin NW: Adjustment of dry weight in hemodialysis patients using intradialytic continuous multifrequency bioimpedance of the calf. Int J Artif Organs 2004;27:104–109.

5   Zhu F, Kaysen G, Sarkar S, Kaitwatcharachai C, Leonard EF, Wang J, Heymsfield S, Greenwood R, Gotch F, Levin NW: Comparison of extracellular and intracellular fluid volumes in hemodialysis patients as measured by segmental and wrist to ankle bioimpedance analysis with magnetic resonance imaging and dilution techniques (abstract). J Am Soc Nephrol 2003;14:55A.

6   Chamney PW, Kramer M, Rode C, Kleinekofort W, Wizemann V: A new technique for establishing dry weight in hemodialysis patients via whole body bioimpedance. Kidney Int 2002;61:2250–2258.

7   Guyton AC: Textbook of Medical Physiology, 10th ed. Philadelphia, Saunders, 2000, p 276.

8   Zhu F, Khilnani R, Enis S, Greenwood R, Ronco C, Kuhlmann MK, Levin NW: Determination of dry weight using intradialytic continuous calf bioimpedance (abstract). J Am Soc Nephrol 2004;15:592A.

Nathan W. Levin, MD
Medical and Research Director
Renal Research Institute, 207 East 94th Street, Suite 303
New York, NY–10128 (USA)
Tel. +1 646 672 4002, Fax +1 212 996 5905, E-Mail nlevin@rriny.com

Ronco C, Brendolan A, Levin NW (eds): Cardiovascular Disorders in Hemodialysis.
Contrib Nephrol. Basel, Karger, 2005, vol 149, pp 168–174

·······················

Cardiovascular Risk in Patients with End-Stage Renal Disease: A Potential Role for Advanced Glycation End Products

Yingjie Wang, Sally M. Marshall, Michael G. Thompson,
Nicholas A. Hoenich

Faculty of Medical Sciences, Medical School, University of Newcastle-upon-Tyne, UK

Abstract

Cardiovascular complications are a major cause of morbidity and mortality in patients with end-stage renal disease (ESRD). Advanced glycation end products (AGE) are elevated in the plasma of such patients and are also found in atherosclerotic plaques. The cellular signalling pathway(s) underlying AGE-induced platelet aggregation have not been elucidated. One pathway currently receiving increased attention is the externalization of the membrane phospholipid, phosphatidylserine (PS), which plays an important role in the activation of clotting factors. In this study, we have investigated ex vivo a possible link between elevated AGE concentration and PS externalization. We observed (i) increased PS externalization in platelets from patients with ESRD, (ii) reconstitution of healthy platelets with serum from patients with ESRD resulted in increased PS externalization and (iii) incubation of platelets with purified human serum albumin (HSA)-AGE elicited PS externalization suggesting a role for AGE.

Introduction

The risk of cardiovascular disease in individuals with end-stage renal disease (ESRD) is markedly elevated. Even after stratification by age, gender, ethnicity and the presence or absence of comorbid conditions such as diabetes, cardiovascular mortality is 20-fold greater than in the general population [1]. Patients with chronic renal disease thus are viewed as one of the groups at greatest risk for cardiovascular events [2]. The mechanisms contributing to this increased risk are not fully understood.

Advanced glycation end products (AGE) are formed by covalent modification of cellular proteins and form a heterogeneous series of compounds [3].

In vivo, the most abundant adduct appears to be N^ε-(carboxymethyl) lysine (CML) [4]. Studies have shown that AGE are greatly increased in both diabetic- and non-diabetic-related renal failure patients compared to the general population [5, 6] and they are not removed during low-flux dialysis which is employed in the majority of clinical settings [7]. The mechanisms by which AGE influence the progression of atherogenesis is not completely understood, however, it is known that they accumulate in atheromatous plaques where they may contribute to disease progression [8]. AGE can also neutralize the processes that protect the vascular bed against the proliferation and aggregation of cells, two key elements in atherogenesis. They may also act directly, or subvert cellular function via specific 'receptors for AGE' [9].

The atherosclerotic process involves interaction between circulating cells and the vessel wall. The human platelet plays a major role in such processes and a previous study has demonstrated AGE-mediated platelet aggregation in vitro [10]. The cell wall of platelets contains phosphatidylserine (PS), a negatively charged phospholipid that resides in the inner leaflet of the plasma membrane in resting cells [11], but following activation, it translocates to the outer leaflet of the plasma membrane where it plays a key role in the activation of clotting factors thus providing a hypercoagulable environment [12]. To characterize the role of AGE on PS externalization and elucidation of the responsible, cell signaling pathway(s) we have performed a series of ex vivo studies.

Materials and Methods

Subjects

Eleven stable ESRD patients (10 males and 1 female) undergoing low flux hemodialysis were studied. The mean age was 59 years (range 38–81 years) and all patients had been undergoing hemodialysis for at least 6 months prior to the study. Patients with diabetes were excluded. Five normal subjects (4 males and 1 female) with an average age of 55 years (range 48–58 years) were used as age-matched controls. The study was approved by the Newcastle and North Tyneside Joint Ethics Committee, and informed consent was obtained from all participants in the study.

Methods

Platelet Rich Plasma Acquisition. Five milliliters of blood was withdrawn from subjects using a standard wide gauge needle (Braun IIG) into a tube with acid citrate dextrose (ACD) and then centrifuged at $120\,g$ for 15 min at 20°C. Platelet rich plasma (PRP) was removed and stored until required.

Platelet Extraction. To inhibit platelet activation, prostaglandin I_2 (PGI$_2$), was added to PRP at a final concentration of 1 μM and centrifuged for 5 min at 20°C to pellet the platelets which were subsequently removed and resuspended in an identical volume of Hepes Glucose Saline containing 1 μM PGI$_2$. Platelets were counted in a hemocytometer and resuspended at a final concentration of 10^7/ml.

Measurement of PS Externalization in PRP/Platelets. This was determined using the binding of the cell membrane-impermeant PS-specific, Annexin V-fluorescein isothiocyanate (FITC) conjugate (Beckman-Coulter) and quantified by a fluorescence-activated cell-scanner (Becton-Dickinson FACScan supporting Lysis II software as described previously) [13].

Measurement of CML Concentration in Plasma. This was performed by ELISA techniques as described previously [14].

Re-Constitution of Control/Uremic Serum with Healthy Platelets. Ten milliliters of blood was obtained from controls/patients and allowed to clot for 90 min at 37°C. Samples were centrifuged for 5 min at 20°C and the serum was used for reconstitution experiments with healthy, resuspended platelets prepared as described above. Platelets (10^7/ml) were incubated with either control or uremic serum for 10 min at 37°C and PS externalization was then determined as described above.

Preparation of Human Serum Albumin-AGE. Human serum albumin (HSA)-AGE was prepared by incubating 1 M glucose with 50 mg/ml HSA in 100 mM PBS (pH 7.4) in the presence of 1.5 mM PMSF, 1 mM EDTA, 100 μg/ml penicillin and 40 μg/ml gentamicin for 6 weeks in the dark at 37°C under sterile conditions. An unmodified protein control was prepared in a parallel incubation in the absence of glucose. After 6 weeks, all proteins were dialyzed extensively against PBS (pH 7.4, 4°C) to remove unreacted sugar and then separated into aliquots and stored at –20°C before use as previously described [15].

Measurement of PS Externalization on Platelets Incubated with HSA-AGE. Cells were incubated with either 100 μg/ml HSA-AGE or appropriate controls for 10 min at 37°C. Samples were then removed and the extent of PS externalization was determined as described above.

Statistical Analysis

Statistical analysis of the data was performed using a standard unpaired Student t test using Minitab 14 (Minitab Inc, State College, Pa., USA). A value of $p < 0.05$ was taken as significant.

Results

PS Externalization on Platelets from Patients with ESRD

Our initial experiments demonstrated that PS externalization in PRP from individuals with ESRD was approximately 7-fold higher than controls [57,009 ± 6,143 vs. 8,613 ± 2,014 binding sites, (mean ± SD) $p < 0.001$].

PS Externalization on Platelets Incubated with Serum from Patients with ESRD

When platelets from normal controls were reconstituted with serum isolated from individuals with ESRD, PS externalization (36,066 ± 4,674, n = 11) increased 5-fold within 15 min when compared to reconstitution with serum from age-matched controls (7,008 ± 1,927, n = 5, $p < 0.001$). Results from an individual experiment are presented in figure 1.

Serum analysis demonstrated an approximate 9-fold increase in CML concentration in those individuals with ESRD compared to normal controls (1,271 ± 108 vs. 145 ± 19 ng/ml, $p < 0.001$) and a significant correlation between PS externalization and CML concentration was noted ($p = 0.02$).

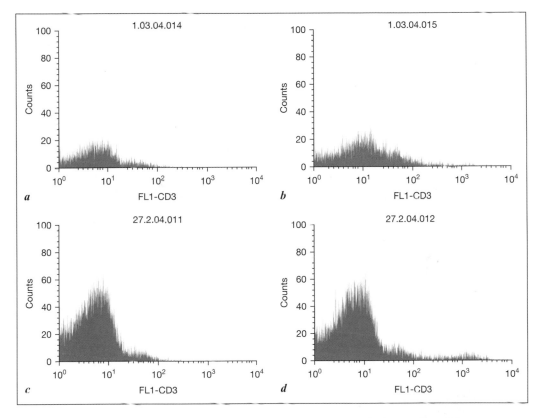

Fig. 1. Effect of serum from uremic patients compared to age-matched controls on PS externalization. Healthy platelets were incubated for 10 min with serum from either individuals with ESRD or healthy controls. PS externalization was then measured by comparing the fluorescence shift in the two samples when incubated in the absence (*a* and *c*) or presence (*b* and *d*) of FITC-labeled Annexin V which specifically binds to PS in the outer leaflet of the cell membrane. The figure shows a greater shift in the population of PS-positive cells in the ESRD patients (*d* vs. *c*) when compared to controls (*b* vs. *a*).

PS Externalization on Platelets Stimulated by HSA-AGE

Our preliminary data demonstrates that incubation of platelets with 100 μg/ ml HSA-AGE for 10 minutes elicited a substantial increase in FITC-Annexin V intensity (fig. 2a/b) demonstrating PS externalization under these conditions. There was no such shift in platelets under control conditions (fig. 2c/d) i.e. using HSA which had been prepared by incubation for 6 weeks in the absence of glucose. Our current work is focused on characterizing this response in detail.

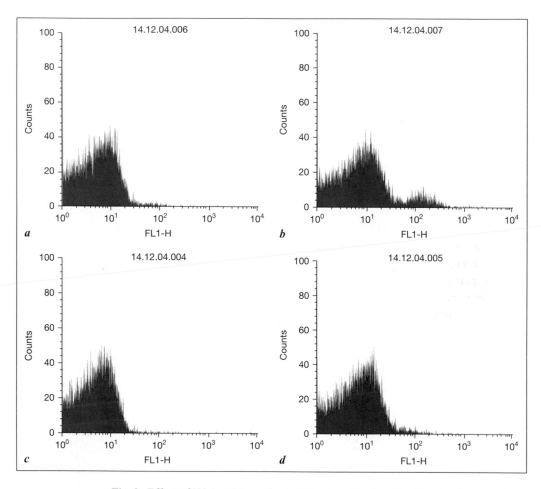

Fig. 2. Effect of HSA-AGE on platelet PS externalization. Healthy platelets were incubated for 10 min with HSA (100 µg/ml) which had been incubated for 6 weeks with or without glucose. PS externalization was then measured by comparing the fluorescence shift in the two samples when incubated in the absence (***a*** and ***c***) or presence (***b*** and ***d***) of FITC-labeled Annexin V which specifically binds to PS in the outer leaflet of the cell membrane. The figure shows a greater shift in the population of PS-positive platelets incubated in HSA prepared with (***b*** vs. ***a***) rather than without (***d*** vs. ***c***) glucose.

Discussion

Under normal circumstances, cells exhibit an asymmetric distribution of PS with this phospholipid being retained in the inner leaflet of the cell membrane. The PS distribution is known to be dependent on the activity of a number of enzymes and is regulated by the activities of three transporter proteins. An ATP-dependent amino phospholipid-specific translocase, which rapidly transports PS and phosphatidylethanolamine from the cell's outer to inner leaflet, an ATP-dependent lipid floppase and a nonspecific lipid scramblase which allows lipids to move randomly between both leaflets [16]. In view of the conditions employed and the speed with which the translocation of PS occurred in our experiments, we postulate that scramblase is the most likely enzyme activity involved. Scramblase exists as a number of isoforms [17] and is activated by a member of the protein kinase C (PKC) family, PKCδ [18]. This enzyme is normally resident in the cell cytosol, but, upon activation, translocates to the plasma membrane where it can phosphorylate and activate its substrate, scramblase [18]. In support of the involvement of scramblase and the role of PKCδ, reconstitution experiments have demonstrated that incubation of healthy platelets with uremic serum, results in the rapid translocation of PKCδ from the cytosol to the membrane.

The loss of PS asymmetry and its appearance on the external face of the platelet plasma membrane is a crucial event promoting coagulation and thrombosis by providing a catalytic surface for the assembly of key regulatory complexes [19]. The binding of activated factor V to surface-exposed PS promotes Ca^{2+}-dependent binding of activated factor X. This results in assembly of the prothrombinase complex that accelerates the generation of thrombin [20] thus promoting a pro-coagulant milieu and raising the likelihood of thrombus formation.

A previous study has demonstrated AGE-mediated platelet aggregation [10] and the data we report in this manuscript represents an important first step in describing a potential link between AGE, PS externalization and increased likelihood of coagulation/thrombosis. There are however, many aspects of this relationship which require further investigation. For example, are the effects of AGE we report on PS externalization mediated by receptors for AGE and is such mediation linked directly/indirectly to the activation of PKCδ and scramblase? Ongoing experiments are directed towards answering these questions.

Acknowledgments

We thank Farzana Shah, Alasdair Mackie and Benjamin Cox for allowing data generated in their experiments to be used, and Dr. Werner Beck for performing the CML analysis.

References

1   Foley RN, Parfrey PS, Sarnak MJ: Clinical epidemiology of cardiovascular disease in chronic renal disease. Am J Kidney Dis 1998;32:S112–S119.
2   Jardine AG: Cardiovascular complications of renal disease. Heart 2001;86:459–466.

3 Horvat S, Jakas A: Peptide and amino acid glycation: New insights into the Maillard reaction. J Pept Sci 2004;10:119–137.

4 Wagner Z, Wittmann I, Mazak I, Schinzel R, Heidland A, Kientsch-Engel R, Nagy J: N(epsilon)-(carboxymethyl) lysine levels in patients with type 2 diabetes: Role of renal function. Am J Kidney Dis 2001;38:785–791.

5 Kilhovd BK, Berg TJ, Birkeland KI, Thorsby P, Hanssen KF: Serum levels of advanced glycation end products are increased in patients with type 2 diabetes and coronary heart disease. Diabetes Care 1998;22:1543–1548.

6 Walter R, Mischak H, Haller H: Haemodialysis, atherosclerosis and inflammation – Identifying molecular mechanisms of chronic vascular disease in ESDR patients. Nephrol Dial Transplant 2002;17:24–29.

7 Agalou S, Ahmed N, Dawnay A, Thornalley PJ: Removal of advanced glycation end products in clinical renal failure by peritoneal dialysis and haemodialysis. Biochem Soc Trans 2003;31: 1394–1396.

8 Sakata N, Imanaga Y, Meng J, Tachikawa Y, Takebayashi S, Nagai R, Horiuchi S: Increased advanced glycation end products in atherosclerotic lesions of patients with end-stage renal disease. Atherosclerosis 1999;142:67–77.

9 Kislinger T, Fu C, Huber B, Qu W, Taguchi A, Du Yan S, Hofmann M, Yan SF, Pischetsrieder M, Stern D, Schmidt AM: N-(carboxymethyl) lysine adducts of proteins are ligands for receptor for advanced glycation end products that activate cell signalling pathways and modulate gene expression. J Biol Chem 1999;274:31740–31749.

10 Hangaishi M, Taguchi J, Miyata T, Ikari Y, Togo M, Hashimoto Y, Watanabe T, Kimura S, Kurokawa K, Ohno M: Increased aggregation of human platelets produced by advanced glycation end products in vitro. Biochem Biophys Res Commun 1998;248:285–292.

11 Daleke DL: Regulation of transbilayer plasma membrane phospholipid asymmetry. J Lipid Res 2003;44:233–242.

12 Lentz BR: Exposure of platelet membrane phosphatidylserine regulates blood coagulation. Prog Lipid Res 2003;42:423–438.

13 Wahid ST, Marshall SM, Thomas TH: Increased platelet and erythrocyte external cell membrane phosphatidylserine in Type I diabetes and microalbuminuria. Diabetes Care 2001;24:2001–2003.

14 Henle T, Deppisch R, Beck W, Hergesell O, Hansch GM, Ritz E: Advanced glycation end products (AGE) during haemodialysis treatment: Discrepant results with different methodologies reflect heterogeneity of AGE compounds. Nephrol Dial Transplant 1999;14:1968–1975.

15 Michelson AD: Platelet activation by thrombin can be directly measured in whole blood through the use of the peptide GPRP and flow cytometry: Methods and clinical applications. Blood Coagul Fibrinolysis 1994;5:121–131.

16 Bevers EM, Comfurius P, Dekkers DW, Zwaal RF: Lipid translocation across the plasma membrane of mammalian cells. Biochim Biophys Acta 1999;1439:317–330.

17 Pastorelli C, Veiga J, Charles N, Voignier E, Moussu H, Monteiro RC, Benhamou M: IgE receptor type I-dependent tyrosine phosphorylation of phospholipid scramblase. J Biol Chem 2001;276: 20407–20412.

18 Heemskerk JW, Siljander PR, Bevers EM, Farndale RW, Lindhout T: Receptors and signalling mechanisms in the procoagulant response of platelets. Platelets 2000;11:301–306.

19 Monroe DM, Hoffman M, Roberts HR: Platelets and thrombin generation. Arterioscler Thromb Vasc Biol 2002;22:1381–1389.

20 Zwaal RF, Schroit AJ: Pathophysiologic implications of membrane phospholipid asymmetry in blood cells. Blood 1997;89:1121–1132.

Dr. Nicholas A. Hoenich
SCMS, Floor 4, William Leech Building
Medical School, University of Newcastle, Framlington Place
Newcastle upon Tyne NE 2 4 HH (UK)
Tel. +44 191 222 6998, Fax +44 191 222 0723, E-Mail nicholas.hoenich@ncl.ac.uk

Ronco C, Brendolan A, Levin NW (eds): Cardiovascular Disorders in Hemodialysis.
Contrib Nephrol. Basel, Karger, 2005, vol 149, pp 175–184

....

Improving Removal of Protein-Bound Retention Solutes

Bert Bammens, Pieter Evenepoel

Department of Medicine, Division of Nephrology,
University Hospital Gasthuisberg, Leuven, Belgium

Abstract

Recent *in vitro* and clinical evidence suggest that protein-bound uremic retention
solutes contribute substantially to the pathophysiology of the uremic syndrome. As compared
to their water-soluble counterparts however, the removal of these molecules by conventional
dialysis techniques is limited. It was the purpose of the present paper to review the existing
data on the dialytic removal of protein-bound solutes and on the potential advantages of
newly developed epuration techniques. Furthermore, the toxicity profile of this group of
molecules is discussed.

Introduction

When the word 'uremia' is entered as a search term for a well-known
online dictionary, the following definition is found: 'accumulation in the blood
of constituents normally eliminated in the urine that produces a severe toxic
condition and usually occurs in severe kidney disease' [1]. It is obvious that this
definition is rather narrow to cover all aspects of the uremic syndrome. Jonas
Bergström [2], one of the pioneers of uremic toxicity research, suggested a
broader definition: 'Uremia is a toxic syndrome caused by severe glomerular
insufficiency, associated with disturbances in tubular and endocrine functions
of the kidney. It is characterized by retention of toxic metabolites, associated
with changes in volume and electrolyte composition of the body fluids and
excess or deficiency of various hormones'. It is clear from both definitions
that, while the exact pathogenetic mechanisms underlying the myriad changes

occurring in the uremic state are not completely understood today, the progressive retention of a large number of compounds which under normal conditions are excreted by the healthy kidneys plays a central role. By convention, these so-called uremic retention solutes are classified according to their molecular size. This subdivision is of course arbitrary, considering that the uremic retention products represent a continuous spectrum of molecular sizes. As a consequence the cutoffs between so-called small, middle and large molecules differ within the literature. In a recent survey by the European Uremic Toxin Work Group (EUTox) the lower cutoff value for the middle molecules was chosen to be 500 daltons (Da) in accordance with the original literature data reporting on this class of uremic solutes [3]. The upper limit for these molecules was defined as 60,000 Da, being about the maximal molecular size that is filterable through the glomerular basement membrane. When uremic retention solutes are listed according to their lipophilicity/hydrophilicity, another characteristic potentially influencing their removal pattern during dialysis or other epuration techniques, a distinct group of solutes appears at the horizon: protein-bound solutes.

Numerous in vitro findings suggest a role for the protein-bound solutes in the biochemical and physiological changes of uremia. Moreover, while still limited in number, recent clinical reports point to the importance of these solutes in an in vivo situation. The recent landmark HEMO and ADEMEX studies failed to show an improvement of patient outcome by increasing the removal of small water-soluble solutes above current standards [4, 5]. Besides several other hypotheses that have been put forward to explain these findings [6, 7], one should consider the limited removal of protein-bound solutes by the dialysis strategies evaluated in these studies. Indeed, the removal of protein-bound solutes during low- and high-flux hemodialysis and peritoneal dialysis was shown to be much lower than that of their water-soluble counterparts [8, 9]. Hence, while the relationship between small water-soluble solute removal and patient outcome may have reached a plateau in modern dialysis practice, efforts should be made to improve the elimination of protein-bound molecules. It was the purpose of this paper to review the existing data on the dialytic removal of protein-bound solutes. First, however, we will briefly discuss the toxicity profile of this group of molecules.

Role of Protein-Bound Uremic Solutes in the Uremic Syndrome

The protein-bound uremic retention solutes listed in the EUTox review cited above [3] consist of the amino acid homocysteine, the urofuranic acid

3-carboxy-4-methyl-5-propyl-2-furanpropionic acid (CMPF) and 23 other molecules belonging to the structural classes of polyamines, peptides, phenols, indoles, hippurates, reactive carbonyl compounds and advanced glycation end-products. Homocysteine, an intermediate of methionine metabolism, has been described as an independent risk factor for vascular disease and atherothrombosis in the general population. Although such a relationship has also been described in end-stage renal disease (ESRD) patients, the evidence is not unequivocal since serum levels of the solute and its relation to cardiovascular outcome is confounded by nutritional and inflammatory status [10]. CMPF, as well as the polyamines spermidine, spermine and putrescine, were shown to be implicated in renal anemia [11, 12]. The latter molecules have also been described in relation to the accelerated atherosclerosis seen in ESRD patients [13]. Leptin is one of the peptides produced by adipocytes and its serum levels were shown to be associated with malnutrition, inflammation and atherosclerosis in patients with renal insufficiency [14, 15]. A recent in vitro study showed inhibition of neutrophil chemotaxis by this molecule, suggesting a potential contribution by leptin to the establishment of infections in ESRD [14]. With regard to the class of phenols most research focused on p-cresol, a 108-Da colonic fermentation metabolite of the amino acid tyrosine. In a study by Vanholder et al. [16] p-cresol depressed the respiratory burst reactivity of granulocytes and monocytes at concentrations encountered in ESRD patients. Later it was shown that the solute inhibits the synthesis of platelet-activating factor, another aspect of leukocyte function [17]. Another evidence for a role of p-cresol in the immunodeficiency of uremia was provided by the finding of its inhibitory effect on cytokine-induced expression of the endothelial adhesion molecules ICAM-1 and VCAM-1 [18]. Furthermore, it was suggested that p-cresol might be implicated in endothelial dysfunction, which is considered to be an important contributor to the burden of cardiovascular atherosclerotic complications in ESRD patients. In vitro studies described an inhibition of the regenerative properties of endothelial cells and an increase of their permeability by uremic concentrations of the solute [19, 20]. Some recent clinical findings on the toxicity of p-cresol are in line with the in vitro data. First, in a study by De Smet et al. [21] serum free levels of the solute were shown to be greater in hemodialysis patients hospitalized for infectious disease. Furthermore, a positive relationship was found between serum total p-cresol levels and a uremic symptom score in patients treated with peritoneal dialysis, whereas a correlation with small water-soluble solutes and the middle molecule β_2-microglobulin was absent [9]. Even more striking, however, are the preliminary findings of a prospective observational study in patients treated with conventional hemodialysis (3×4 h/week) indicating that the accumulation of p-cresol is a risk factor for overall mortality [Bammens et al., unpubl. data]. Furthermore, the data show a significantly positive correlation between serum levels of the solute and the Malnutrition-Inflammation Score, reflecting both inflammation and nutritional status. Like *p*-cresol, indoxyl sulfate was found to inhibit endothelial proliferation and wound repair [19]. Moreover, the molecule was shown to induce free radical production by renal tubular cells and activate NF-κB, a central mediator of atherosclerosis [22]. Earlier studies by Niwa [23] showed that the molecule constitutes one of the factors responsible for the progression of glomerular sclerosis in rats and humans. Recently another protein-bound uremic compound, phenylacetic acid, was identified to have an inhibitory impact on inducible nitric oxide synthase, an enzyme with a protective potential on endothelial function [24]. Also carbonyl compounds and advanced glycation end-products have been shown to be related to oxidative stress, progression of atherosclerosis and furthermore to peritoneal membrane dysfunction in patients treated with peritoneal dialysis [25, 26]. The protein binding of these molecules (e.g. fructoselysine, glyoxal, methylglyoxal, N^ε-carboxymethyllysine, pentosidine and 3-deoxyglucosone), however, should be considered to be rather tight and long-lived in contrast to the other ligands mentioned above which display a continuous and dynamic competition among each other and with drugs for the binding sites. In the following paragraphs, only solutes with the latter type of protein binding will be discussed.

Removal of Protein-Bound Solutes

In a study by Lesaffer et al. [8] the removal of protein-bound solutes during conventional low-flux hemodialysis was found to be much lower than that of their water-soluble counterparts. In contrast to a 77.8% reduction ratio of urea nitrogen, serum levels of indoxyl sulfate and *p*-cresol were reduced only by 41.5 and 32.9%, respectively, while CMPF levels did not decrease at all. Furthermore, the data indicated that high-flux hemodialysis has no advantage over low-flux hemodialysis with regard to protein-bound solute removal. Protein losses into the dialysate were limited and seemed to have no major impact on solute removal. This could best be interpreted from the data regarding CMPF, which is virtually 100% bound to protein. CMPF was not detectable in the dialysate. For all other protein-bound solutes, known to have a substantial free fraction, concentrations in the dialysate were higher. These results pointed to the most important mechanism for the removal of protein-bound solutes through non-protein-leaking membranes: elimination of their unbound fraction.

Of course, protein-leaking membranes can also be used to enhance the elimination of protein-bound solutes. Six months of hemodialysis treatment using a polymethylmethacrylate-based dialyzer with an albumin-sieving coefficient of 0.03 resulted in a reduction of serum total homocysteine from 25.3 ± 5.9 µmol/l to 17.2 ± 4.2 µmol/l in a study by Galli et al. [27]. While removal of CMPF is nonexistent during conventional dialysis [8], its elimination was better with the use of a protein-leaking membrane resulting in a better control of renal anemia [11]. The widespread application of protein-leaking membranes in hemodialysis settings, however, is hampered by the risks of hypoalbuminemia. As a consequence, the most feasible approach to enhance the elimination of protein-bound solutes is by improving the removal of their unbound fraction.

In this regard, mathematical modeling by Meyer et al. [28] showed that substantially higher clearances of protein-bound solutes could be achieved by combined increases of the dialyzer mass transfer area coefficient (K_0A) and the dialysate flow rate (Q_d). In accordance, preliminary data suggest that the use of super-flux dialyzers (high K_0A) may result in a reduction of serum levels of indoxyl sulfate. Again, this effect seems not to be entirely attributable to protein losses, but may be related to a better elimination of the free fraction [De Smet et al., J Am Soc Nephrol 2004;15:363A]. Also with regard to homocysteine, beneficial effects of super-flux dialyzers have been shown [29, 30]. However, the results of these studies suggest that improved removal of uremic toxins with inhibitory effects on metabolizing enzymes, rather than better dialytic clearance of homocysteine itself, may explain the findings.

Limited data are available on the elimination of protein-bound solutes during alternative dialysis time schedules. In a study by Fagugli et al. [31] lower serum levels of indoxyl sulfate and *p*-cresol were found after 6 months of daily 2-hour dialysis as compared to an equal period on three-times-a-week 4-hour dialysis. Theoretically, the same effect can be expected from long-duration (daily or alternate day nocturnal) hemodialysis. Although the amount of free molecules removed per unit of time will not be higher (or even lower) than during conventional hemodialysis, the total solute removal at the end of the session might be greater. Moreover, since rebound can be expected to be less than during conventional dialysis, the influence of long-duration dialysis on serum levels of protein-bound solutes might be favorable. Further investigation is warranted, however, to confirm this hypothesis.

The application of convection is another way to enhance the transmembrane transport of the unbound fraction of protein-bound solutes. Six months of predilution hemofiltration had a small, but significant, lowering effect on the serum levels of several protein-bound molecules [Meert et al. J Am Soc Nephrol 2004;15:363A]. As suggested by in vitro measurements, higher clearances can be achieved by combining convective and diffusive transport in the setting of hemodiafiltration [Meyer et al. J Am Soc Nephrol 2004;15:363A]. This is illustrated by the results of a recent randomized cross-over study including 14 patients,

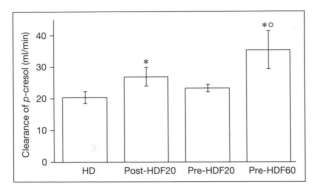

Fig. 1. Dialytic clearances of the protein-bound solute *p*-cresol during hemodialysis (HD), postdilution hemodiafiltration with a substitution flow rate of 87.0 ml/min (post-HDF20), predilution hemodiafiltration with a substitution flow rate of 87.0 ml/min (pre-HDF20) and predilution hemodiafiltration with a substitution flow rate of 260.9 ml/min (pre-HDF60). Actual blood flow rates were approximately 320 ml/min and did not differ among the treatments. The figures of the substitution flow rates are the result of a total substitution volume of 20 l (post-HDF20, pre-HDF20) and 60 l (pre-HDF60) during an actual dialysis time of 230 min per session. Means ± SEM are illustrated; * $p < 0.05$ versus HD; ° $p < 0.05$ versus pre-HDF20. Adapted from reference [32], with permission.

which investigated the effect of hemodiafiltration on the removal of the protein-bound solute *p*-cresol [32]. For this purpose the newly developed high-flux FX80 dialyzer was used [33]. Main characteristics of this device are the reduced diameter of the capillary fibers (185 μm), the nanocontrolled spinning of the membrane and its reduced wall thickness (35 μm). It was shown earlier that these features result in an increased sieving coefficient of the middle molecule β_2-microglobulin and a higher K_0A for urea nitrogen [33]. Figure 1 illustrates the mean clearances (ml/min) of the protein-bound solute *p*-cresol by the FX80 dialyzer calculated from total dialysate collections during hemodialysis, postdilution hemodiafiltration with a substitution flow rate of 87.0 ml/min, predilution hemodiafiltration with a substitution flow rate of 87.0 ml/min and predilution hemodiafiltration with a substitution flow rate of 260.9 ml/min, respectively. As can be seen, the application of convection resulted in a significant increase of *p*-cresol removal. Three observations from the study suggested that a higher elimination of unbound *p*-cresol underlies this finding. First, using rate nephelometry as a sensitive detection method, albumin levels were found to be lower than the detection limit of 6.17 mg/l in all dialysate samples. Hence, considering the total collected dialysate volumes, the amount of albumin lost per session never exceeded 1.236 ± 0.011 g. If *p*-cresol removal was due to albumin loss, at least 167 ± 20 g would be expected (calculations based on the molar ratio of serum

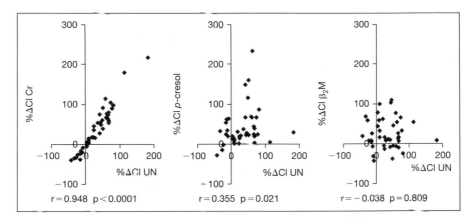

Fig. 2. Percentage of change of clearance versus hemodialysis; pooled data from the three hemodiafiltration periods (post-HDF20, pre-HDF20, pre-HDF60). The three panels show percentage of change of clearance of creatinine (%ΔCl Cr), *p*-cresol (%ΔCl *p*-cresol), and β2-microglobulin (%ΔCl β₂M) versus percentage of change of clearance of urea nitrogen (%ΔCl UN). Spearman's correlation coefficients (r) and p are indicated. Reprinted from reference [32], with permission.

p-cresol to albumin and the total amount of *p*-cresol removed during the session). Second, identical concentrations of total *p*-cresol were found in native dialysate samples as well as after passage through a filter with a cutoff of 30,000 Da. Third, from a correlation analysis evaluating the changes in solute clearances due to the application of convection, significant relationships between urea nitrogen, creatinine, and *p*-cresol were found (fig. 2). The latter finding suggests that free *p*-cresol presumably behaves like other small molecules such as urea nitrogen and creatinine. By the combination of convection and diffusion, a constant shift in the bound-unbound equilibrium allows more free *p*-cresol to pass through the membrane.

Besides diffusion and convection, adsorption of albumin and its ligands to the membrane may theoretically account for some protein-bound solute removal [34]. Furthermore, the adsorptive capacity of albumin itself can be exploited to improve the removal of protein-bound molecules. This is the case in so-called albumin-dialysis techniques where human serum albumin is added to the dialysate as a molecular adsorbent. In this setting unbound solutes cross the high-flux membrane used, bind to the albumin on the other side and are carried away. In single pass albumin-dialysis, the albumin-rich dialysate is discharged after circulation through the dialyzer, while in the molecular adsorbent recirculating system (MARS®) it is regenerated in a separate circuit by using low-flux dialysis and different adsorbers to wash away or to bind the molecules unloaded from albumin. This 'regenerated' albumin is then used again for the detoxification process. To overcome the need for large amounts of commercial human serum albumin and related safety and cost issues, a method called fractionated plasma separation and adsorption has been developed. In this system, a special albumin-permeable filter with a cutoff of approximately 250,000 Da is used. Thus albumin passes through the membrane together with bound solutes, which are directly removed by special adsorbers within the secondary circuit. The native albumin is subsequently returned to the patient. In clinical practice, fractionated plasma separation and adsorption is combined with hemodialysis to eliminate water-soluble solutes (Prometheus®). Albumin-dialysis techniques have been shown to enable the removal of protein-bound solutes in the setting of liver failure [35; Stange et al., ASAIO J 1995;41:11]. To date, however, information on their performance in ESRD patients is limited and studies on this issue are warranted [Mitzner et al., Nephrol Dial Transplant 1994; 14: A201].

Conclusions

In view of the existing and emerging data on protein-bound uremic retention solutes as important contributors to the pathophysiology of the uremic syndrome, efforts should be made to improve their elimination. Next to dialysis with protein-leaking membranes and more sophisticated therapies including albumin-dialysis and adsorptive strategies, the more feasible application of convective therapy seems promising in this regard. Data on the use of alternative dialysis time schedules are limited to date. Nevertheless, it seems worthwhile to investigate their ability to remove protein-bound solutes, since it can be expected that long and/or daily dialysis result in a better elimination of the unbound fraction of these molecules.

References

1   Merriam-Webster Online. Version current 1 January 2005. Internet: http://www.webster.com/cgi-bin/dictionary (accessed 13 January 2005).
2   Bergström J: Uremic toxicity; in Kopple JD, Massry SG (eds): Nutritional Management of Renal Disease. Baltimore, Williams & Wilkins, 1997, pp 97–190.
3   Vanholder R, De Smet R, Glorieux R, Argilés A, Baurmeister U, Brunet P, Clark W, Cohen G, De Deyn PP, Deppisch R, Descamps-Latscha B, Henle T, Jörres A, Lemke HD, Massy ZA, Passlick-Deetjen J, Rodriguez M, Stegmayr B, Stenvinkel P, Tetta C, Wanner C, Zidek W for the European Uremic Toxin Work Group (EUTox): Review on uremic toxins: Classification, concentration, and interindividual variability. Kidney Int 2003;63:1934–1943.
4   Eknoyan G, Beck G, Cheung A, Daugirdas JT, Greene T, Kusek JW, Allon M, Bailey J, Delmez JA, Depner TA, Dwyer JT, Levey AS, Levin NW, Milford E, Ornt DB, Rocco MV, Schulman G,

Schwab SJ, Teehan BP, Toto R for the Hemodialysis (HEMO) Study Group: Effect of dialysis dose and membrane flux in maintenance hemodialysis. N Engl J Med 2002;347:2010–2019.

5   Paniagua R, Amato D, Vonesh E, Correa-Rotter R, Ramos A, Moran J, Mujais S: Effects of increased peritoneal clearances on mortality rates in peritoneal dialysis: ADEMEX, a prospective, randomized, controlled trial. J Am Soc Nephrol 2002;13:1307–1320.

6   Levin N, Greenwood R: Reflections on the HEMO study: The American viewpoint. Nephrol Dial Transplant 2003;18:1059–1060.

7   Locatelli F: Dose of dialysis, convection and haemodialysis patients outcome – What the HEMO study doen't tell us: The European viewpoint. Nephrol Dial Transplant 2003;18:1061–1065.

8   Lesaffer G, De Smet R, Lameire N, Dhondt A, Duym P, Vanholder R: Intradialytic removal of protein-bound uraemic toxins: Role of solute characteristics and of dialyser membrane. Nephrol Dial Transplant 2000;15:50–57.

9   Bammens B, Evenepoel P, Verbeke K, Vanrenterghem Y: Removal of middle molecules and protein-bound solutes by peritoneal dialysis and relation with uremic symptoms. Kidney Int 2003;64:2238–2243.

10 Suliman ME, Stenvinkel P, Qureshi AR, Barany P, Heimburger O, Anderstam B, Alvestrand A, Lindholm B: Hyperhomocysteinemia in relation to plasma free amino acids, biomarkers of inflammation and mortality in patients with chronic kidney disease starting dialysis therapy. Am J Kidney Dis 2004;44:455–465.

11 Niwa T, Asada H, Tsutsui S, Miyazaki T: Efficient removal of albumin-bound furancarboxylic acid by protein-leaking hemodialysis. Am J Nephrol 1995;15:463–467.

12 Kushner D, Beckman B, Nguyen L, Chen S, Della Santina C, Husserl F, Rice J, Fisher JW: Polyamines in the anemia of end-stage renal disease. Kidney Int 1991;39:725–732.

13 Bagdade JD, Subbaiah PV, Bartos D, Bartos F, Campbell RA: Polyamines: An unrecognised cardiovascular risk factor in chronic dialysis. Lancet 1979;1:412–413.

14 Ottonello L, Gnerre P, Bertolotto M, Mancini M, Dapino P, Russo R, Garibotto G, Barreca T, Dallegri F: Leptin as a uremic toxin interferes with neutrophil chemotaxis. J Am Soc Nephrol 2004;15:2366–2372.

15 Pecoits-Filho R, Nordfors L, Heimburger O, Lindholm B, Anderstam B, Marchlewska A, Stenvinkel P: Soluble leptin receptors and serum leptin in end-stage renal disease: Relationship with inflammation and body composition. Eur J Clin Invest 2002;32:811–817.

16 Vanholder R, De Smet R, Waterloos MA, Van Landschoot N, Vogeleere P, Hoste E, Ringoir S: Mechanisms of uremic inhibition of phagocyte reactive species production: Characterization of the role of p-cresol. Kidney Int 1995;47:510–517.

17 Wratten ML, Tetta C, De Smet R, Neri R, Sereni L, Camussi G, Vanholder R: Uremic ultrafiltrate inhibits platelet-activating factor synthesis. Blood Purif 1999;17:134–141.

18 Dou L, Cerini C, Brunet P, Guilianelli C, Moal V, Grau G, De Smet R, Vanholder R, Sampol J, Berland Y: P-cresol, a uremic toxin, decreases endothelial cell response to inflammatory cytokines. Kidney Int 2002;62:1999–2009.

19 Dou L, Bertrand E, Cerini C, Faure V, Sampol J, Vanholder R, Berland Y, Brunet P: The uremic solutes p-cresol and indoxyl sulfate inhibit endothelial proliferation and wound repair. Kidney Int 2004;65:442–451.

20 Cerini C, Dou L, Anfosso F, Sabatier F, Moal V, Glorieux G, De Smet R, Vanholder R, Dignat-George F, Sampol J, Berland Y, Brunet P: P-cresol, a uremic retention solute, alters the endothelial barrier function in vitro. Thromb Haemost 2004;92:140–150.

21 De Smet R, Van Kaer J, Van Vlem B, De Cubber A, Brunet P, Lameire N, Vanholder R: Toxicity of free p-cresol: A prospective and cross-sectional analysis. Clin Chem 2003;49:470–478.

22 Motojima M, Hosokawa A, Yamato H, Muraki T, Yoshioka T: Uremic toxins of organic anions upregulate PAI-1 expression by induction of NF-κB and free radical in proximal tubular cells. Kidney Int 2003;63:1671–1680.

23 Niwa T: Organic acids and the uremic syndrome: Protein metabolite hypothesis in the progression of chronic renal failure. Semin Nephrol 1996;16:167–182.

24 Jankowski J, van der Giet M, Jankowski V, Schmidt S, Hemeier M, Mahn B, Giebing G, Tolle M, Luftmann H, Schluter H, Zidek W, Tepel M: Increased plasma phenylacetic acid in patients with end-stage renal failure inhibits iNOS expression. J Clin Invest 2003;112:256–264.

25   Rashid G, Benchetrit S, Fishman D, Bernheim J: Effect of advanced glycation end-products on gene expression and synthesis of TNF-alpha and endothelial nitric oxide synthase by endothelial cells. Kidney Int 2004;66:1099–1106.

26   Honda K, Nitta K, Horita S, Yumura W, Nihei H, Nagai R, Ikeda K, Horiuchi S: Accumulation of advanced glycation end products in the peritoneal vasculature of continuous ambulatory peritoneal dialysis patients with low ultra-filtration. Nephrol Dial Transplant 1999;14:1541–1549.

27   Galli F, Benedetti S, Buoncristiani U, Piroddi M, Conte C, Canestrari F, Buoncristiani E, Floridi A: The effect of PMMA-based protein-leaking dialyzers on plasma homocysteine levels. Kidney Int 2003;64:748–755.

28   Meyer TW, Leeper EC, Bartlett DW, Depner TA, Lit YZ, Robertson CR, Hostetter TH: Increasing dialysate flow and dialyzer mass transfer area coefficient to increase the clearance of protein-bound solutes. J Am Soc Nephrol 2004;15:1927–1935.

29   van Tellingen A, Grooteman MP, Bartels PC, van Limbeek J, van Guldener C, ter Wee PM, Nubé MJ: Long-term reduction of plasma homocysteine levels by super-flux dialyzers in hemodialysis patients. Kidney Int 2001;59:342–347.

30   De Vriese AS, Langlois M, Bernard D, Geerolf I, Stevens L, Boelaert JR, Schurgers M, Matthys E: Effect of dialyser membrane pore size on plasma homocysteine levels in haemodialysis patients. Nephrol Dial Transplant 2003;18:2596–2600.

31   Fagugli RM, De Smet R, Buoncristiani U, Lameire N, Vanholder R: Behavior of non-protein-bound and protein-bound solutes during daily hemodialysis. Am J Kidney Dis 2002;40:339–347.

32   Bammens B, Evenepoel P, Verbeke K, Vanrenterghem Y: Removal of the protein-bound solute p-cresol by convective transport: A randomized crossover study. Am J Kidney Dis 2004;44:278–285.

33   Ronco C, Bowry SK, Brendolan A, Crepaldi C, Soffiati G, Fortunato A, Bordoni V, Granziero A, Torsello G, La Greca G: Hemodialyzer: From macro-design to membrane nanostructure: The case of the FX-class of hemodialyzers. Kidney Int 2002;61(suppl 80):S126–S142.

34   Clark WR, Macias WL, Molitoris BA, Wang NHL: Plasma protein adsorption to highly permeable hemodialysis membranes. Kidney Int 1995;48:481–488.

35   Evenepoel P, Maes B, Wilmer A, Nevens F, Fevery J, Kuypers D, Bammens B, Vanrenterghem Y: Detoxifying capacity and kinetics of the Molecular Adsorbent Recycling System Contribution of the different inbuilt filters. Blood Purif 2003;21:244–252.

36   Mitzner S, Stange J, Dillmonn A, Winkler RE, Michelson A, Knippel M, Schmidt R: Removal of protein-bound uremic toxins by albumin dialysis: In vivo results (abstract). Nephrol Dial Transplant 1999;14:A201.

Pieter Evenepoel, MD, PhD
Department of Medicine, Division of Nephrology
University Hospital Gasthuisberg, Herestraat 49
BE–3000 Leuven (Belgium)
Tel. +32 16 344591, Fax +32 16 344599, E-Mail Pieter.Evenepoel@uz.kuleuven.ac.be

Ronco C, Brendolan A, Levin NW (eds): Cardiovascular Disorders in Hemodialysis.
Contrib Nephrol. Basel, Karger, 2005, vol 149, pp 185–199

....

Inflammation in End-Stage Renal Disease – A Fire that Burns within

Peter Stenvinkel

Division of Renal Medicine, Department of Clinical Science,
Karolinska University Hospital, Karolinska Institutet, Stockholm, Sweden

Abstract

Cardiovascular disease (CVD) remains the major cause of morbidity and mortality in end-stage renal disease (ESRD) patients. As traditional risk factors cannot alone explain the unacceptable high prevalence and incidence of CVD in this population, inflammation (which is interrelated to insulin resistance, oxidative stress, wasting and endothelial dysfunction) has been suggested to be a significant contributor. Indeed, several different inflammatory biomarkers, such as high sensitivity C-reactive protein (hs-CRP), have been shown to independently predict mortality in ESRD patients. As CRP is so strongly associated with vascular disease it has been suggested that this hepatic-derived protein is not only a marker, but also a mediator of vascular disease. Indeed, recent in vitro data from studies on endothelial cells, monocytes-macrophages and smooth muscle cells support a direct role for CRP in atherogenesis. The causes of the highly prevalent state of inflammation in ESRD are multiple, including decreased renal function, volume overload, comorbidity and intercurrent clinical events, factors associated with the dialysis procedure and genetic factors. Recent evidence suggests that several cytokine DNA polymorphisms may affect the inflammatory state as well as outcome in ESRD patients. As interventions directed towards traditional risk factors have, so far, not proven to be very effective, controlled studies are needed to evaluate if various pharmacological as well as non-pharmacological anti-inflammatory treatment strategies, alone or in combination, may be an option to affect the unacceptable high cardiovascular mortality rate in this patient group.

Inflammation as a Risk Factor for Vascular Disease

The lifespan of end-stage renal disease (ESRD) patients is markedly reduced, and cardiovascular disease (CVD) accounts for a premature death in more than 50% of patients from Western Europe and North America undergoing regular dialysis [1]. Despite many recent improvements in dialysis technology, the majority of maintenance dialysis patients die within a 5-year period, a survival worser than that of the majority of patients with malignancies. By extrapolating data from the general population, nephrologists have mostly been focusing on conventional (i.e. Framingham) risk factors, such as hypertension and dyslipidemia. However, although the prevalence of traditional Framingham risk factors is high in most ESRD patients, the extent and severity of cardiovascular complications is clearly disproportionate to the underlying risk factor profile [2], and survival has not improved very much in the last two decades. Moreover, recent multicenter clinical trials, such as ADEMEX [3] and HEMO [4], have failed to show any survival advantage of increased dialysis dosage. In the light of these disappointing findings, recent interest has, therefore, focused on the impact of non-traditional risk factors, one of which being chronic inflammation, a putative 'secret killer' that has been hypothesized to promote atherosclerosis [5].

According to the inflammation hypothesis of atherosclerosis [6], local inflammatory stimuli, such as oxidatively modified products, advanced glycation endproducts (AGEs) or various persistent infectious processes may change the milieu of the arterial wall. The change in milieu promotes the production of pro-atherogenic adhesion molecules, growth factors and chemokines, all of which play important roles in the atherogenic process. In addition, an increased production of interleukin (IL)-6 is the chief stimulator of various pro-coagulants, such as fibrinogen, and also the most important downstream marker of inflammation, C-reactive protein (CRP). It should be noted that although the correlation between circulating IL-6 and CRP levels is very strong, the circulating levels of CRP can vary 100-fold for any given IL-6 level (fig. 1). Clearly, the pattern of cytokine production and the acute phase response differ in various inflammatory conditions. Thus, as uremic inflammation may be different in etiology from inflammation under various non-renal conditions, results obtained in non-renal populations cannot automatically be transferred to the ESRD population. Although the concept that inflammation plays a central role in the pathophysiology of atherosclerosis has gained a lot of recent interest, we do not know, yet, whether inflammation is just a reflection of vascular injury or instead a promotor of vascular injury. However, the lack of association between the extent of atherosclerosis and CRP levels argues against CRP elevation being purely a consequence of atherosclerosis development [7].

Evidence that CRP per se Promote Vascular Disease

While CRP initially was felt to be only a marker of inflammation, recent evidences suggest that CRP also may be a direct mediator of vascular disease.

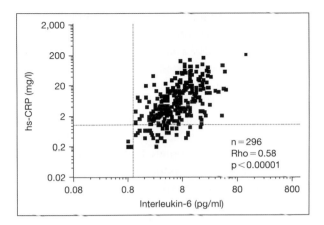

Fig. 1. Correlation between interleukin-6 and high sensitivity C-reactive protein (hs-CRP) in 296 incident end-stage renal disease patients starting renal replacement therapy.

As any inflammatory stimuli that would prompt the release of pro-atherogenic cytokines, such as IL-1, IL-6 and TNF-α, also stimulate hepatic CRP production, it has been argued that the association between vascular damage and CRP may be merely indirect and that pro-inflammatory cytokines, such as IL-6, are the main culprit. Moreover, as strong associations between CRP and other cardiovascular risk factors, such as insulin resistance, oxidative stress endothelial dysfunction and vascular calcification have been documented [5, 8, 9], this may also support the concept of an indirect association between CRP and vascular disease. However, abundant in vitro data have recently emerged from the studies on endothelial cells, monocytes-macrophages and smooth muscle cells, supporting a direct role for CRP in atherogenesis. First, as CRP has significant effects on monocytes and macrophages and activates the production of monocyte chemotactic protein, it may be an activator of the entire recruitment cascade [10]. Secondly, using cultures of umbilical vein endothelial cells, Pasceri et al. [11] showed that CRP activates endothelial cells to express adhesion molecules. Other documented effects of CRP on endothelial cells include inhibition of endothelial progenitor cell differentiation and function [12], decreased eNOS expression and activity in human aortic endothelial cells [13], activation of NF-κB [14] and release of the potent endothelial-derived contracting factor endothelin-1 [15]. CRP seems also to exert direct pro-atherogenic effects on vascular smooth muscles, including an upregulation of iNOS and NF-κB [16] and angiotensin type-1 receptors [17]; the latter finding links inflammation to the renin-angiotensin system. Finally, as CRP and complement are known to colocalize in human atherosclerotic lesions it could be suggested that by activating complement CRP may be an active participant in the atherosclerotic development [18]. Despite the numerous reports of potential pro-atherogenic properties of CRP in vitro, there are no direct evidences available suggesting that CRP promotes atherosclerosis in humans. However, a recent study in Apo E-deficient mice showed that human transgene expression caused accelerated aortic atherosclerosis [19]. Moreover, in this study, CRP detected in the atherosclerotic lesions was associated with increased complement deposition and increased angiotensin type-1 receptors and adhesion molecule expression. Finally, it should be mentioned that inflammation might promote both dyslipidemia [20] and hypertension [17], two traditional risk factors associated with the metabolic syndrome. Whether or not the uremic milieu may affect the atherogenic potential of CRP has, to the best of our knowledge, not yet been tested. However, as the pro-atherogenic effect of CRP is exaggerated in a hyperglycemic milieu [21], the in vitro effects of CRP (in serum concentration observed in ESRD patients) should be tested on monocytes-macrophages, endothelial cells, smooth muscle cells and in the uremic milieu.

Application of CRP in Detection and Prevention of CVD in the General Population

Every year about 1.3 million of American patients have an acute coronary event. As only half of these patients would be identified on the basis of traditional Framingham risk factors, additional tools to screen for CVD risk have been urgently needed. It is, therefore, of interest that the relationship between the baseline level of CRP and subsequent risk of vascular events has been remarkably consistent in more than 20 prospective epidemiological studies [22]. Moreover, as several cohort studies have confirmed that hs-CRP evaluation adds prognostic information beyond that available from traditional Framingham risk factors [22], it has been suggested that hs-CRP should be used as a predictive tool in selected risk populations [23]. As CRP levels are not closely associated with subclinical atherosclerosis (as measured by cardiac catheterization, intima-media thickness and ankle-brachial index), it has been suggested that inflammation is associated with plaque vulnerability and rupture rather than the total plaque burden [24]. Indeed, clinical trials have shown that a CRP evaluation at the time of admission of an acute coronary event identifies the high cardiovascular risk patients [25]. It should be mentioned that a recent study [26] has called into question the magnitude of the independent effect of CRP in the general population. In this large study (2,459 patients from Reykjavik, Iceland) the relative risk of elevated CRP ($>2\,mg/l$) was reduced after adjustment for established risk factors, such as smoking, blood pressure, body mass index and total cholesterol. Nevertheless, a highly significant fully adjusted odds ratio of 1.5 was still observed, an odds ratio similar to that observed for hypertension and smoking. Although an abundance of epidemiological data support the hypothesis that inflammation is an independent predictor of vascular events, studies linking a greater reduction in CRP to reduced rate of vascular disease have so far been lacking. However, two studies recently demonstrated that reducing the inflammatory component of CVD through the use of statins improve outcome independent of any reduction in lipoprotein levels. First, Ridker et al. [27] demonstrated in 3,745 patients with acute coronary syndromes assigned to either 80 mg of atorvastatin or 40 mg of pravastatin that patients with low CRP levels after statin therapy had better outcome than those with higher CRP levels. Secondly, Nissen et al. [28] have performed intravascular ultrasonography in 502 patients with angiographically documented coronary disease and repeated the investigation following either moderate (40 mg pravastatin) or intensive (80 mg atorvastatin) statin treatment for 18 months. It is notable that even after the adjustment for reduction in lipid levels the observed decrease in CRP levels was independently and significantly correlated with the rate of atheroma progression. Taken together, available evidences from the general population support the hypothesis that CRP has pro-atherogenic properties.

CRP Predicts Outcome also in ESRD Patients

Considering the fact that the median CRP level is about 1.5 mg/l in the general population [29], the majority of ESRD patients could be considered to be inflamed. In a survey of 663 European ESRD patients we have found that almost 75% of this patient group have CRP level above 3.4 mg/l [30]. Evidence from the CRIB study [31] demonstrates that kidney disease is associated with inflammation even in patients with moderate renal impairment. As chronic inflammation is such a common phenomenon in European [30] and American [32] ESRD populations, its role as an atherosclerotic mediator and prognostic indicator has been the area of much recent interest. The documented lower prevalence of inflammation in East Asian countries may depend on genetic factors as well as cultural habits, such as food intake [33]. In accordance with the findings in the general population several studies have shown that elevated CRP predicts both all-cause and cardiovascular mortality in hemodialysis (HD) [34–36] and peritoneal dialysis (PD) [37–39] patients. Moreover, elevated CRP concentrations observed after a HD session is associated with both cardiac hypertrophy [40] and a higher mortality risk [41]. In accordance, in PD patients an elevated CRP was shown to be an independent predictor of non-fatal myocardial infarction [42] and increased incidence of CVD [38]. Data from the so-called MDRD study show that, after adjusting for traditional CVD risk factors, the odds of CVD were 1.7 times greater in CKD patients with high CRP level [32]. In addition, in CKD stage 3–4 patients' elevations in CRP, IL-6 and TNF-α were significantly associated with future coronary events in the women health study [43]. It is noteworthy that a Korean study demonstrated that a persistent elevation of CRP during 6 months in PD patients was strongly associated with ischemic heart disease [44]. Further support linking inflammation to poor outcome in ESRD comes from two recent large studies of HD patients, showing a direct association between the neutrophil count and mortality [45, 46]. Also other less commonly measured inflammatory markers, such as IL-6 [47–49], hyaluronan [50] and fibrinogen [51], have been shown to predict mortality in dialysis patients. Further support for the role of inflammation in atherogenesis comes from an evaluation of a historical cohort of 393,451 US dialysis patients, which demonstrated that septicemia was associated with increased cardiovascular death risk [52].

Many Causes of Inflammation in ESRD

Clearly, both dialysis-related and dialysis-unrelated factors may contribute to the high prevalence of inflammation in ESRD patients. Not surprisingly, a prospective study by van Tellingen et al. [53] in 74 HD patients showed that an intercurrent clinical event was the most important factor predicting CRP alterations in this patient group. Evidence suggests that chronic infections with *Chlamydia pneumoniae* and *Helicobacter pylori* [54–56] and peridontitis [57] are likely to contribute. Moreover, the putative role of biofilm formation on subclinical inflammation was recently discussed by Cappeli et al. [58]. Obesity is another factor that may contribute to a state of subclinical inflammation in ESRD. In a recent study we found that truncal (i.e. visceral) fat mass (estimated by DEXA) was significantly associated to circulating IL-6 levels [59]. Moreover, as recent studies suggest that a reduction of kidney function per se may be associated with an inflammatory response both in mild [60] and advanced [61] kidney failure, differences in residual renal function may also contribute. As the failing heart produces large quantities of pro-inflammatory cytokines [62], volume overload and/or congestive heart failure may also link inflammation to reduced residual renal function. Indeed, a recent study by Wang et al. [63] has documented strong interrelations between inflammation, residual renal function and cardiac hypertrophy in PD patients. It should be emphasized that also processes associated with the HD procedure, such as clotted access grafts [64], contact of blood with artificial dialysis membranes [65] and catheter infections [66] contribute to an inflammatory process. Moreover, a study by

Lopez-Gomez et al. [67] have demonstrated that failed kidney transplants may be yet another reason of inflammation in dialysis patients.

Genetic Factors Affect the Prevalence of Inflammation in ESRD

Although multiple, both dialysis-related and dialysis-unrelated, factors might be related to chronic inflammation in ESRD patients [68], it should be emphasized that not all ESRD patients show signs of inflammation. Indeed, as there is a great inter- and intra-individual variability in the prevalence of inflammation that cannot be explained only by dialysis treatment or other clinical inflammatory events, genetic factors may account for these differences. The identification of genetic variations in genes related to the inflammation-malnutrition axis in particular individuals more susceptible to chronic inflammation might be an interesting approach for tracking ESRD patients at high risk who may benefit from more aggressive treatment. In fact, Pankow et al. [69] found evidence of substantial heritability (35–40%) for CRP levels as well as for leukocytes and albumin levels and a prospective evaluation by Tsirpanlis et al. [70] demonstrated that individual factors significantly influence the levels of inflammatory markers in dialysis patients. No doubt, several candidate genes may affect the prevalence of inflammation in ESRD, which in turn may affect the risk of vascular complications and outcome.

Given the strong associations between circulating IL-6 levels and outcome in ESRD patients [47–49], single nucleotide polymorphisms within the IL-6 gene should be of major interest to study. A recent study demonstrated that healthy subjects carrying the -174 G-allele had significantly stronger inflammatory IL-6 response to vaccination against *Salmonella typhii*, compared to individuals who were C homozygous [71]. Few studies have investigated the impact of IL-6 single nucleotide polymorphisms on plasma IL-6 levels and/or the phenotype in ESRD patients. Balakrishnan et al. [72] found a significantly higher plasma IL-6 level as well as comorbidity score in -174 G-allele (high producers) carriers in 183 ESRD patients, whereas another group [73] investigated 161 ESRD patients and found that individuals with the G/C and C/C phenotype had higher diastolic blood pressure and left ventricular mass than G/G homozygotes. As inter-individual differences in TNF-α release have been described, a TNF-α G$\rightarrow$A transition at position -308 may be another important functional polymorphism [74] that affects both the prevalence of inflammation and its associated complications in ESRD. It has been reported that HD patients who were A-allele carriers had increased comorbidity and lower S-albumin levels [72]. Although CRP polymorphisms ($+1059$ G/C and -286 C/T/A) have been reported to be associated with circulating CRP levels [75, 76] there are, to the best of our knowledge, not yet any published reports on the association between CRP polymorphisms and inflammatory biomarkers and/or risk for vascular complications in ESRD patients. The peroxisome proliferator-activated receptor-γ (PPAR-γ) is an important regulator of adipocyte differentiation and intracellular insulin-signaling events that may have an inflammatory impact. In a recent prospective evaluation of 229 ESRD patients, we found that the presence of the Ala12 allele of PPAR-γ2 was associated not only with lower levels of inflammatory biomarkers but also better survival [77]. IL-10 is a product of monocytes and lymphocytes that has been regarded as one of the most important anti-inflammatory and anti-atherogenic immune-regulating cytokines as it effectively downregulates pro-inflammatory cytokines, such as IL-1, IL-6 and TNF-α [9]. A strong heritability for IL-10 production appears to exist [78] and similar to most other cytokine genes, polymorphic sequences have been described in the IL-10 promotor (-1082 [A$\rightarrow$G], -819 [C$\rightarrow$T] and -592 [C$\rightarrow$A]) [79]. Findings by Girndt et al. [80] suggest that genetic variations in the IL-10 gene may have important clinical implications in this patient group and demonstrated that HD patients with the genotype associated with higher circulating IL-10 levels have a better immuneresponse to the hepatitis B vaccination, less inflammatory activity as well as a lower risk of cardiovascular death [81].

Inflammation in Renal Disease – Can It Be Treated by Statins?

Despite the fact that interventions to treat traditional risk factors in ESRD have been used since several years, the survival in this patient group has not improved substantially in the last two decades. Thus, new treatment strategies should be considered and various pharmacological and non-pharmacological strategies to reduce the high prevalence of inflammation and wasting have been discussed [82]. Of course, all manipulation of inflammation in ESRD should first include identification of the probable cause of inflammation and, if possible, targeted treatment. However, in our experience specific causes of inflammation may not always be found in ESRD and unspecific anti-inflammatory treatment strategies could, therefore, be discussed. In this respect several drugs, such as statins, angiotensin-converting enzyme inhibitors, angiotensin receptor blockers, PPAR-γ activators, N-acetylcysteine (and other antioxidants) may be of interest to study. Lifestyle modification may be another important component in normalizing a dysregulated cytokine system activity in ESRD [82]. Although not designed to lower CRP the use of angiotensin-converting enzyme inhibitors has been reported to be associated with lower CRP levels both in

Table 1. Studies in ESRD patients in which the effect of short-term statin therapy on CRP levels have been evaluated

Authors	Country	Number of patients	Statin and dose	Study period	Change in median CRP (mg/l)
Chang et al. [88]	Korea	58 HD	Simvastatin 20 mg/day	8 weeks	2.3–1.2 (p < 0.01)
Ichihara et al. [90]	Japan	22 HD	Fluvastatin 20 mg/day	6 months	9.7–2.6 (p < 0.05)
van der Akker et al. [92]	The Netherlands	28 HD	Atorvastatin (10–40 mg/day) or simvastatin (10–40 mg/day)	18 weeks	20.0–17.7 (NS)
Vernaglione et al. [89]	Italy	35 HD	Atorvastatin 10 mg/day	6 months	9.5–5.0 (p < 0.01)
Ikejiri et al. [91]	Japan	35 HD	Atorvastatin 10 mg/day	3 months	8.2–4.2 (p < 0.05)

HD = Hemodialysis; NS = non significant.

non-renal [83] and renal [84] patients. Among other drugs having primary or secondary anti-inflammatory properties HMG-CoA reductase inhibitors (statins) may be one of the most interesting, as it is evident that statins not only inhibit cholesterol synthesis but also have important pleiotropic anti-inflammatory and immuno-protective effects. Indeed, the list of disorders for which statins might prove beneficial is growing and today includes diseases such as multiple sclerosis and rheumatoid arthritis. However, it should be noted that although there are compelling evidences that statins lower CRP in the non-renal population [27, 28, 85, 86], some studies [86, 87] show no effect of this drug on circulating IL-6 levels. This may suggest that statins interfere with the generation and/or release of CRP from the liver rather than modulating inflammatory processes at the site of the vessel wall [87]. Although most [88–91], but not all [92], short-term studies in ESRD patients have showed that statins lower CRP levels, the effect of statins on IL-6 levels in ESRD has, to the best of our knowledge, not yet been evaluated (table 1). In contrast to the clear-cut beneficial effects of statins on vascular mortality in the general population the data suggesting a beneficial effect of this class of drug in ESRD are highly controversial. Data from DOPPS study suggest that statin prescription is associated with a reduced mortality in HD patients [93]. Moreover, a study including patients with mildly impaired renal function suggests that statin therapy is associated with better outcome [94]. In fact, given the high risk associated with chronic kidney disease the absolute benefit from the use of statins was recently proposed to be greater than in those with normal renal function [95]. On the other hand, the double-blind placebo-controlled 4-dimensional trial performed in 178 dialysis centers failed to show improved cardiovascular outcome among 1,255 type-2 diabetic HD patients receiving 20 mg of atorvastatin [Wanner et al., presented at ASN 2004]. Considering the putative anti-inflammatory effect of statins the negative results in the 4-dimensional study may suggest that anti-inflammatory treatment strategies may not be beneficial in the ESRD patient population. However, as Ridker et al. [27] recently showed that non-renal patients receiving 80-mg atorvastatin was more likely to reach a CRP level <2 mg/l, it could be speculated that 20 mg of atorvastatin was not a sufficient high dose to achieve any significant anti-inflammatory effect in this inflamed high-risk patient population. Hopefully, results from other ongoing statin trials in ESRD, such as CREATE, SHARP and AURORA, will provide some insight into the clinical value of this anti-inflammatory class of drug.

Conclusion

In summary, the significance of inflammation and elevated circulating concentrations of CRP in both non-renal and renal patient populations has increased during the last couple of years. The available data suggest that the prevalence and magnitude of inflammation increases as renal function declines. However, as not all ESRD patients have activated acute phase response this suggests that genetic factors play an important role in the propensity of inflammation in this patient group. Indeed, although the impact of various DNA polymorphisms in inflammation has only started to be investigated there is evidence suggesting that genetic variation may affect the inflammatory phenotype of this patient group. Although the clinical usefulness of regular CRP monitoring has not yet been fully validated careful search for infectious processes, correction of volume status as well as the use of biocompatible dialysis membranes and ultrapure water can be recommended. As data regarding the value of statin therapy in ESRD has been conflicting, further studies are needed to evaluate the potential anti-inflammatory potential and usefulness of this class of drug.

Acknowledgment

Söderbergs Foundation presented the present study.

References

1 Foley RN, Parfrey PS, Sarnak MJ: Clinical epidemiology of cardiovascular disease in chronic renal failure. Am J Kidney Dis 1998;32 (suppl 5):S112–S119.

2 Cheung AK, Sarnak MJ, Yan G, Dwyer JT, Heyka RJ, Rocco MV, et al: Atherosclerotic cardiovascular disease risks in chronic hemodialysis patients. Kidney Int 2000;58:353–362.

3 Paniagua R, Amato D, Vonesh E, Correa-Rotter R, Ramos A, Moran J, et al: Effects of increased peritoneal clearances on mortality rates in peritoneal dialysis: ADEMEX, a prospective, randomized, controlled trial. J Am Soc Nephrol 2002;13:1307–1320.

4 Eknoyan G, Beck GJ, Cheung AK, Daugirdas JT, Greene T, Kusek JW, et al: Effect of dialysis dose and membrane flux in maintenance hemodialysis. N Engl J Med 2002;347:2010–2019.

5 Himmelfarb J, Stenvinkel P, Ikizler TA, Hakim RM: The elephant of uremia: Oxidative stress as a unifying concept of cardiovascular disease in uremia. Kidney Int 2002;62:1524–1538.

6 Ross R: Atherosclerosis: An inflammatory disease. N Engl J Med 1999;340:115–126.

7 Pearson TA, Mensah GA, Alexander RW, Anderson JL, Cannon RO, Criqui M, et al: Markers of inflammation and cardiovascular disease: Application to clinical and public health practice: A statement for healthcare professionals from the Centers of Disease Control and Prevention and the American Heart Association. Circulation 2003;107:499–511.

8 Stenvinkel P: Interactions between inflammation, oxidative stress and endothelial dysfunction in end-stage renal disease. J Ren Nutr 2003;13:144–148.

9 Stenvinkel P, Ketteler M, Johnson RJ, Lindholm B, Pecoits-Filho R, Riella M, et al: Interleukin-10, IL-6 and TNF-a: Important factors in the altered cytokine network of end-stage renal disease – The good, the bad and the ugly. Kidney Int 2004;in press.

10 Pasceri V, Cheng JS, Willerson JT, Yeh ET, Chang JW: Modulation of C-reactive protein-mediated monocyte chemoattractant protein-1 induction in human endothelial cells by anti-atherosclerosis drugs. Circulation 2001;103:2531–2534.

11 Pasceri V, Willerson JT, Yeh ET: Direct proinflammatory effect of C-reactive protein on human endothelial cells. Circulation 2000;102:2165–2168.

12 Verma S, Kuliszewski MA, Li SH, Szmitko PE, Zucco L, Wang CH, et al: C-reactive protein attenuates endothelial progenitor cell survival, differentiation, and function: Further evidence of a mechanistic link between C-reactive protein and cardiovascular disease. Circulation 2004;109:2058–2067.

13 Venugopal SK, Devaraj S, Yuhanna I, Shaul P, Jialal I: Demonstration that C-reactive protein decreases eNOS expression and bioactivity in human aortic endothelial cells. Circulation 2002;106:1439–1441.

14 Verma S, Badiwala MV, Weisel RD, Li SH, Wang CH, Fedak PW, et al: C-reactive protein activates the nuclear factor-kappaB signal transduction pathway in saphenous vein endothelial cells: Implications for atherosclerosis and restenosis. J Thorac Cardiovasc Surg 2003;126:1886–1891.

15 Verma S, Li SH, Badiwala MV, Weisel RD, Fedak PW, Li RK, et al: Endothelin antagonism and interleukin-6 inhibition attenuate the proatherogenic effects of C-reactive protein. Circulation 2002;105:1890–1896.

16 Hattori Y, Matsumura M, Kasai K: Vascular smooth muscle cell activation by C-reactive protein. Cardiovasc Res 2003;58:186–195.

17 Wang CH, Li SH, Weisel RD, Fedak PW, Dumont AS, Szmitko P, et al: C-reactive protein upregulates angiotensin type 1 receptors in vascular smooth muscle. Circulation 2003;107:1783–1790.

18 Torzewski J, Torzewski M, Bowyer DE, Fröhlich M, Koenig W, Waltenberger J, et al: C-reactive protein frequently colocalizes with the terminal complement complex in the intima of early atherosclerotic lesions of human coronary arteries. Arterioscler Thromb Vasc Biol 1998;18:1386–1392.

19 Paul A, Kerry WS, Li L, Yechoor V, McCrory MA, Szalai A, et al: C-reactive protein accelerates the progression of atherosclerosis in apolipoprotein E-deficient mice. Circulation 2004;109:647–655.

20 Fernandez-Real JM, Ricart W: Insulin resistance and chronic cardiovascular inflammatory syndrome. Endocr Rev 2003;24:278–301.

21 Verma S, Wang CH, Weisel RD, Badiwala MV, Li SH, Fedak PW, et al: Hyperglycemia potentiates the proatherogenic effects of C-reactive protein: Reversal with rosiglitazone. J Mol Cell Cardiol 2003;35:417–419.

22   Ridker PM: Clinical application of C-reactive protein for cardiovascular disease detection and prevention. Circulation 2003;107:363–369.

23   Pearson TA, Mensah GA, Alexander RW, Anderson JL, Cannon RO, Criqui M, et al: Markers of inflammation and cardiovascular disease: Application to clinical and public health practice: A Statement for Healthcare Professionals From the Centers for Disease Control and Prevention and the American Heart Association. Circulation 2003;107:499–511.

24   Ridker PM, Wilson PW, Grundy SM: Should C-reactive protein be added to metabolic syndrome and to assesment of global cardiovascular risk? Circulation 2004;109:2818–2825.

25   Suleiman M, Aronson D, Resiner SA, Kapeliovich MR, Markiewicz W, Levy Y, Hammerman H: Admission C-reactive protein levels and 30-day mortality in patients with acute myocardial infarction. Am J Med 2003;115:695–701.

26   Danesh J, Wheeler JG, Hirschfield GM, Eda S, Eiriksdottir G, Rumley A, et al: C-reactive protein and other circulating markers of inflammation in the prediction of coronary heart disease. N Engl J Med 2004;350:1387–1397.

27   Ridker PM, Cannon CP, Morrow D, Rifai N, Rose LM, McCabe CH, et al: C-reactive protein levels and outcome after statin therapy. N Engl J Med 2005;352:20–28.

28   Nissen SE, Tuzcu EM, Schoenhagen P, Crowe T, Sasiela WJ, Tasi J, et al: Statin therapy, LDL cholesterol, C-reactive protein, and coronary heart disease. N Engl J Med 2005;352:29–38.

29   Ridker PM, Rifai N, Rose L, Buring JE, Cook NR: Comparison of C-reactive protein and low-density lipoprotein cholesterol levels in the prediction of first cardiovascular events. N Engl J Med 2002;347:1615–1617.

30   Stenvinkel P, Wanner C, Metzger T, Heimbürger O, Mallamaci F, Tripepi G, et al: Inflammation and outcome in end-stage renal failure: Does female gender constitute a survival advantage? Kidney Int 2002;62:1791–1798.

31   Landray MJ, Wheeler DC, Lip GY, Newman DJ, Blann AD, McGlynn FJ, et al: Inflammation, endothelial dysfunction, and platelet activation in patients with chronic kidney disease: The chronic renal impairment in Birmingham (CRIB) study. Am J Kidney Dis 2004;43:244–253.

32   Menon V, Wang X, Greene T, Beck GJ, Kusek JW, Marcovina SM, et al: Relationship between C-reactive protein, albumin, and cardiovascular disease in patients with chronic kidney disease. Am J Kidney Dis 2003;42:44–52.

33   Nascimento MM, Pecoits-Filho R, Lindholm B, Riella MC, Stenvinkel P: Inflammation, malnutrition and atherosclerosis in end-stage renal disease: A global perspective. Blood Purif 2002;20:454–458.

34   Yeun JY, Levine RA, Mantadilok V, Kaysen GA: C-reactive protein predicts all-cause and cardiovascular mortality in hemodialysis patients. Am J Kidney Dis 2000;35:469–476.

35   Zimmermann J, Herrlinger S, Pruy A, Metzger T, Wanner C: Inflammation enhances cardiovascular risk and mortality in hemodialysis patients. Kidney Int 1999;55:648–658.

36   Iseki K, Tozawa M, Yoshi S, Fukiyama K: Serum C-reactive (CRP) and risk of death in chronic dialysis patients. Nephrol Dial Transpl 1999;14:1956–1960.

37   Noh H, Lee SW, Kang SW, Shin SK, Choi KH, Lee HY, et al: Serum C-reactive protein: A predictor of mortality in continuous ambulatory peritoneal dialysis patients. Nephrol Dial Transplant 1998;18:387–394.

38   Ducloux D, Bresson-Vautrin C, Kribs M, Abdelfatah A, Chalopin J-M: C-reactive protein and cardiovascular disease in peritoneal dialysis patients. Kidney Int 2002;62:1417–1422.

39   Wang A, Woo J, Wai Kei C, Wang M, Man.Sei M, Lui S-F, et al: Is a single time-point C-reactive protein predictive of outcome in peritoneal dialysis patients? J Am Soc Nephrol 2003;14:1871–1879.

40   Park CW, Shin YS, Kim CM, Lee SY, Yu SE, Kim SY, et al: Increased C-reactive protein following hemodialysis predicts cardiac hypertrophy in chronic hemodialysis patients. Am J Kidney Dis 2002;40:1230–1239.

41   Korevaar JC, van Manen JG, Dekker FD, de Waart DR, Boeschoten EW, Krediet RT: Effect of an increase in CRP level during a hemodialysis session on mortality. J Am Soc Nephrol 2004;15:2916–2922.

42   Herzig KA, Purdie DM, Chang W, Brown AM, Hawley CM, Campbell SB, et al: Is C-reactive protein a useful predictor of outcome in peritoneal dialysis patients? J Am Soc Nephrol 2001;12:814–821.

Stenvinkel

43   Knight EL, Rimm EB, Pai JK, Rexrode KM, Cannuscio CC, Manson JE, et al: Kidney dysfunction, inflammation and coronary events: A prospective study. J Am Soc Nephrol 2004;15: 1897–1903.

44   Kim SB, Min WK, Lee SK, Park JS, Hong CD, Yang WS: Persistent elevation of C-reactive protein and ischemic heart disease in patients with continuous ambulatory peritoneal dialysis. Am J Kidney Dis 2002;39:342–346.

45   Pifer TB, McCullogh KP, Port FK, Goodkin DA, Maroni BJ, Held PJ, et al: Mortality risk in hemodialysis patients and changes in nutritional indicators: DOPPS. Kidney Int 2002;62:2238–2245.

46   Reddan DN, Klassen PS, Szczech LA, Coladonato JA, O'Shea S, Owen WF, et al: White blood cells as a novel predictor in haemodialysis patients. Nephrol Dial Transpl 2003;18:1167–1173.

47   Bologa RM, Levine DM, Parker TS, Cheigh JS, Seur D, Stenzel KH, et al: Interleukin-6 predicts hypoalbuminemia, hypocholesterolemia, and mortality in hemodialysis patients. Am J Kidney Dis 1998;32:107–114.

48   Pecoits-Filho R, Barany B, Lindholm B, Heimbürger O, Stenvinkel P: Interleukin-6 and its receptor is an independent predictor of mortality in patients starting dialysis treatment. Nephrol Dial Transpl 2002;17:1684–1688.

49   Panichi V, Maggiore U, Taccola D, Migliori M, Rizza GM, Consani C, et al: Interleukin-6 is a stronger predictor of total and cardiovascular mortality than C-reactive protein in haemodialysis patients. Nephrol Dial Transpl 2004;19:1154–1160.

50   Stenvinkel P, Heimburger O, Wang T, Lindholm B, Bergström J, Elinder CG: High serum hyaluronan indicates poor survival in renal replacement therapy. Am J Kidney Dis 1999;34:1083–1088.

51   Zoccali C, Mallamaci F, Tripepi G, Cutrupi S, Parlongo S, Malatino LS, et al: Fibrinogen, mortality and incident cardiovascular complications in end-stage renal failure. J Intern Med 2003;254: 132–139.

52   Foley RN, Guo H, Snyder JJ, Gilbertson DT, Collins AJ: Septicemia in the United States dialysis population 1991–1999. J Am Soc Nephrol 2004;15:1038–1045.

53   van Tellingen A, Grooteman MPC, Schoorl M, Bartels PCM, Schoorl M, van der Ploeg T, et al: Intercurrent clinical events are predicitive of plasma C-reactive protein levels in hemodialysis patients. Kidney Int 2002;62:632–638.

54   Stenvinkel P, Heimbürger O, Jogestrand T: Elevated interleukin-6 predicts progressive carotid atherosclerosis in dialysis patients: Association to Chlamydia pneumoniae seropositivity. Am J Kidney Dis 2002;39:274–282.

55   Haubitz M, Brunkhorst R: C-reactive protein and chronic Chlamydia pneumonia infection – Long term predictors of cardiovascular disease survival in patients on peritoneal dialysis. Nephrol Dial Transplant 2001;16:809–815.

56   Sezer S, Ibis A, Ozdemir BH, Ozdemir FN, Kulah E, Boyacioglu S, et al: Association of Helicobacter pylori infection with nutritional status in hemodialysis patients. Transplant Proc 2004;36:47–49.

57   Craig RG, Spittle MA, Levin NM: Importance of peridontal disease in the kidney patient. Blood Purif 2002;20:113–119.

58   Capelli G, Tetta C, Canaud B: Is biofilm a cause of silent chronic inflammation in haemodialysis patients? A fascinating working hypothesis. Nephrol Dial Transpl 2005;20:266–270.

59   Axelsson J, Qureshi AR, Suliman ME, Honda H, Pecoits-Filho R, Heimbürger O, et al: Truncal fat mass as a contributor to inflammation in end-stage renal disease. Am J Clin Nutr 2004;80:1222–1229.

60   Shlipak MG, Fried LF, Crump C, Bleyer AJ, Manolio TA, Tracy RP, et al: Elevations of inflammatory and procoagulant biomarkers in elderly persons with renal insufficiency. Circulation 2003;107:87–92.

61   Pecoits-Filho R, Heimbürger O, Bárány P, Suliman M, Fehrman-Ekholm I, Lindholm B, et al: Associations between circulating inflammatory markers and residual renal function in CRF patients. Am J Kidney Dis 2003;41:1212–1218.

62   Feldman AM, Combes A, Wagner D, Kadakomi T, Kubota T, Li YY, et al: The role of tumor necrosis factor in the pathophysiology of heart failure. J Am Coll Cardiol 2000;35:537–544.

63   Wang AY, Wang M, Woo J, Lam CW, Lui SF, Li PK, et al: Inflammation, residual kidney function, and cardiac hypertrophy are interrelated and combine adversely to enhance mortality and cardiovascular death risk of peritoneal dialysis patients. J Am Soc Nephrol 2004;15:2186–2194.

64 Ayus JC, Sheikh-Hamad D: Silent infection in clotted hemodialysis access grafts. J Am Soc Nephrol 1998;9:1314–1321.

65 Caglar K, Peng Y, Pupim LB, Flakoll PJ, Levenhagen D, Hakim RM, et al: Inflammatory signals associated with hemodialysis. Kidney Int 2002;62:1408–1416.

66 Allon M, Depner TA, Radeva M, Bailey J, Beddhu S, Butterly D, et al: Impact of dialysis dose and membrane on infection-related hospitalization and death: Results of the HEMO study. J Am Soc Nephrol 2003;14:1863–1870.

67 Lopez-Gomez JM, Perez-Flores I, Jofre R, Carretero D, Rodriguez-Beniter P, Villaverde M, et al: Presence of a failed kidney transplant in patients who are on hemodialysis is associated with chronic inflammatory state and erytropoietin resistance. J Am Soc Nephrol 2004;15:2494–2501.

68 Yao Q, Lindholm B, Stenvinkel P: Inflammation as a cause of malnutrition, atherosclerotic cardiovascular disease, and poor outcome in hemodialysis patients. Hemodial Int 2004;8:118–129.

69 Pankow JS, Folsom AR, Cushman M, Borecki IB, Hopkins PN, Eckfeldt JH, et al: Familial and genetic determinants of systemic markers of inflammation: The NHLBI family heart study. Atherosclerosis 2001;154:681–689.

70 Tsirpanlis G, Bagos P, Ioannou D, Bleta A, Marinou I, Lagouranis A, et al: The variability and accurate assessment of microinflammation in haemodialysis patients. Nephrol Dial Transplant 2004;19:150–157.

71 Bennermo M, Held C, Stemme S, Ericsson CG, Silveira A, Green F, et al: Genetic predisposition of the interleukin-6 response to inflammation: Implications for a variety of major diseases? Clin Chim Acta 2004;50:2136–2140.

72 Balakrishnan VS, Guo D, Rao M, Jaberm BL, Tighiouart H, Freeman RL, et al: Cytokine gene polymorphisms in hemodialysis patients: Association with comorbidity, functionality, and serum albumin. Kidney Int 2004;65:1449–1460.

73 Losito A, Kalidas K, Santoni S, Jeffery S: Association of interleukin-6 -174G/C promoter polymorphism with hypertension and left ventricular hypertrophy in dialysis patients. Kidney Int 2003;64:616–622.

74 Wilson AG, de Vries N, Pociot F, di Giovine FS, van der Putte LB, Duff GW: An allelic polymorphism within the human tumor necrosis factor alpha promoter region is strongly associated with HLA A1, B8, and DR3 alleles. J Exp Med 1993;177:557–560.

75 Zee RY, Ridker PM: Polymorphism in the human C-reactive protein (CRP) gene, plasma concentrations of CRP, and the risk of future arterial thrombosis. Atherosclerosis 2002;162:217–219.

76 Kovacs A, Green F, Hansson LO, Lundman P, Samnegård A, Boquist S, et al: A novel common single nucleotide polymorphism in the promotor region of the C-reactive protein gene associated with the plasma concentration of C-reactive protein. Atherosclerosis 2005;178:193–198.

77 Yao Q, Axelsson J, Nordfors L, Heimbürger O, Qureshi AR, Barany P, et al: Peroxisome proliferator-activated receptor gamma polymorphisms in relation to inflammation and survival in end-stage renal disease patients starting renal replacement therapy. Atherosclerosis 2005;in press.

78 Westendorp RG, Langermans JA, Huizinga TW, Elouali AH, Verweij CL, Boomsma DI, et al: Genetic influence on cytokine production and fatal meningococcal disease. Lancet 1997;349:170–173.

79 Eskdale J, Gallagher G: A polymorphic dinucleotide repeat in the human IL-10 promoter. Immunogenetics 1995;42:444–445.

80 Girndt M, Sester U, Sester M, Deman E, Ulrich C, Kaul H, et al: The interleukin-10 promotor genotype determines clinical immune function in hemodialysis patients. Kidney Int 2001;60:2385–2391.

81 Girndt M, Kaul H, Sester U, Ulrich C, Sester M, Georg T, et al: Anti-inflammatory interleukin-10 genotype protects dialysis patients from cardiovascular events. Kidney Int 2002;62:949–955.

82 Stenvinkel P, Lindholm B, Heimbürger O: Novel approaches in an integrated therapy of inflammatory-associated wasting in end-stage renal disease. Semin Dial 2004;17:505–515.

83 Di Napoli M, Papa F: Angiotensin-converting enzyme inhibitor use is associated with reduced plasma concentration of C-reactive protein in patients with first-ever ischemic stroke. Stroke 2003;34:2922–2929.

84 Stenvinkel P, Andersson A, Wang T, Lindholm B, Bergström J, Palmblad J, et al: Do ACE-inhibitors suppress tumor necrosis factor-α production in advanced chronic renal failure? J Intern Med 1999;246:503–507.

Stenvinkel

85 Jialal I, Stein D, Balis D, Grundy SM, Adams-Huet B, Devaraj S: Effect of hydroxymethyl glutaryl coenzyme A reductase inhibitor therapy on high sensitive C-reactive protein levels. Circulation 2001;103:1933–1935.

86 Albert MA, Danielsson E, Rifai N, Ridker PM, Investigators P: Effect of statin therapy on C-reactive protein levels: The pravastatin inflammation/CRP evaluation (PRINCE): A randomized trial and cohort study. JAMA 2001;286:64–70.

87 März W, Winkler K, Nauck M, Böhm BO, Winkelmann BR: Effects of statins on C-reactive protein and interleukin-6 (The Ludwigshafen Risk and Cardiovascular health Study). Am J Cardiol 2003;92:305–308.

88 Chang JW, Yang WS, Min WK, Lee SK, Park JS, Kim SB: Effects of simvastatin on high-sensitivity C-reactive protein and serum albumin in hemodialysis patients. Am J Kidney Dis 2002;39: 1213–1217.

89 Vernaglione L, Cristofano C, Muscogiuri P, Chimienti S: Does atorvastatin influence serum C-reactive protein levels in patients on long-term hemodialysis? Am J Kidney Dis 2004;43: 471–478.

90 Ichihara A, Hayashi M, Ryuzaki M, Handa M, Furukawa T, Saruta T: Fluvastatin prevents development of arterial stiffness in haemodialysis patients with type 2 diabetes mellitus. Nephrol Dial Transpl 2002;17:1513–1517.

91 Ikejiri A, Hirano T, Muryama S, Yoshino G, Gushijen N, Hyodo T, et al: Effects of atorvastatin on triglyceride-rich lipoproteins, low-density lipoprotein subclass, and C-reactive protein in hemodialysis patients. Metabolism 2004;53:1113–1117.

92 van der Akker JM, Bredie SJ, Diepenveen SH, van Tits LJ, Stalenhoef AF, van Leusen R: Atorvastatin and simvastatin in patients on hemodialysis: Effects on lipoproteins, C-reactive protein and in vivo oxidized LDL. J Nephrol 2003;16:238–244.

93 Mason NA, Bailie GR, Satayathum S, Bragg-Gresham JL, Akiba T, Akizawa T, et al: HMG-coenzyme A reductase inhibitor use is associated with mortality reduction in hemodialysis patients. Am J Kidney Dis 2005;45:119–126.

94 Tonelli M, Sacks FM, Kiberd B, Curhan GC: Pravastatin for secondary prevention of cardiovascular events in persons with mild chronic renal insufficiency. Ann Intern Med 2003;138:98–104.

95 Tonelli M, Isles C, Curhan GC, Tonkin A, Pfeffer MA, Shepherd J, et al: Effect of pravastatin on cardiovascular events in people with chronic kidney disease. Circulation 2004;110:1557–1563.

Peter Stenvinkel, MD, PhD
Department of Renal Medicine K56, Karolinska Institutet
Karolinska University Hospital at Huddinge
SE–141 86 Stockholm (Sweden)
Tel. +46 8 585 82532, Fax +46 8 7114742, E-Mail peter.stenvinkel@klinvet.ki.se

Ronco C, Brendolan A, Levin NW (eds): Cardiovascular Disorders in Hemodialysis.
Contrib Nephrol. Basel, Karger, 2005, vol 149, pp 200–207

....

Interaction between Nutrition and Inflammation in Hemodialysis Patients

Martin K. Kuhlmann, Nathan W. Levin

Renal Research Institute, New York, N.Y., USA

Abstract

The excessive cardiovascular mortality of dialysis patients is at least in part related to chronic inflammation, which is associated with the occurrence of malnutrition. The negative effects of chronic inflammation on nutritional status are mediated by proinflammatory cytokines leading to a reduction in appetite and increased muscle catabolism. However, dietary behavior itself may also independently affect inflammation. Reduced dietary supply of vitamins C, B6, B12 and folate, as well as regular coffee consumption and increased intake of dietary advanced glycation end products may trigger chronic inflammation. On the other hand, a Mediterranean dietary pattern and regular soy intake both have been shown to attenuate chronic inflammation. Dietary interventions aiming to attenuate the chronic inflammatory status in dialysis patients need further exploration.

Introduction

The triad of protein-energy malnutrition, inflammation and atherosclerosis (MIA syndrome) is common in long-term hemodialysis (HD) patients and is strongly associated with outcome. This malicious combination was originally described by Stenvinkel et al. [1] in a study on 109 pre-end-stage renal disease (pre-ESRD) patients. Patients with clinical signs of malnutrition had significantly higher C-reactive protein (CRP) levels and a higher prevalence of atherosclerosis than their well-nourished counterparts. At the same time patients with CRP levels >10 mg/l showed significantly lower serum albumin levels as well as a higher prevalence of protein-energy malnutrition and atherosclerosis than patients with CRP levels <10 mg/l. The combination of all three MIA entities was found in 22% of their patients, whereas malnutrition alone (4%) and inflammation alone (<1%) were rare findings. The full triad of this syndrome is associated with the highest mortality risk [2]. It has been hypothesized that inflammation is the cause of both, malnutrition and atherosclerosis, and a link between inflammation and atherosclerosis has been established in recent years. On the other hand, the relationship between chronic inflammation and malnutrition appears to be bi-directional, with inflammation affecting nutritional status and dietary factors influencing the state of inflammation.

Pathogenesis of Inflammation

The hallmarks of chronic inflammation are increased plasma levels of pro-inflammatory cytokines (IL-1, IL-6, TNF-α, etc.) released from activated monocytes and macrophages. These inflammatory mediators induce the hepatic synthesis of acute phase proteins (e.g. CRP), and inhibit the hepatic generation of negative acute phase reactants, such as albumin. There are multiple potential causes for inflammation in chronic kidney disease patients, both dialysis-related and non-dialysis-related. The most frequent dialysis-related causes are bio-incompatibility, suboptimal water quality linked to dialysate backfiltration, and vascular access infections. Non-dialysis-related causes include decreased renal cytokine clearance, insulin resistance, increased body fat mass, chronic heart failure, coronary heart disease, and accumulation of advanced glycation endproducts (AGEs). It is now well accepted that chronic inflammation contributes to atherosclerosis not only in the general population but also in patients with chronic kidney disease, where cardiovascular events are the main causes of death. In ESRD patients pro-inflammatory cytokine levels are correlated with severity and progression of carotid atherosclerosis [3].

Effect of Inflammation on Nutritional State

Besides their effects on atherosclerosis pro-inflammatory cytokines also have a profound effect on nutritional state. In animal studies they do interfere with the satiety center inducing loss of appetite and also induce catabolism of skeletal muscle protein ultimately leading to loss of body mass [4]. Furthermore, inflammatory mediators are associated with chronic fatigue, depression, weight loss, decreased gastric emptying and intestinal bleeding. In dialysis patients mid-arm muscle area is lower and loss of body weight higher when IL-6 levels are

Table 1. Dietary factors influencing inflammation in HD patients

Dietary factor	Effect on inflammation
Reduced vitamin C intake	Impaired antioxidant capacity
Reduced vitmin B and folate intake	Oxidative stress (hyperhomocysteinemia)
Dietary AGEs	Increased plasma AGEs, oxidative stress
Regular coffee consumption	Inflammatory stimulus (?)
Mediterranean diet	Attenuation of inflammation
Soy intake (isoflavones)	Attenuation of inflammation (?)

increased [5]. Hand grip strength, a functional marker of muscle mass and also a predictor of mortality risk, is inversely correlated with IL-6 levels in HD patients [6]. Leptin is involved in appetite regulation and may be involved in the development of malnutrition in HD patients. Leptin and its receptors share structural and functional similarities with members of the cytokine family, and an inverse relationship between leptin levels and spontaneous dietary energy intake has been reported. In uremic children increasing leptin levels are associated with decreasing lean body mass [7]. Taken together, there are many data indicating a direct link between chronic inflammation and the development of malnutrition in ESRD patients.

Effects of Dietary Factors on Inflammation

There is the saying 'we are what we eat', but does this also apply to the field of dialysis and inflammation? Actually, there are data indicating that the inflammatory status can either be triggered or attenuated by diet modification (table 1).

AGEs are pro-oxidant and pro-inflammatory compounds, which are commonly elevated in patients with diabetes or chronic kidney failure. Traditionally, it was assumed that AGEs are mainly produced endogenously and retained in kidney failure; however, it has recently been shown that increased dietary AGE intake also contributes to the elevated plasma AGE levels in dialysis patients.

Dietary AGEs are formed during heating of common foods due to spontaneous reactions between reducing sugars and proteins or lipids. These so-called 'dietary glycotoxins' are generated in high amounts during roasting, broiling and oven frying, and considerably less during boiling, poaching, stewing and steaming of foods. The amount of AGEs ingested with a conventional diet is much higher than the total amount of AGEs in plasma and tissue. Roughly 10% of the dietary AGE amount ingested is absorbed and out of this fraction two thirds are retained in tissues in a bioactive form [8].

In HD as well as in peritoneal dialysis patients serum levels of carboxymethyllysine, a representative AGE, correlate well with dietary AGE intake as estimated from 3-day food records or dietary questionnaires [9]. In a small prospective short-term study over 4 weeks on 18 peritoneal dialysis patients it was further demonstrated that circulating AGE levels significantly decreased in patients randomized to a diet low in AGE content, whereas it increased in those patients randomized to the high AGE diet [10]. The changes in serum AGE levels in response to dietary AGE restriction were accompanied by significant parallel changes in serum CRP levels [11]. In conjunction with similar observations reported in patients with diabetes and normal renal function [9], these data indicate a potential direct link between dietary AGE intake and inflammation.

In HD patients plasma isoprostanes, markers of oxidative lipid peroxidation, are significantly higher compared to healthy controls and are directly correlated with serum CRP levels, indicating a link between oxidative stress and inflammation [12]. A disruption of the natural balance between the generation of oxidants and the activity of antioxidant systems has been documented. Oxidative stress is commonly attributed to the activation of polymorphonuclear neutrophils and monocytes and subsequent generation of highly reactive oxygen species during dialysis. On the other hand, the natural defense system against oxidative stress, measured as plasma antioxidant capacity, has been reported to be impaired in dialysis patients. The antioxidant capacity of human plasma is composed of vitamins, endogenous antioxidants (e.g. glutathione) and natural antioxidants mainly derived from fruits, such as polyphenolic flavones and catechins. In dialysis patients, however, the consumption of fresh fruits is restricted due to high potassium content. An either absolute or functional vitamin deficit has been documented in ESRD patients. For example, low vitamin C levels, even within the scorbutic range, have been found in a large percentage of dialysis patients [13]. Since vitamin C constitutes one of the most important water-soluble antioxidants in plasma and serves as the primary intracellular antioxidant in concert with glutathione, a severe depletion of this compound may well contribute to reduced antioxidant status and subsequently to inflammation.

A deficit in vitamins B_{12}, B_6 and folic acid in association with hyperhomocysteinemia is frequently found in dialysis patients. Besides being considered a non-traditional risk factor for cardiovascular disease, homocysteine (Hcy) is also viewed as an oxidant and as an indicator of oxidative stress in dialysis patients. We have recently shown that normalization of Hcy levels can be achieved in over 90% of HD patients by administering folic acid, vitamin B_6 and B_{12} intravenously during dialysis [14]. This supports the notion that a lack of important antioxidant vitamins occurs in dialysis patients and that higher doses of those vitamins are required to reduce oxidative stress and restore antioxidant capacity in dialysis patients. Whether normalization of Hcy levels also reduced the levels of pro-inflammatory cytokines and acute phase proteins remains to be studied.

Prescription of multivitamin is generally recommended for dialysis patients, but in practice this is not consistently the case. New information is available from the DOPPS study, which recently reported on the relationship between patterns of water-soluble vitamin use and outcomes in HD patients [15]. Among 308 representative dialysis facilities in France, Germany, Italy, Japan, Spain, the United Kingdom, and the United States the percentage of patients administered water-soluble vitamins varied largely. The fraction of patients on multivitamins was considerably higher in the United States (71.9%) than in Japan (5.6%) and in Europe where it ranged from 3.7% in UK to 37.9% in Spain. Multivariate analysis revealed that the use of water-soluble vitamins was independently associated with a substantial 16% reduction of relative risk for mortality. These data do not definitely prove that water-soluble vitamins improve outcomes, but they should be considered as important hint that this low cost intervention may be beneficial in this population. So far, no information is available from this study on the effects of multivitamins on inflammatory markers.

Another interesting aspect on diet and inflammation comes from the cross-sectional ATTICA study involving more than 3,000 healthy subjects from Greece. In this study a positive association was found between coffee consumption and inflammatory markers [16]. Men who consumed >200 ml coffee/day had 50% higher IL-6, 30% higher CRP and 28% higher TNF-α levels than coffee non-drinkers. Similar significant data were found for women, even after controlling for the interactions between coffee consumption and age, sex, smoking, body mass index, and physical activity status. There was also a positive association of coffee consumption with blood pressure and hypercholesterolemia. An association between coffee consumption and cardiovascular mortality is known from earlier studies. It is not sure whether the observed effect can be solely explained by caffeine intake, or whether other, so far unknown factors are also involved. If caffeine should be the culprit, it may be hypothesized that exposure to caffeine and its metabolites may be higher in dialysis patients due to lack of renal clearance.

Dietary patterns differ considerably between the Unites States, Europe and Japan, as do mortality rates from cardiovascular disease. It is known for some time that a Mediterranean diet is associated with a significantly lower mortality risk [17]. A typical Mediterranean diet is characterized by high intakes of vegetables, legumes, fruits and nuts, cereals, fish, and olive oil and relatively low intakes of dairy products and meat. In this regard another recent report from the ATTICA study deserves attention. Adherence to a Mediterranean diet, scored from 1 (low) to 9 (high) was associated with lower levels of inflammatory markers [18]. Patients in the highest score tertile had on average 20% lower CRP, 17% lower IL-6 and 15% lower Hcy levels compared with the lowest tertile. This association with inflammation may at least partially explain the beneficial effects of a Mediterranean diet on cardiovascular outcome.

Similar data have been reported from the Nurses' Health Study I cohort, where dietary intake was documented twice, 1986 and 1990, in 732 women aged 43–69 years [19]. A so-called 'prudent dietary pattern', characterized by higher intake in fruit, vegetables, legumes, fish, poultry, and whole grains was inversely associated with CRP levels. In contrast, a 'Western dietary pattern', composed of higher consumption of red and processed meats, sweets, dessert, French fries and refined grains was positively correlated with CRP and IL-6 levels.

Finally, soy foods have received increasing attention because of their nutritional properties and high isoflavone content. In a recent study the relationship between habitual soy intake on blood isoflavone levels in clinically stable HD patients from the US, Thailand and Japan was assessed. Soy intake was higher in Japanese than in Thai patients, and negligible in the US patients. Blood levels of the isoflavone genistein correlated with soy intake and were significantly higher in the Japanese patients as compared with Thai and the US patients. Other dietary antioxidants, including tocopherols, carotenoids and retinol, differed only minimally. ESRD patients appear to accumulate isoflavones as a function of dietary soy intake. Isoflavones have well-documented antioxidant and anti-inflammatory potency and high levels of isoflavones in subjects habitually consuming soy products may be associated with lower levels of inflammation. Preliminary data from a small-randomized study on the effects of soy-isoflavone containing supplements on inflammatory parameters in American HD patients reported an inverse correlation between plasma-isoflavone concentration after 4 weeks of treatment and serum CRP levels [20].

Taken together, data published so far clearly document a link between chronic inflammation and deterioration of nutritional status. On the other hand, dietary patterns may have an important impact on the extent of inflammation and 'prudent' changes of dietary behavior may even attenuate the inflammatory state. When searching for new ways to reduce chronic inflammation in HD patients, dietary interventions should also be considered.

References

1   Stenvinkel P, Heimburger O, Paultre F, Diczfalusy U, Wang T, Berglund L, Jogestrand T: Strong association between malnutrition, inflammation, and atherosclerosis in chronic renal failure. Kidney Int 1999;55:1899–1911.
2   Pecoits-Filho R, Lindholm B, Stenvinkel P: The malnutrition, inflammation, and atherosclerosis syndrome – The heart of the matter. Nephrol Dial Transplant 2002;17:28–31.
3   Stenvinkel P, Heimburger O, Jogestrand T: Elevated interleukin-6 predicts progressive carotid artery atherosclerosis in dialysis patients: Association with *Chlamydia pneumoniae* seropositivity. Am J Kidney Dis 2002;39:274–282.
4   Ikizler TA: Role of nutrition for cardiovascular risk reduction in chronic kidney disease patients. Advances in Chronic Kidney Disease 2004;11:162–171.
5   Kaizu Y, Ohkawa S, Odamaki M, Ikegaya N, Hibi I, Miyaji K: Association between inflammatory mediators and muscle mass in long-term hemodialysis patients. Am J Kidney Dis 2003;42: 295–302.
6   Heimburger O, Qureshi AR, Blaner WS, Berglund L, Stenvinkel P: Hand-grip muscle strength, lean body mass, and plasma proteins as markers of nutritional status in patients with chronic renal failure close to start of dialysis therapy. Am J Kidney Dis 2000;36:1213–1225.
7   Stenvinkel P, Pecoits-Filho R, Lindholm B: Leptin, ghrelin, and proinflammatory cytokines: Compounds with nutritional impact in chronic kidney disease? Adv Ren Replace Ther 2003;10: 155–169.
8   Vlassara H, Cai W, Crandall J, Goldberg T, Oberstein R, Daradaine V, Peppa M, Rayfield EJ: Inflammatory mediators are induced by dietary glycotoxins, a major risk factor for diabetic angiopathy. PNAS 2002;99:15596–15601.
9   Uribarri J, Peppa M, Cai W, Goldberg T, Lu M, Baliga S, Vassalotti JA, Vlassara H: Dietary glyco-toxins correlate with circulating advanced glycation end product levels in renal failure patients. Am J Kidney Dis 2003;42:532–538.
10 Uribarri J, Peppa M, Cai W, Goldberg T, Lu M, He C, Vlassara H: Restriction of dietary glycotox-ins reduces excessive advanced glycation end products in renal failure patients. J Am Soc Nephrol 2003;14:728–731.
11 Peppa M, Uribarri J, Cai W, Lu M, Vlassara H: Glycoxidation and inflammation in renal failure patients. Am J Kidney Dis 2004;43:690–695.
12 Spittle MA, Hoenich NA, Handelman GJ, Adhikarla R, Homel P, Lewin NW: Oxidative stress and inflammation in hemodialysis patients. Am J Kidney Dis 2001;38:1408–1413.
13 Wang S, Eide TC, Sogn EM, Berg KJ, Sund RB: Plasma ascorbic acid in patients undergoing chronic hemodialysis. Eur J Clin Pharmacol 1999;55:527–532.
14 Obeid R, Kuhlmann MK, Kohler H, Herrmann W: Response of homocysteine, cystathionine, and methylmalonic acid to vitamin treatment in dialysis patients. Clin Chem 2005;51:196–201.
15 Fissell RB, Bragg-Gresham JL, Gillespie BW, Goodkin DA, Bommer J, Saito A, Akiba T, Port FK, Young EW: International variation in vitamin prescription and association with mortality in the Dialysis Outcomes and Practice Patterns Study (DOPPS). Am J Kidney Dis 2004;44: 293–299.
16 Zampelas A, Panagiotakos DB, Pitsavos C, Chrysohoou C, Stefanadis C: Associations between coffee consumption and inflammatory markers in healthy persons: The ATTICA Study. Am J Clin Nutr 2004;80:862–867.
17 Trichopoulou A, Costacou T, Bamia C, Trichopoulos D: Adherence to a Mediterranean diet and survival in a Greek population. N Engl J Med 2003;348:2599–2608.

18   Chrysohoou C, Panagiotakos DB, Pitsavos C, Das UN, Stefanadis C: Adherence to the Mediterranean diet attenuates inflammation and coagulation process in healthy adults: The ATTICA Study. J Am Coll Cardiol 2004;44:152–158.
19   Lopez-Garcia E, Schulze MB, Fung TT, Meigs JB, Rifai N, Manson JE, Hu FB: Major dietary patterns are related to plasma concentrations of markers of inflammation and endothelial dysfunction. Am J Clin Nutr 2004;80:1029–1035.
20   Fanti P, Asmis R, Fraer M, et al: Improvement of inflammation and malnutrition correlates with the blood levels of soy isoflavones following dietary intervention in Hemodialysis patients (abstract). J Am Soc Nephrol 2004;15:169A.

Martin K. Kuhlmann, MD
Research Laboratory Director, Renal Research Institute
207 East 94th Street, Suite 303
New York, NY–10128 (USA)
Tel. +1 646 672 4042, Fax +1 646 672 4174, E-Mail mkuhlmann@rriny.com

Ronco C, Brendolan A, Levin NW (eds): Cardiovascular Disorders in Hemodialysis.
Contrib Nephrol. Basel, Karger, 2005, vol 149, pp 208–218

....

Monocytes from Dialysis Patients Exhibit Characteristics of Senescent Cells: Does It Really Mean Inflammation?

Julia Carracedo, Rafael Ramirez, Sagrario Soriano,
Maria Antonia Alvarez de Lara, Mariano Rodriguez,
Alejandro Martin-Malo, Pedro Aljama

Unidad de Investigación, Servicio de Nefrología. Hospital Universitario Reina Sofía, Córdoba, Spain

Abstract

Hemodialysis treatment induces mononuclear cell activation particularly if cellulosic hemodialysis membrane is used. In normal cells, repeated activation induce a process of accelerate cellular senescence. The aim of the present study was to evaluate whether the mononuclear cell activation associated to hemodialysis with cellulosic membranes favors a process of accelerate senescence in mononuclear cells. Our results show that mononuclear cells from patients dialyzed with cellulosic membranes, exhibit: decrease telomere length, increase percentage of cells CD14dim/CD16bright and increase production of IL-1β, IL1Ra and IL6 cytokines. After culture in vitro, these cells shown increase susceptibility to undergoing spontaneous apoptosis, that is enhanced by IL-4 and prevented by IL-1β or LPS. All of these characteristics have been reported associated to senescence of monocytes, and not are observed in cells from controls subjects or patients dialyzed with non-cellulosic membranes, suggesting that hemodialysis with cellulosic membranes induce a process of senescence in mononuclear cells.

Hemodialysis (HD) therapy in patients with chronic renal failure has been associated with impairment in the immune response. Particularly, cellulosic membranes affect mononuclear cells that exhibit characteristics of activated cells such as: increase production of pro-inflammatory cytokines [1–4], upregulation of cell-activation molecules (adhesion molecules or the CD14 receptor), and phosphorylation of surface molecules. In addition, we have demonstrated that cellulosic HD membranes may induce increased spontaneous apoptosis in normal mononuclear cells and mononuclear cells from hemodialyzed patients.

In normal, somatic cells have been demonstrated as a process of replicative senescence that occur associated to their division potential. Thus, after certain number of population doublings normal somatic cells become exhausted and enter into a post-replicative state characterized by the progressive decline in cell function that eventually results in the cessation of cell growth. It has been proposed that telomere shortening could act as the molecular clock that signals eventual growth arrest in this process of replicative senescence. Human telomeres are simple repeats of the sequence TTAGGG [5–8]. It was shown that telomeres shorten with each division in human somatic cells. This shortening leads to the suggestion that telomeres might act as a mitotic clock, counting the number of divisions and eventually activating replicative senescence as an ultimate DNA damage checkpoint. In several cell types, but particularly in immune cells, it has been documented that cellular senescence is associated to a progressive decline in cell function with the cessation of proliferative capacity. Previous reports suggest that telomere protects the ends of the chromosome against degradation and prevents ligation of the ends of DNA by DNA-repair enzymes. The activation of a ribonucleoprotein complex termed telomerase prevents telomere shortening during cell division. Telomerase synthesized, with enzymatic activity, the DNA repeated sequence of telomeres. In the absence of telomerase, the failure of DNA polymerase to synthesize DNA leads to chromosomal shortening with each round of replication, and entering into a post-replicative senescent state [7, 8].

Although the pivotal role of monocytes in immune regulation by their production of pro-inflammatory and inhibitory cytokines is acknowledged, limited information is available on monocyte changes with senescence. A recent report shows that monocytes of aged individuals exhibit in vivo activation with increased production of pro-inflammatory cytokines and CD14dim/CD16bright phenotype. In younger, the increased life span of monocytes are also associated to cell activation with the production of pro-inflammatory cytokines. Additionally, as occurs in neutrophils, senescent monocytes undergo spontaneous apoptosis after culture in vitro in a process that is upregulated by anti-inflammatory cytokines such as interleukin (IL)-4 and downregulated by pro-inflammatory cytokines such as IL-1 [9, 10].

A process of replicative senescence has been postulated to explain the finite proliferative capacity of human T-cell clones in long-term culture [11] or the inability of the CD8$^+$CD28-T cells to respond in HIV disease [7]. In the present work, we postulate that the repeated in vivo stimulation during the HD session, particularly with the use of cellulosic HD membranes, induces a process of replicative senescence in mononuclear cells.

Methods

Patients

The present study includes HD patients: patients dialyzed with cellulosic membranes (GFS 20 Plus-Alwall; Gambro, Germany), and with non-cellulosic membranes (HF80S; Fresenius, Germany). The mean age was 46.9 ± 15.9 years (range, 20–67) in the cellulosic group and 50.4 ± 15.8 years (range, 23–64) in the non-cellulosic group. All the patients were dialyzed in the same dialysis unit three times per week through arteriovenous fistulae; 12 were native and only 2 were polytetrafluoroethylene fistulae. Bicarbonate dialysate solutions were used in all treatments; the blood flow was 300–400 ml/min and the duration of dialysis was individually adjusted to maintain a Kt/V above 1.2. Both groups of patients were not significantly different regarding gender time on dialysis, erythropoietin therapy and etiology of chronic renal failure. The type of membrane was unchanged during at least 2 months prior to the study. During the period of the study the analysis of the dialysis system always revealed the absence of bacteria (less than 100 cfu/ml) or bacteriological contaminant products (endotoxin levels below 0.25 EU). Criteria for patient selection included the absence of: acute or chronic infection, autoimmune disease, hepatic insufficiency, diabetes or malignancy. The patients were not on anti-inflammatory medication and/or immunosuppressive therapy. Thirty-two healthy volunteers were included as controls. Informed consent was obtained from all the patients after institutional approval.

Isolation of Human Mononuclear Cells

Peripheral blood leukocytes were obtained from 20 ml of heparinized whole blood drawn immediately before a HD session. Buffy coat cells were separated by differential centrifugation gradient (Ficoll-Hypaque; Pharmacia LKB, Uppsala, Sweden). Thereafter, cells were washed and seeded in 12-well culture plates with a complete culture medium described below. Adherent mononuclear cells, containing mainly monocytes, were isolated after adherence to plates; a purity of >75% of monocytes was demonstrated by staining with anti-CD14 monoclonal antibody (Mab) (M5E2; Pharmingen, Calif., USA). Contamination with $CD3^+$ and $CD19^+$ (Leu-4 and Leu-12; Becton Dickinson, Calif., USA) lymphocytes was less than 8%.

Considering the possibility that other leukocyte adherent to plastic may contaminate the monocyte population, representative control experiments were performed in cells isolated by flow cytometry and sorted using a Mab against the CD14 molecule (AB383; R&D system, UK). In these experiments, the forward and side light scatter analysis in the flow cytometer identified a gate in which monocytes were grouped, and other leukocytes were gated out according to forward and side scatter characteristics.

Cell Culture

Monocytes were cultured at 37°C in complete culture medium containing RPMI 1640 supplemented with L-glutamine (2 mM), Hepes (20 mM), sodium pyruvate (1 mM), streptomycin (50 ng/ml), penicillin (100 UI/ml), and 10% human autologous serum (BioWhittaker,

Walkersville, Md., USA). Serum was heated (at 56°C for 60 min period) to inactivate complement. Cells were seeded in 96-well microtiter plates (Falcon; Becton Dickinson and Company, Paramus, N.J., USA) at 2×10^5 cells per well.

In some experiments lymphokines and/or lipopolysaccharide (LPS) were added to the culture. LPS from *Escherichia coli*, strain 0127:B8 (Sigma Chemical Co., Poole, UK) was added at 1 ng/ml. IL-4 and IL-1β were purchased from R&D system (R&D system, Minn., USA) and were used at 100 ng/ml and 500 U/ml, respectively.

Flow Cytometric Detection of Telomere FISH

Telomere length measurement of mononuclear cells was performed by flow cytometric detection of telomere FISH (flow-FISH) as previously described. Briefly, cells (1×10^6) were washed and placed in 1.5 ml Eppendorf tubes. After centrifugation, 30 s at 13,000 rpm, the supernatant was removed and cell pellets were resuspended in a hybridization buffer containing 70% deionized formamide (Sigma Chemical Co., St Louis, Mo., USA), 10 mM Tris, pH 7.0, 10% FCS and 0.3 μg/ml of the telomere specific FITC-conjugated probe (FITC-O-CCCATAACTAAACAC-NH2). DNA from samples was denatured by heating for 10 min at 80°C in a Thermomixer 5436 (Eppendorf-Netheler, Germany) followed by hybridization for 2 h at room temperature. Thereafter cells were centrifuged (30 s at 13,000 rpm), the supernatant was removed, and the cells were washed in washing buffer containing 70% deionized formamide (Sigma Chemical Co.), 10 mM Tris, pH 7.0, FCS, 0.1% Tween 20 (Sigma Chemical Co.). After incubation for 1 h at room temperature, cells were washed and resuspended in PBS, 10% FCS, RNAse A at 10 μg/ml (Boehringer Mannheim, Laval, Canada) and propidium iodide, incubated for 1 h at room temperature, washed and analyzed on a FACS-Cam flow cytometer. Daily, the flow cytometer was aligned using FITC-labeled fluorescence Sphero microparticles (Pharmingen). Green fluorescence was measured on a linear scale and results were expressed in kilomolecular equivalents of soluble fluorochrome units. The relative telomere length value was calculated comparing the kilomolecular equivalents of soluble fluorochrome values to those obtained using different cell lines as telomere length (U937, Jurkat, Molt4 and HL60) controls.

Measurement of Cytokines

IL-1β, IL-1Ra, IFN-γ, IL-6 and IL-10 were measured according to the manufacturer's instructions using ELISA kits purchased from R&D Systems (Minn., USA). Soluble cytokines were measured in the supernatant of cultured mononuclear cells from aged-matched controls (n = 10) or HD patients. For this purpose, 10^5 cells/ml were cultured in complete medium plus 20 ng/ml of Phorbol-12-myristate-13 acetate from Sigma.

Evaluation of CD14 and CD16 Monocyte Expression

Cells (10^5/ml) were incubated for 30 min at 4°C with the Mabs M5E2 against the molecule CD14 or 3G8 against the molecule CD16. Both Mabs were phycoerythrin-conjugated Mabs and were purchased from Becton Dickinson. For the study, cells were washed with PBS buffer (PBS/0.1% NaN$_3$, 5% human autologous serum) and incubated for 30 min at 4°C with 20 μl of the appropriate Mab. Thereafter, cells were washed in cold PBS buffer and analyzed in the flow cytometer. Background fluorescence was determined by phycoerythrin-conjugated mouse immunoglobulins. Cytofluorometric analysis was performed with a FACScan cytometer (Becton Dickinson).

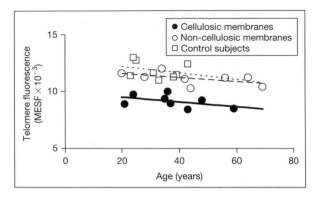

Fig. 1. Telomere length determination by flow cytometry (Flow-FISH). Lineage-specific telomere shortening in human mononuclear cells with age-matched normal individuals or hemodialyzed patients.

Cell Apoptosis

An FITC-Annexin-V probe was used to determine the percentage of cells undergoing apoptosis. Cells were washed in PBS and adjusted to 5×10^5/ml in binding buffer [10 mM Hepes/NaOH, pH 7.4, 140 mM NaCl, 2.5 mM CaCl$_2$ (Bender MedSystems, Vienna, Austria) filtered through 0.2 μm filter]. Five microliters of Annexin-V FITC (Bender MedSystems) were added to 195-μl cell suspension, and after 10 min of incubation in the dark, cells were washed and resuspended in 190-μl binding buffer and 10-μl propidium iodide stock solution (20 μg/ml). The degree of apoptosis was assessed by flow cytometry: live cells exclude both dyes, dead cells were positive for both fluorochomes, and cells positive for Annexin-V and excluding propidium iodide were considered as apoptotic cells. Background fluorescence was determined by FITC-conjugated mouse immunoglobulins.

Results

Mononuclear Cells from Patients Hemodialyzed with Cellulosic Membranes Exhibit Short Telomere

Flow-FISH was performed in mononuclear cells from control individuals and patients dialyzed with cellulosic or non-cellulosic membranes. Although, there was a large variation in telomere fluorescence values in cells from either age-matched normal individuals or hemodialyzed patients (fig. 1), the fluorescence intensity from labeled telomere probe decrease correlates with aging, indicating that cells from aged subjects have shorter telomeres than cells from young individuals. When we compared these data to the obtained using mononuclear cells from hemodialyzed patients, cells from patient hemodialyzed with non-cellulosic membranes have a similar telomere length to the observed in age-matched controls. However, mononuclear cells from patient hemodialyzed with cellulosic membranes show shorter telomeres than the controls or patients hemodialyzed with non-cellulosic membranes. The simple histogram analysis indicates that shortening in telomere occur in a percentage of cells that vary between the subjects studied.

Cell Expression of CD14 and CD16 in Mononuclear Cells

In previous reports, we have observed that senescent monocytes that exhibit short telomeres have a phenotype CD14dim/CD16bright, in contrast to the monocytes with long telomeres that were CD14bright/CD16dim. A $32 \pm 11\%$ of mononuclear cells from cellulosic hemodialyzed patients were CD14dim/CD16bright. In contrast, the percentage of CD14bright/CD16dim cells in patients hemodialyzed with non-cellulosic membranes was $8 \pm 3\%$, similar to the observed in cells from age-matched controls ($5 \pm 3\%$).

Cytokine Production in Mononuclear Cell Cultures

Production of IL-1β, IL-1Ra, IFN-γ, IL-6 and IL-10 by mononuclear cells was tested at 24 h of culture. Increased levels of IL-1β (232 ± 49 pg/ml), IL-1Ra (473 ± 98 pg/ml) and IL-6 (96 ± 29 pg/ml) were measured in the culture supernatants of cells from HD patients, to the observed in controls IL-1β (115 ± 37 pg/ml), IL-1Ra (208 ± 67 pg/ml) and IL-6 (34 ± 15 pg/ml) or non-cellulosic HD patients IL-1β (149 ± 73 pg/ml), IL-1Ra (247 ± 96 pg/ml) and IL-6 (38 ± 11 pg/ml). The levels of IFN-γ and IL-10 were similar in controls and HD patients.

Spontaneous Apoptosis in vitro of Mononuclear Cells from Cellulosic Hemodialyzed Patients

Freshly isolated mononuclear cells from controls or hemodialyzed patients did not exhibit apoptosis. After 48 h in culture there was $5.2 \pm 2.3\%$ of apoptosis in cells from controls, 14.7 ± 4 in mononuclear cells from non-cellulosic hemodialyzed patients and as high as $32.2 \pm 5.2\%$ in mononuclear cells from cellulosic hemodialyzed patients (fig. 2).

Effect of Lymphokines and LPS in Mononuclear Cell Phenotype and Apoptosis

Addition of either LPS or IL-1β inhibited mononuclear cell apoptosis in cells from cellulosic-treated patients. In addition both cytokines prevent the presence of CD14dim/CD16bright cells. In cells for non-cellulosic patients or in control groups, neither LPS nor IL-1β has a significant effect. In contrast, to the observed with LPS or IL-1β, IL-4 increases apoptosis in cells from

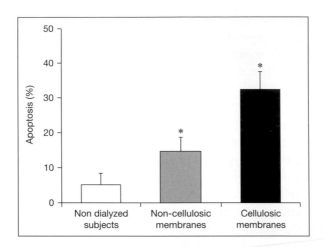

Fig. 2. Apoptosis in mononuclear cells cultured for 48 h. Results are expressed as mean percentage of cell apoptosis in healthy subjects, and patients on hemodialysis with non-cellulosic or cellulosic membranes.

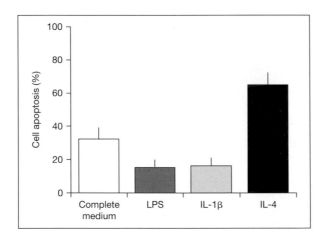

Fig. 3. Determination of apoptosis in mononuclear cells from patients on hemodialysis with cellulosic membranes, cultured for 48 h in complete medium alone or with the addition of LPS (1 ng/ml), IL-1β (100 ng/ ml), or IL-4 (500 U/ml).

cellulosic-treated patients (from $32.3 \pm 5.2\%$ to $65 \pm 5.9\%$, $p < 0.05$), although did not modify the results obtained in cells from non-cellulosic patients or in control groups (fig. 3).

Discussion

The results of the present study show that freshly isolated mononuclear cells from patients hemodialyzed with cellulosic membranes have characteristics of senescent monocytes such as: short telomeres, phenotype CD14dim/CD16bright and increased production of IL-1β, IL-1Ra and IL-6 cytokines. In addition, these cells underwent spontaneous apoptosis after culture in vitro, apoptosis that is increased after the addition of IL-4 and is prevented by IL-1β or LPS. These characteristics are not observed in mononuclear cells from patients dialyzed with non-cellulosic membranes or in control subjects, suggesting that dialysis with cellulosic membranes induce a process of replicative senescence in mononuclear cells.

Mononuclear cells from patients dialyzed with cellulosic membranes have shown shorter telomeres as compared with the age-matched controls subjects or patients dialyzed with non-cellulosic membranes. Normal somatic cells die after a finite number of cell divisions, a phenomenon described as cellular senescence. Shortening of telomeres has been implied as the molecular mechanism that controls the life span in somatic cells and in consequence their senescence [12-14]. It is of interest that the CD14 and CD32 receptors have increased expression as compared with cells from non-cellulosic HD patients and control subjects [15]. Other authors have reported increased expression of CD14 and CD32 molecules in senescent of mononuclear cells [16-19]. These results suggest that HD with cellulosic membranes favors the accumulation of senescent mononuclear cells. By contrast HD with non-cellulosic membrane, a more biocompatible membrane, does not seem to be associated with increased expression of CD14 and CD32 molecules.

After 48 h in culture, a large percentage of mononuclear cells from patients dialyzed with cellulosic membrances underwent spontaneous apoptosis. These results confirm and extend previous results by Hendenreich et al. [14] and our own observations [13]. The rate of apoptosis is also increased in senescent monocyte cultured in vitro [20–23]. The increased rate of spontaneous apoptosis together with the increased expression of CD14 and CD32 molecules suggest that monocytes from patients dialyzed with cellulosic membranes are senescents [24-26].

The mechanism(s) of senescence in monocytes are not well understood, however, it is well established that inflammatory mediators and/or lymphokines may regulate in vivo and in vitro the activation of monocytes and consequently their life span and the accumulation of senescent cells [26–28]. During HD therapy mononuclear cells may be activated through different mechanisms including: interaction between cell surface molecules and dialysis membrane [12, 29]; release of pro-inflammatory lymphokines during HD therapy or bacterial LPS contamination from the dialysate [30]. In vivo activation of mononuclear cells from patients dialyzed with cellulosic membranes is supported by the fact that in these cells IL-4 downregulates the expression of CD14 and increases apoptosis; and LPS but not IL-1β suppressed the pro-apoptotic effect of IL-4. Mangan et al. [28] have demonstrated that IL-4 only induces downregulation in CD14 expression and increase apoptosis when monocytes are activated, suggesting that the anti-inflammatory effect of some cytokines such as IL-4 is mediated in part by reducing survival of stimulated monocytes in chronic lesions [27–28].

In summary, mononuclear cells from patients in HD therapy with cellulosic but not with non-cellulosic membranes exhibit characteristics of senescent cells. In addition these cells, lymphokines and bacterial products regulate apoptosis in a manner which is similar to what has been described in senescent monocytes. Our study suggests that in patients hemodialyzed with cellulosic membranes there is an accumulation of senescent cells which may result in abnormal immune response. These results have important clinical implication as described in elderly subjects, an abnormal immune response is implicated in the increased incidence of infectious, neoplasias and degenerative diseases.

Aknowledgments

This work was supported by grants from Fondo de Investigaciones Científicas de la Seguridad Social (FIS 02/0154, 03/0946), Junta de Andalucia (57/03), and Fundación Nefrologica. J. Carracedo is supported by FIS.

References

1 Lindner A, Farewell VT, Sherrard DJ: High incidence of neoplasia in uremic patients receiving long-term dialysis. Cancer and long-term dialysis. Nephron 1981;27:292–296.
2 Vanholder R, Ringoir S: Infectious morbidity and defects of phagocytic function in end-stage renal disease: A review. J Am Soc Nephrol 1993;3:1541–1554.
3 Vanholder RC, Ringoir S: Adequacy of dialysis: A critical analysis. Kidney Int 1992;42: 540–558.
4 Cohen G, Haag-Webbr M, Horl WH: Immune dysfunction in uremia. Kidney Int 1997;62(suppl): S79–S82.

5   Pawelec G, Effros G, Effros RB, Caruso C, Remarque E, Barnett Y, Solana R: T cells and aging. Front Biosci 1997;13:69–77.

6   Solana R, Alonso MC, Pena J: Natural killer cells in healthy aging. Exp Gerontol 1999;34: 435–443.

7   Chakravarti B, Abraham GN: Aging and T-cell-mediated immunity. Mech Ageing Dev 1999;108: 183–206.

8   Girndt M, Heisel LO, Kohler H: Influence of dialysis with polyamide vs haemophan haemodialysers on monokines and complement activation during a 4-month long-term study. Nephrol Dial Transplant 1999;14:676–682.

9   Ando M, Lundkvist I, Bergstrom J, Lindholm B: Enhanced scavenger receptor expression in monocyte-macrophages in dialysis patients. Kidney Int 1996;49:773–780.

10 Coli L, Tumietto F, De Pascalis A, La Manna G, Zanchelli F, Isola F, Perna C, Raimondil C, Dee Sanctis LB, Marseglia CD, Costigliola P, Stefoni S: Effects of dialysis membrane nature on intradialytic phagocytizing activity. Int J Artif Organs 1999;22:74–80.

11 Descamps-Latsscha B, Hervellin A, Nguyen AT, De Groote D, Chauveau P, Verger C, Jungers P, Zingraff J: Soluble CD23 as an effector of immune dysregulation in chronic uremia and dialysis. Kidney Int 1994;14:253–260.

12 Carracedo J, Ramirez R, Pintado O, Gomez-Villamandos JC, Martin-Malo A, Rodriguez M, Aljama P: Cell aggregation and apoptosis induced by hemodialysis membranes. J Am Soc Nephrol 1995;6:1586–1591.

13 Martin-Malo A, Carracedo J, Ramirez R, Rodriguez-Benot A, Soriano S, Rodriguez M, Aljama P: Effect of uremia and dialysis modality on mononuclear cell apoptosis. J Am Soc Nephrol 2000;11: 936–942.

14 Heidenreich S, Schmidt M, Bachmann J, Harrach B: Apoptosis of monocytes cultured from long-term hemodialysis patients. Kidney Int 1996;49:792–799.

15 Williams MA, Newland AC, Kelsey SM: The potential for monocyte-mediated immunotherapy during infection and malignancy. Part I: aApoptosis induction and cytotoxic mechanisms. Leuk Lymphoma 1999;34:1–23.

16 Heidenreich S: Monocyte CD14. A multifunctional receptor engaged in apoptosis from both sides. J Leukoc Biol 1999;65:737–743.

17 Heidenreich S, Schmidt M, August C, Cullen P, Rademaekers A, Pauels HG: Regulation of human monocyte apoptosis by the CD14 molecule. J Immunol 1997;159:3178–3188.

18 Kizaki T, Ookawara T, Oh-Ishi S, Itoh Y, Iwabuchi K, Onoe K, Day NK, Good RA, Ohno H: An increase in basal glucocorticoid concentration with age induces suppressor macrophages with high-density Fc gamma RII/III. Immunology 1998;93:409–414.

19 Stohlawetz P, Hahn P, Koller M, Hauer J, Resch H, Smolen J, Pietschmannp P: Immunophenotypic characteristics of monocytes in elderly subjects. Scand J Immunol 1998;48:324–326.

20 Schneider EL, Braunnschweiger K, Mitsui Y: The effect of serum batch on the in vitro lifespans of cell cultures derived from old and young human donors. Exp Cell Res 1978;115:47–52.

21 Smith JR, Pereira-Smith OM, Schneider EL: Colony size distributions as a measure of in vivo and in vitro aging. Proc Natl Acad Sci USA 1978;75:1353–1356.

22 Schneider EL: In vivo versus in vitro cellular aging. Birth Defects Orig Artic Ser 1978;14: 159–169.

23 Goldstein S, Moerman EJ, Soeldner JS, Gleason RE, Barnett DM: Chronologic and physiologic age affect replicative life-span of fibroblasts from diabetic, prediabetic, and normal donors. Science 1978;199:781–782.

24 Rabatic S, Sabioncello A, Dekaris D, Kardumi I: Age-related changes in functions of peripheral blood phagocytes. Mech Ageing Dev 1988;45:223–229.

25 Mege JL, Capo C, Michel B, Gastaut JL, Bongrand P: Phagocytic cell function in aged subjects. Neurobiol Aging 1988;9:217–220.

26 Mangan DF, Wahl SM: Differential regulation of human monocyte programmed cell death (apoptosis) by chemotactic factors and pro-inflammatory cytokines. J Immunol 1991;147:3408–3412.

27 Mangan DF, Wekch GR, Wahl SM: Lipopolysaccharide, tumor necrosis factor-alpha, and IL-1 beta prevent programmed cell death (apoptosis) in human peripheral blood monocytes. J Immunol 1991;146:1541–1546.

28   Mangan DF, Robertson B, Wahl SM: IL-4 enhances programmed cell death (apoptosis) in stimulated human monocytes. J Immunol 1992;148:1812–1816.
29   Carracedo J, Ramirez R, Martin-Malo A, Rodriguez M, Aljama P: Nonbiocompatible hemodialysis membranes induce apoptosis in mononuclear cells: The role of G-proteins. J Am Soc Nephrol 1998;9:46–53.
30   Girndt M, Sester U, Kaul H, Kohler H: Production of proinflammatory and regulatory monokines in hemodialysis patients shown at a single-cell level. J Am Soc Nephrol 1998;9:1689–1696.

Dr. Pedro Aljama
Servicio de Nefrología
Hospital Universitario Reina Sofía, Avda Menendez Pidal, 1
ES–14004 Córdoba (Spain)
Tel. +34 957 010440, Fax +34 957 010 307, E-Mail pedro.aljama.sspa@juntadeandalucia.es

Ronco C, Brendolan A, Levin NW (eds): Cardiovascular Disorders in Hemodialysis.
Contrib Nephrol. Basel, Karger, 2005, vol 149, pp 219–229

....

Modifiable Risk Factors for Cardiovascular Disease in CKD Patients

Eric Seibert, Martin K. Kuhlmann, Nathan W. Levin

Renal Research Institute, New York, N.Y., USA

Abstract

Risk factors for cardiovascular disease (CVD) have been studied extensively in CKD patients. It can be differentiated between modifiable, potentially-modifiable and non-modifiable risk factors. Nonetheless, even for easily modifiable risk factors there is still a lack of data demonstrating the benefit of common interventions, such as statin treatment for dyslipidemia, improvement of HbA1c levels in diabetic patients, implementation of physical exercise, normalization of Hgb and achievement of adequate dry weight in dialysis patients. This article gives an overview of modifiable and potentially modifiable risk factors and available modification strategies.

Introduction

The study of risk factors for cardiovascular disease (CVD) in dialysis patients has been a medical literature growth industry [1a]. There is clearly good reason to go further than the division of these factors into 'traditional' and 'non-traditional' since this partition adds nothing to understanding. Much better is the division into non-modifiable, modifiable and potentially modifiable risk factors which provide the opportunities for defining the truth of the classification by evaluating the outcomes following attempts at modification.

As noted above, there is a plethora of studies relating individual or groups of factors as predictors for subsequent cardiac outcome. Unfortunately, it is difficult to compare the relative value of specific risk factors since varying qualities, both medical and those inherent in the population under examination, interfere with such comparisons. An example of this is the presence of diabetes, which in the white population in the United States has such a marked effect on the severity of atherosclerotic disease while with a similar incidence in Japan diabetes is associated with a much lower coronary artery disease prevalence. Statistical adjustments for covariates, a staple of clinical reports, may be equally inadequate when populations are compared in which the effect of the covariates is unequal.

Another series of phenomena which makes the interpretation of risk factors more difficult is exemplified in the term 'reverse epidemiology' (the term itself has little meaning since epidemiology cannot be reversed). What is meant is a factor apparently being causative in a direction opposite to that which is usually reported. Some examples are low blood pressure (BP), low cholesterol, increased obesity, and higher creatinine, all of which in dialysis patients are usually associated with better outcomes. Analysis of this reverse causation shows that in each case a specific explanation affecting only a fraction of the patients accounts for the apparent unexpected effect. For example low BP associated with increased mortality is explained by severe cardiomyopathy.

This paper discusses modifiable and potentially modifiable cardiovascular risk factors with an attempt to categorize the status of measures used to influence those factors.

Of course there is a number of risk factors which cannot be modified, such as age, gender, race, menopause, family history, low-molecular-weight apolipoprotein (Apo) (a) phenotype and Apo (E) polymorphism. Nevertheless, there is a long list of modifiable and potentially modifiable risk factors (tables 1 and 2) which will be discussed in the following.

Modifiable Risk Factors

Smoking
Data from the USRDS Dialysis Morbidity and Mortality Wave 2 study show that nearly half the dialysis population smoked during lifetime and about 14% are current smokers [1b]. Smoking is highly associated with peripheral vascular disease, congestive heart failure and mortality and this emphasizes the importance of discouraging patients from smoking. Current smokers have a hazard ratio of 1.59 for new-onset congestive heart failure in comparison to non-smokers. Former smokers who had quit for more than one year showed similar event risks as non-smokers, suggesting that quitting even at a late disease stage can still improve outcomes. Furthermore, smoking cessation may slow the progression of chronic kidney disease (CKD) [2] and prolong the dialysis-free interval. Smoking is a relatively easily modifiable risk factor,

Table 1. Modifiable cardiovascular risk factors

Risk factor	Modification strategies
Tobacco use	Further efforts in smoking reduction, counseling by dialysis team; nicotine dermal patches; tobacco restriction groups
Hypertension	Should be treated despite 'reverse epidemiology'; new K/DOQI-Guidelines in Hypertension (May 2004) and CVD (April 2005)
ECV overload	May be a major reason for hypertension and LV damage; importance of accurate determination of dry weight
Dyslipidemia	Statins?, ezetimibe?
Diabetes mellitus	Tight control, (insulin inhalers in the future)
BMI, body size, obesity	Encourage lifestyle changes; small patients might need more frequent dialysis
Diet	Fruits and vegetables, increase omega-3 fatty acids
Alcohol consumption	Small amounts especially in low high-density lipoprotein patients may be protective
Physical inactivity	Physical exercise during and between dialysis treatments
Psychological stress	Check with BDI (Beck Depression Inventory) for depression; improve patient education and staff awareness of psychological influences
Hypo- and hyperparathyroidism	DOQI-Guidelines; vitamin D analogs, calcimimetics, PTH between 150 and 300 pg/ml
Abnormal calcium and/or phosphorus metabolism	Phosphate binders, dietary counseling; dialysis dose and frequency
Malnutrition	Dietary supplementation during and between dialysis treatments; check protein catabolic rate (in stable patients)
Anemia	Erythropoietin; increased Hb useful, but may be harmful at high hematocrit with high UF rates and hypotension

provided that all members of the dialysis team cooperate and all efforts are made to encourage patients, including psychological counseling and support with nicotine dermal patches.

Hypertension

Everyday clinical practice as well as several studies show that the BP control in dialysis patients can be improved. The main causes for hypertension in the dialysis population are high interdialytic weight gain in conjunction with an

Table 2. Potentially modifiable cardiovascular risk factors

Risk factor	Modification strategies
Infection and inflammation (CRP, cytokines)	'Fistula first', ultrapure water, reduce usage of catheters, check for acute infections
PVD (elevated ankle-brachial index, pulse wave velocity, pulse pressure)	Stop smoking, statins
RAS-activity	ACE-inhibitors
Thrombogenic factors	Low-dose aspirin
Oxidative stress	Vitamin D, E supplements with questionable effects, prevention
Elevated homocysteine	Folic acid, Vitamin B_6 + B_{12} intravenous

inappropriate dietary sodium intake and the clinical difficulty to determine dry weight [3]. Recently a risk factor paradox for BP in CKD stage 5 patients (low-normal BP as a negative prognostic marker) gained attention [4]. These observational results must be questioned, as the used study design was open to potential confounding by disease (reverse causality bias). Pre-existing diseases leading to low BP itself, such as cardiomyopathy, may explain the observed association between low BP and poor cardiovascular outcome [5]. Long-term observational studies with adequately controlled baseline cardiac parameters show, that patients with normal BP have improved cardiovascular outcome in comparison with hypertensive patients. Therefore, all dialysis patients should be treated aggressively to decrease BP to the lowest tolerable value in the absence of intradialytic hypotension [3].

ECV Overload

Fluid overload initiated predominantly by increased sodium intake and followed by thirst is the usual state of affairs in dialysis patients. Prescription of dry weight is still a major problem in clinical practice. To date, there is no simple technique available to determine the body hydration state. Therefore, clinical assessment which is highly dependable on clinical experience is the only method for assessment of dry weight. As both over- and underestimation of dry weight have severe consequences in terms of quality of dialysis treatment and patients' quality of life, other more objective strategies need to be developed.

Upon a variety of methods such as biochemical markers (ANP, BNP, cGMP), inferior vena cava ultrasound, continuous blood volume monitoring during dialysis and bioimpedance measurements, segmental bioimpedance of the calf seems to be the most promising technique. Albeit further validation and simplification of its interpretation is needed, its clinical implementation is on the horizon (6, see other paper by Levin NW et al. in this issue).

Dyslipidemia and Lipoprotein (a)

Atherogenic lipid abnormalities in dialysis patients are common. Classically, low-density lipoprotein (LDL) levels in hemodialysis (HD) patients are normal, triglycerides and VLDL are increased and high-density lipoprotein (HDL) is decreased. In the non-dialysis population, treatment of dyslipidemia improves cardiovascular outcomes significantly. Unfortunately, there are only very few data for the renal population. Retrospective analysis from the USRDS Dialysis Morbidity and Mortality Wave 2 study shows a risk reduction for all-cause death (relative risk = 0.68) and cardiovascular mortality (relative risk = 0.64) in CKD patients with statin therapy [7]. In contrast, latest results from the 4-dimensional study, a randomized, placebo-controlled study in 1,255 type 2 diabetic patients on chronic HD, showed no significant difference in the primary end point (cardiac death, non-fatal myocardial infarction and fatal or non-fatal stroke) between the placebo group and patients treated with 20 mg atorvastatin daily [Wanner ASN 2004, unpubl. data]. Even though statins are safe and effective in reducing LDL levels, at this time *general treatment* of HD patients with statins cannot be recommended due to lacking evidence from prospective randomized trials. Nonetheless, given the strong evidence of risk reduction and benefits of lipid-lowering therapy in the general population, it is still in dispute whether dialysis patients should be treated for dyslipidemia to an LDL goal <100 mg/dl, as recommended in the K/DOQI-Guidelines from 2003. Albeit an additional important effect of lipid-lowering therapy with statins might be in lowering microinflammatory state and endothelial dysfunction associated with uremia, there are no data available that associate this with improved cardiovascular outcome.

Lipoprotein (a) serum concentrations are increased in dialysis patients compared to the general population. Nonetheless, treatments with adenocorticotropic hormone, D-thyroxin or nandrolone, which could potentially reduce lipoprotein (a) levels may not be recommended at this time, due to lacking evidence for cardiovascular benefit.

A meta-analysis clarified lately the conflicting results of former studies and identified Apo E4 genotype as a significant risk factor for CVD [8]. Carriers of the Apo E4 allele have a 42% higher risk for coronary heart disease. In contrast, Apo E2 genotype seems to have no such effect. So far, there are no ways to modify Apo E4 levels.

Recently, several studies demonstrated that endothelial progenitor cells which play a pivotal role in vascular repair mechanisms are numerically and functionally impaired in CKD patients. This may lead to accelerated atherosclerosis and reduced angiogenesis which is observed regularly in CKD. In the future, endothelial progenitor cells action may be amenable to therapeutic intervention [9, 10].

Diabetes Mellitus

To date, there is no evidence that diabetic dialysis patients benefit from tight glycemic control in comparison with standard therapy. Nevertheless, current guidelines recommend tight control. It has to be kept in mind that tight control of blood glucose levels may raise potential problems for dialysis patients. Reduced dietary intake and nausea as well as increased half-life of insulin may increase the risk for hypoglycemia. Therefore, reduced insulin doses may be appropriate at the onset of dialysis treatment. Patients with tight blood glucose control may experience gradual increase of body weight, which must be considered in adjustment of prescribed dry weight. Glycosylated hemoglobin (HbA_{1c}) may not precisely reflect the glycemic control in HD patients, due to decreased metabolism, anemia and shorter life of red cells [11]. An accurate target value for HbA_{1c} in dialysis patients has not yet been established. The use of new insulin regimens and preparations with properties closer to physiological conditions should be favored and oral hypoglycemic agents should be used with caution in dialysis patients. Metformin implies the possibility of severe lactic acidosis and sulfonylurea class of drugs create a high risk of persistent hypoglycemia, due to decreased clearance of these drug types.

Body Mass Index, Obesity and Dialysis Dose

Obesity has reached epidemic status in western countries within the last decades. Growing concern about renal complications of obesity has developed as it seems to play a major role in the genesis of hypertension, CKD and renal cell carcinoma. Still, the impact on outcome of chronic HD patients and renal transplant patients remains uncertain [12]. Patients with a higher body mass index (BMI) tend to have a lower mortality whereas small body size despite adequate dialysis dose is viewed as an independent risk factor for mortality [13]. However, concerning adequate dialysis dose, there is no evidence that smaller patients require proportionally lower dialysis dose, as one could intuitionally guess. In contrast, it seems that patients with a lower body size in terms of a lower volume might require more frequent dialysis than patients with higher volumes to achieve comparable outcomes [14]. Decreased Kt/V and late referral of patients can generally be considered as contributors to CVD.

Diet and Alcohol Consumption

The relationship between diet and cardiovascular mortality in the general population has been studied extensively. Unfortunately, the optimal diet for prevention of CVD in CKD patients has not yet been investigated thoroughly. Nevertheless, some conclusions from studies in the general population can be transferred to renal patients. A 'cardio-protective diet' should be low in saturated fat and cholesterol, include non-hydrogenated unsaturated fatty acids or monosaturated fatty acids as the predominant form of dietary fat, whole grains as the main form of carbohydrates and regular intake of fruits and vegetables. Data from several studies suggest that adequate amounts of omega-3 fatty acids can offer additional protection against CVD [15–18]. Therefore, diet should also be high in omega-3 fatty acid content, which can be achieved by the use of flaxseed, canola, olive oil, soybean or fish oil.

Regarding the widespread recommendation for the general population that regular alcohol consumption in small amounts might be a protective factor, especially for patients with low serum high-density lipoprotein, there are no data so far addressing this fact in CKD patients. Most studies that identified regular alcohol consumption as a protective factor did mainly focus on cardiovascular outcome and there are little data on the effects of regular alcohol consumption, even in small amounts, on the incidence of liver disease and induction of alcoholism.

Physical Inactivity

Positive effects on functional capacity, anemia, psychosocial problems and cardiovascular risk factors can be achieved by physical exercise. Unfortunately, only few patients participate in exercise programs in an outpatient setting. Therefore, a new strategy, exercise during dialysis, was recently examined by Daul et al. [19]. The training program was performed with a bed bicycle ergometer and supported by gymnastics and relaxation techniques. It was demonstrated in several studies that this type of exercise has effects similar to outpatient training programs. Furthermore, exercise during dialysis supports fluid removal and dialysis efficiency. Even elderly patients are able to participate in this kind of programs and benefit in terms of activity for daily living. As specialized physiotherapists survey the patients in these programs, no serious adverse effects have been reported so far and nearly all patients can participate in training programs during dialysis, we recommend the more frequent implementation of these programs in HD units. Nonetheless, careful selection of patients is required because of the risk of myocardial infarction and data on outcome, including nutritional status and inflammation are urgently needed.

Psychological Stress and Depression

The psychosocial factors that might affect outcome in CKD patients have not yet been studied extensively except for depression, for which some data regarding its role as a predictor of mortality are available. Kimmel et al. [20] reported in 2000 that higher levels of depressive affect in HD patients are associated with increased mortality. Interestingly, crude baseline Beck Depression Inventory (BDI) values were not significantly associated with mortality at a relatively long mean follow-up (38.6 months). Using BDI as a time-varying covariate, it predicted survival with an estimated relative risk of 1.24, indicating that a 1 SD increment in BDI scores (8.1 points) was associated with a 24% higher risk of death. A similar albeit smaller effect was reported for Cognitive Depression Inventory scores. Both BDI and Cognitive Depression Inventory scores remained significant when including other variables such as age, serum albumin, Kt/V and protein catabolic rate in the analysis. However, the mechanisms linking depression and survival remain obscure and must be further investigated. Nonetheless, high levels of self-reported depression in HD patients should be taken seriously and interventions should include cognitive or psychopharmacological therapy.

Calcium and Phosphate Metabolism Disorders and Parathyroid Hormone

Elevated phosphate levels and increased calcium-phosphorus product are associated with vascular and valvular calcifications, arterial hypertension, left ventricular hypertrophy (LVH) and increased risk of cardiovascular death. The aims of therapeutic measures are phosphorus (P) levels <5.5 mg/dl and calcium (Ca) levels <10 mg/dl (Ca $\times$ P <55 mg^2/dl^2). This can be achieved by a phosphate-restricted diet, phosphate binders and dialysis [3]. In most patients the use of phosphate binders is unavoidable, but should be restricted to the use of calcium-free agents such as sevelamer or lanthanum carbonate in order to avoid hypercalcemia and its consequences.

Interestingly, both hypo- and hyperparathyroidism pose an increased risk in patients on HD. Parathyroid hormone (PTH) is involved in smooth muscle cell proliferation, LVH, hypertension, calcium deposition in vessel walls and contributes to arterial stiffness and thus to increased LV afterload [21]. PTH values should be assessed every 3 months using an intact PTH assay and maintained between 150 and 300 pg/ml. This can be achieved by different strategies using vitamin D analogs and calcimimetics.

Uremic Malnutrition

Protein-energy malnutrition is an important predictor of mortality and hospitalization events in HD patients. Although several markers correlate with nutritional status, there is no single unequivocal predictor of outcome among them. Serum albumin is a strong indicator for cardiovascular outcome, but may not indicate solely the nutritional status as it is affected by several non-nutritional factors (inflammation, comorbidity, external protein loss). To date, simultaneous assessment of several parameters (e.g. BMI, handgrip strength, serum albumin, serum creatinine, Subjective Global Assessment) is considered to deliver the best estimation of nutritional status. Recent studies show in a clear-cut fashion that HD itself induces catabolism which can be compensated by intradialytic parenteral nutrition or oral food intake during dialysis. Additionally, patients can even develop anabolism by adequate protein intake and sufficient presence of insulin which emphasizes the demand on dialysis quality and adequate protein supplementation [22].

Anemia

Consequences of renal anemia in the dialysis population are well described. Its high prevalence and association with poor cardiovascular outcome emphasize the demand for adequate therapy. There is a close association between low Hb levels, LV dilatation, cardiac failure and mortality [23]. Erythropoietin treatment is widely available, but the most appropriate Hb value for optimum cardiovascular risk reduction is still a matter of debate. A recent meta-analysis of randomized controlled trials emphasizes that Hb values <12 g/dl are associated with a lower all-cause mortality than values >13 g/dl, whereas values <10 g/dl are associated with increased risk of seizures. Therefore, in CKD patients with CVD, the benefits associated with higher Hb targets (reduced seizures) are outweighed by potential harmful effects (increased risk of hypertension and death). It was concluded that the preferred target Hb should be <12 g/dl [24]. Higher Hb levels have long been considered as beneficial regarding quality of life. Nevertheless, potential harms (thromboembolic events) especially in combination with high UF rates and hypotension must be considered until additional well-designed trials that address safety and quality of life are available. For CKD patients without CVD and the pre-dialysis population, there are still insufficient data regarding the optimal Hb level.

Potentially Modifiable Risk Factors

Inflammation and Infection

Recently, inflammatory response has gained much attention as a risk factor for CVD in CKD patients, especially in those undergoing HD treatment. Albeit, it is not definitely known whether inflammation is a marker or a cause of vascular injury, latest data suggest that inflammatory markers such as C-reactive protein (CRP) and interleukin-6 are mediators of vascular disease. In the general population, high-sensitivity CRP seems to be the best inflammatory marker to assess cardiovascular risk. Its usefulness in CKD patients needs to be further investigated. Nevertheless, it can serve to monitor and identify inflammatory response in CKD patients on a case-by-case basis [25]. There is no consensus with regard to the optimal CRP level yet, but it seems prudent to seek sources of inflammation when CRP levels are elevated above 5–10 mg/l, including infectious processes (vascular access, catheter use, periodontitis, gastritis, other covert infections), impure dialysate (endotoxin or bacterial contamination), bioincompatible dialyzers and backfiltration. Use of catheters should in general be reduced to the smallest necessary degree, as it is a major contributor to infection and inflammation and, therefore, can be considered as indirect risk factor.

References

1a   Yeo FE, Villines TC, Bucci JR, Taylor AJ, Abbott KC: Cardiovascular risk in stage 4 and 5 nephropathy. Adv Chronic Kidney Dis 2004;11:116–133.
1b   Foley RN, Herzog CA, Collins AJ: Smoking and cardiovascular outcomes in dialysis patients: The United States Renal Data System Wave 2 Study. Kidney Int 2003;63:1462–1467.
2   Schiffl H, Lang SM, Fischer R: Stopping smoking slows accelerated progression of renal failure in primary renal disease. J Nephrol 2002;15:270–274.
3   Covic A, Gusbeth-Tatomir P, Goldsmith DJA: The challenge of cardiovascular risk factors in end-stage renal disease. J Nephrol 2003;16:476–486.
4   Kalantar-Zadeh K, Block G, Humphreys MH, Kopple JD: Reverse epidemiology of cardiovascular risk factors in maintenance dialysis patients. Kidney Int 2003;63:793–808.
5   Culleton BF, Hemmelgarn B: Inadequate treatment of cardiovascular disease and cardiovascular disease risk factors in dialysis patients: A commentary. Semin Dial 2004;17:342–345.
6   Zhu F, Kuhlmann MK, Sarkar S, Kaitwatcharachai C, Khilnani R, Leonard EF, Greenwood R, Levin NW: Adjustment of dry weight in hemodialysis patients using intradialytic continuous multifrequency bioimpedance of the calf. Int J Artif Organs 2004;27:104–109.
7   Seliger SL, Weiss NS, Gillen DL: HMG-CoA reductase inhibitors are associated with reduced mortality in ESRD patients. Kidney Int 2002;61:297–304.
8   Song Y, Stampfer MJ, Liu S: Meta-analysis: Apolipoprotein E genotypes and risk for coronary heart disease. Ann Intern Med 2004;141:137–147.
9   Woywodt A, Bahlmann FH, De Groot K, Haller H, Haubitz M: Circulating endothelial cells: Life, death, detachment and repair of the endothelial cell layer. Nephrol Dial Transplant 2002;17:1728–1730.
10   Choi JH, Kim KL, Huh W, Kim B, Byun J, Suh W, Sung J, Jeon ES, Oh HY, Kim DK: Decreased number and impaired angiogenic function of endothelial progenitor cells in patients with chronic renal failure. Arterioscler Thromb Vasc Biol 2004;24:1246–1252.
11   Joy MS, Cefalu WT, Hogan SL, Nachman PH: Long-term glycemic control measurements in diabetic patients receiving hemodialysis. Am J Kidney Dis 2002;39:297–307.
12   Saxena AK, Chopra R: Renal risks of an emerging 'Epidemic' of Obesity: The role of adipocyte-derived factors. Dial Transplant 2004;33:11–20.
13   Port FK, Ashby VB, Dhingra RK, Roys EC, Wolfe RA: Dialysis dose and body mass index are strongly associated with survival in hemodialysis patients. J Am Soc Nephrol 2002;13:1061–1066.
14   Lowrie EG, Li Z, Ofsthun N, Lazarus JM: Body size, dialysis dose and death risk relationships among hemodialysis patients. Kidney Int 2002;62:1891–1897.

15 Burr ML, Fehily AM, Gilbert JF Rogers S, Holliday RM, Sweetnam PM, Elwood PC, Deadman NM: Effects of changes in fat, fish, and fibre intakes on death and myocardial reinfarction: Diet and Reinfarction Trial (DART). Lancet 1989;2:757–761.

16 GISSI-Prevenzione Investigators: Dietary supplementation with n-3 polyunsaturated fatty acids and vitamin E after myocardial infarction: Results from the GISSI-Prevenzione trial. Lancet 1999;354:447–455.

17 Marchioli R, Barzi F, Bomba E, Chieffo C, Di Gregorio D, Di Mascio R, Franzosi MG, Geraci E, Levantesi G, Maggioni AP, Mantini L, Marfisi RM, Mastrogiuseppe G, Mininni N, Nicolosi GL, Santini M, Schweiger C, Tavazzi L, Tognoni G, Tucci C, Valagussa F, GISSI-Prevenzione Investigators: Early protection against sudden death by n-3 polyunsaturated fatty acids after myocardial infarction: Time-course analysis of the results of the Gruppo Italiano per lo Studio della Sopravvivenza nell'Infarto Miocardico (GISSI)-Prevenzione. Circulation 2002;23;105:1897–1903.

18 Singh RB, Niaz MA, Sharma JP, Kumar R, Rastogi V, Moshiri M: Randomized, double-blind, placebo-controlled trial of fish oil and mustard oil in patients with suspected acute myocardial infarction: The Indian experiment of infarct survival-4. Cardiovasc Drugs Ther 1997;11:485–491.

19 Daul AE, Schafers RF, Daul K, Philipp T: Exercise during hemodialysis. Clin Nephrol 2004;61(suppl 1):S26–S30.

20 Kimmel PL, Peterson RA, Weihs KL, Simmens SJ, Alleyne S, Cruz I, Veis JH: Multiple measurements of depression predict mortality in a longitudinal study of chronic hemodialysis outpatients. Kidney Int 2000;57:2093–2098.

21 Levin A: Cardiac disease in chronic kidney disease: Current understandings and opportunities for change. Blood Purif 2004;22:21–27.

22 Ikizler TA: Protein and energy: Recommended intake and nutrient supplementation in chronic dialysis patients. Semin Dial 2004;17:471–478.

23 Foley RN, Parfrey PS, Harnett JD, Kent GM, Murray DC, Barre PE: The impact of anemia on cardiomyopathy, morbidity, and mortality in end-stage renal disease. Am J Kidney Dis 1996;28: 53–61.

24 Strippoli GF, Craig JC, Manno C, Schena FP: Hemoglobin targets for the anemia of chronic kidney disease: A meta-analysis of randomized, controlled trials. J Am Soc Nephrol 2004;15:3154–3165.

25 Lacson E Jr, Levin NW: C-reactive protein and end-stage renal disease. Semin Dial 2004;17: 438–448.

Nathan W. Levin, MD
Medical and Research Director
Renal Research Institute
207 East 94th Street, Suite 303
New York, NY–10128 (USA)
Tel. +1 646 672 4002, Fax +1 212 996 5905, E-Mail nlevin@rrny.com

Ronco C, Brendolan A, Levin NW (eds): Cardiovascular Disorders in Hemodialysis.
Contrib Nephrol. Basel, Karger, 2005, vol 149, pp 230–239

....

The Cardiovascular Burden of the Dialysis Patient: The Impact of Dialysis Technology

Sudhir K Bowry[a], Ute Kuchinke-Kiehn[a], Claudio Ronco[b]

[a]Fresenius Medical Care, Scientific Affairs, Bad Homburg, Germany;
[b]Department of Nephrology, St. Bortolo Hospital, Vicenza, Italy

Abstract

There is widespread recognition that the poor survival rates of dialysis patients, attributed predominantly to cardiovascular disease, need to be addressed and improved. In this paper, we relate diverse aspects of modern dialysis technology with factors that are considered to contribute towards increased mortality and morbidity in the dialysis population. Firstly, we assess the overall cardiovascular burden of the dialysis patient: it is the sum of uraemia-related risk factors (URRF), traditional risk factors and dialysis-therapy-related factors. Secondly, we describe how key components of the dialysis procedure may be directly related to the more common URRF: the dialyser and the membrane, microbiological quality of water and dialysate, treatment modality and online monitory equipment. The judicious selection and application of these components may collectively help improve patient outcomes in hemodialysis therapy.

Copyright © 2005 S. Karger AG, Basel

Introduction

Patients with chronic kidney disease (CKD), irrespective of diagnosis, are at increased risk of cardiovascular disease (CVD), including coronary heart disease, cerebrovascular disease, peripheral vascular disease, and heart failure [1]. Even mild renal impairment adds to the cardiovascular risk [2]. Furthermore, a wealth of data suggests that mortality rates are higher in the dialysis patients compared to the general population or even to patients suffering from other diseases [3, 4]. About half the deaths of patients on dialysis are attributed to cardiovascular causes [5].

Thus, there is widespread recognition today that the poor survival rates of dialysis patients, attributed predominantly to cardiovascular death, need to be addressed and improved [6]. However, there is also the general concern that poor patient outcomes still occur 'despite significant improvements in dialysis technology' [7], and that technological advancements over the last 25 years have essentially done little to abate the continuing dilemma of the dialysis patients worldwide.

We therefore consider it pertinent to examine the validity of such statements and viewpoints and attempt, in this paper, to relate diverse aspects of modern dialysis technology with factors that are considered to contribute towards increased mortality and morbidity in the dialysis population. In view of the complexity of such an exercise, and to make such a task more transparent, we restricted our considerations to factors related to cardiovascular issues since the majority of the dialysis patients are at some stage afflicted with cardiovascular complications. Before any strategies are undertaken to improve CVD-related outcomes of dialysis patients, we considered it essential to first assess the overall cardiovascular burden encountered by dialysis patients.

Assessing the Total Cardiovascular Risk for the Dialysis Patient

There has been a recent spate of publications dealing with 'cardiovascular risk factors' related to CKD patients [8–11]. It has become customary to deal with these 'risk factors' under the categories of 'traditional' risk factors (TRF) pertaining to the general population, and 'uremia-related' risk factors (URRF, or 'nontraditional factors') afflicting patients with CKD [12, 13]. We have examined several publications dealing with the different risk factors in both categories [9–18] and have found considerable inconsistencies and variations in listing of the risk factors in the various publications. These so-called risk 'factors' include:
- Common diseases, e.g. hypertension, diabetes, anemia
- General biological systems or pathways, e.g. inflammation, oxidative stress
- Broad group of functional substances, e.g. advanced glycation end-products (AGEs)
- Specific molecules, e.g. fibrinogen, C-reactive protein
- Lifestyle-associated aspects, e.g. smoking, obesity, inactivity
- Miscellaneous e.g. gender, age

CVD risk factors discussed in the literature in relation to CKD thus constitute a very broad and general group of 'factors' ranging from complex, multifactorial

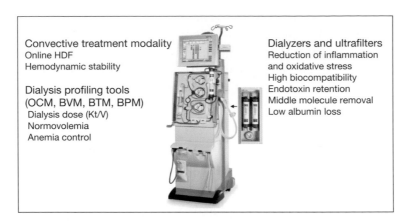

Convective treatment modality
Online HDF
Hemodynamic stability

Dialysis profiling tools
(OCM, BVM, BTM, BPM)
Dialysis dose (Kt/V)
Normovolemia
Anemia control

Dialyzers and ultrafilters
Reduction of inflammation
and oxidative stress
High biocompatibility
Endotoxin retention
Middle molecule removal
Low albumin loss

Fig. 1. Modern hemodialysis technology provides an integrated system that is geared towards the reduction of the overall cardiovascular burden of the dialysis patient. In addition advanced treatment modalities such as online-HDF and patient surveillance tools further support the control of the uremia-related risk factors. OCM = Online clearance monitor; BVM = blood volume monitor; BTM = blood temperature monitor; BPM = blood pressure monitor; HDF = hemodiafiltration.

biological systems, defined biological entities that function in a specific biochemical reaction or process, or even, to habits. In our attempts to ascertain more specifically the overall risk a dialysis patient has from CVD (to enable a more systematic approach to the management of CVD in this group of patients), we surmised that the total cardiovascular burden of the hemodialysis (HD) patient arises from three different sources (table 1).

Firstly, as already mentioned, there appears to be a substantial body of evidence now indicating that uremia, per se, contributes to the bulk of the risk of CVD to the HD patient. In table 1 we have therefore listed what we consider to be the key, inter-related contributing 'factors' pertaining to URRF: no attempt has been made to compile an exhaustive list of factors that belong to this category as the literature showed no clear consensus on the composition of risk factors in this category.

Secondly, most dialysis patients suffer from one or a number of comorbid conditions, i.e. have a high prevalence of TRF such hypertension and diabetes that exacerbate uremia and vice versa. It is not uncommon to have dialysis patients who suffer from malnutrition, anemia and diabetes and are aged, smokers and/or are obese. Clearly, the cardiovascular burden carried by dialysis patients is both multifactorial and can be extraordinarily large when uremia-related and TRF (affecting the general population) are considered together.

Table 1. The overall cardiovascular burden experienced by a dialysis patient is the sum of the uremia-related risk factors (URRF) specific to CKD, plus the traditional risk factors afflicting the general population as well as dialysis therapy related factors. It should be noted that dialysis-related factors could have a beneficial or detrimental effect on the URRF

URRF	TRF	Dialysis-therapy-related factors (influencing URRF)
Inflammation	Hypertension	Membrane bioincompatibility
Oxidative stress	Diabetes	Dialyzer solute removal
Dyslipidemia	Hyperlipidemia	Endotoxins (water quality)
Salt/water overload	Obesity	Dialysis treatment modality
Anemia	Smoking	
Malnutrition	Old age	
Calcium-phosphate product	Gender	
Left ventricular hypertrophy		

A third source of contribution to the overall cardiovascular burden encountered by the dialysis patient could be due to the inadvertent and adverse influence of certain features related to the dialysis procedure itself on either the URRF or the TRF. As dialysis is unquestionably a highly beneficial, life-saving therapy, we refrain from addressing these factors as 'dialysis-related risk factors' as, by implication, any factor that is perceived to represent a risk should not then be also applicable for therapeutic purposes. Nevertheless, these treatment-related factors that are 'administered' repeatedly each time a patient undergoes dialysis have the potential of making an already bad situation worse and are to the detriment of the long-term well-being of the patient. Hence the overall cardiovascular burden could thus be summarized as follows:

$$CVD_{Risk\ for\ dialysis\ patient} = URRF + TRF + dialysis\ therapy\text{-}related\ factors.$$

In the following section we consider some of the evidence that is now beginning to emerge showing how key individual components of the dialysis therapy may have a direct bearing on the more common URRF. Additionally, we deliberate upon the central dialysis-related technological features and modalities, which if implemented, have the potential of favorably modifying the overall cardiovascular burden that has to be tolerated by the dialysis patient. Again, no attempt has been made to present a comprehensive review of how each and every component of the dialysis procedure could influence the cardiovascular status of the dialysis patient: we deal here only with the key components of the dialysis procedure.

Components of the Dialysis Procedure Linked to Cardiovascular Risk Factors

Dialyzer and Membrane

While it is well recognized that various commercial dialyzers vary significantly in their ability to remove different-sized uremic retention solutes, some recent publications have demonstrated differences between dialyzers (flux and membrane material effects) in terms of parameters implicated in biological mechanisms contributing to cardiovascular events:

(a) p-cresol, an end-product of protein breakdown is involved in uremic toxicity and inhibits endothelial proliferation; being essentially protein bound, it is eliminated by high-flux membranes. A dialyzer-dependent enhanced removal of p-cresol has been reported with the new FX-class of dialyzers (containing the advanced polysulfone membrane, Helixone) showing superior removal than conventional dialyzers [19].

(b) Leptin is a peptide hormone (16 kDa) that is an index of the nutritional status, accumulates and is elevated in serum of HD patients leading to a reduced appetite and in parallel results to weight loss; being larger than β_2-microgobulin it is also classified as a middle molecule (MM). Its clearance kinetics and removal depends on the type of high-flux membranes used, with polysulfone membranes being better than triacetate and polyarylethersulfone membranes [20].

(c) Dyslipidemia, together with inflammation and oxidative stress, are key risk factors that aggravate CVD in HD patients. Wanner et al. [21], in a randomized crossover study (36 patients) showed the influence of dialyzer flux and membrane material on the three CVD risk factors. Patients showed improved lipid and apolipoprotein profiles after 6 weeks of high-flux treatment. In particular, levels of oxidized LDL were significantly reduced when treated with Helixone than with cellulose triacetate membrane.

(d) Bioincompatibility of dialysis membranes has long been considered to have detrimental effects on the long-term outcome of patients although conclusive data, from controlled, prospective studies has not been forthcoming. Complement activation, in conjunction with leukocyte stimulation, has been shown to be dependent upon the membrane material type, with differences also apparent between different types of synthetic dialysis membranes [22]. The resultant proinflammatory response and oxidative stress over an extended period of time is considered to add to the cardiovascular burden of the dialysis patient.

(e) Anemia is an important predictor of mortality and morbidity in patients with CKD undergoing dialysis. Ayli et al. [23] showed beneficial effects of high-flux dialysis (compared to low-flux dialysis) in controlling renal anemia and reducing the cost of the therapy, an effect achieved probably by the efficient removal of moderate- and high-molecular-weight erythropoiesis inhibitors.

Contaminated Dialysis Fluid (Water Quality)

Biofilm formation, involving the proliferation of bacteria on surfaces and conduits of the extracorporeal circuit could occur easily in most dialysis units [24] in the absence of efficient water treatment systems and disinfection procedures. During bacterial growth and lysis, endotoxins are produced and are now accepted to be the cause of silent chronic inflammation in HD patients. A number of papers have shown that endotoxin elimination from dialysis fluids necessitates the usage of efficient water-treatment systems as well as dialyzer membranes having a high endotoxin-retention capacity.

(a) Two recent publications have highlighted the membrane material-dependent retention of endotoxins from contaminated dialysate: Firstly, Weber et al. [25] showed that polysulfone-based membranes (Fresenius Polysulfone and Helixone) had a significantly higher endotoxin retention compared to two polyarylethersulfone-based membranes. Likewise, Schindler et al. [26] demonstrated differences in the permeability of high-flux dialyzer membranes for bacterial pyrogens, concluding that dialyzers that leak cytokine-inducing substances should not be used unless dialysate has passed through an ultrafilter (e.g. the DIASAFEplus ultrafilters).

(b) AGEs are inflammatory molecules that are thought to contribute to the pathogenesis of atherosclerosis in HD patients. Reznikov et al. [27] describe the contribution of AGEs to elevate cytokine levels; significantly, they demonstrated that AGEs act synergistically with endotoxins and the cytokine-inducing effect of AGEs plus endotoxin is more potent than the effect of either alone.

(c) An association between ultrapure dialysate and iron utilization and erythropoietin response in chronic hemodialysis patients has recently been published [28]. Through endotoxin reduction, ultrapure dialysate in dialysis patients manifested a reduced inflammatory parameter (C-reactive protein), reduced erythropoietin dose and improved iron utilization.

Selection of the Dialytic Treatment Modality
(e.g. Online-Hemodiafiltration)

Treatment modalities that involve convection, rather than diffusion, as the predominant mechanism of uremic toxin removal are increasingly being acknowledged as more relevant from the standpoint that, like the natural kidney, a broader spectrum of uremic retention solutes are removed. Further, it is now becoming apparent that significant additional advantages can be attributed to therapy modalities such as hemodiafiltration (HDF) which point towards improved patient outcomes.

(a) Elimination of MM during dialysis is increasingly gaining significance: the European Best Practice Guidelines [29] states that 'to enhance MM removal, synthetic high-flux membranes should be used' and 'additional strategies, such as adding a convective component, or increasing time or frequency, should be used to maximise MM removal'. Removal of p-cresol (see above) as well as β_2-microglobulin was shown by Bammens et al. [30] to be better with HDF (and increased with greater filtration volumes) than with conventional HD.

(b) Dialysis membranes designed specifically to enhance MM removal necessitate changing key membrane parameters. Mostly, β_2-microglobulin (a surrogate for MM) removal is addressed by an increase of the membrane sieving coefficient for this molecule. However, the increased removal of MM is associated with high leakage of essential substances like albumin from the patient's blood, thereby compromising the nutritional status of the patient. Ahrenholz et al. [31], in a clinical study examining the β_2-microglobulin/ albumin relationship, showed considerable differences between eight high-flux membranes in terms of removal of these two molecules during postdilution HDF carried out at various fluid filtration/substitution rates.

(c) Bonomini et al. [32] have reported the effect of certain uremic retention solutes that cause an abnormal exposure of phosphatidylserine on the surface of erythrocytes; cells modified in this way then contribute to anemia by providing a signal for recognition (and removal) by macrophages, or, contribute to vascular damage by adhering strongly to the vascular endothelium, thereby accelerating atherosclerosis. Both HDF and Helixone were found to be more efficient than HD and conventional polysulfone, respectively, in reducing the phosphatidylserine exposure on uremic erythrocytes.

Online Monitoring Equipment

One of the major issues of concern in contemporary dialysis therapy is how to cope with severe fluctuations of physiological processes during a single dialysis session as well as in the intradialytic interval. Dialysis profiling is now being recognized as highly useful towards providing quality assurance through monitoring of the therapy during each and every dialysis session. Further, it also reduces the incidence and occurrence of severe unphysiological events, thereby ensuring safety to the dialysis care team as well as the patient and possibly favorably influence patient outcomes [33].

Treatment-related variations relate either to changes in the concentration of uremic retention solutes (e.g. urea) as well as the volume status/dry weight of the patient:

(a) Online clearance monitoring (OCM) relies on the measurement of the ionic strength of the dialysate and is then correlated to urea clearance; this is required for the estimation of the dose of dialysis (Kt/V), an index that is associated with patient outcomes.

(b) Normovolemia ('dry weight'). The uremic population is confronted in general with vascular impairment attributed to arterial stiffness, making it harder to regulate the fluid status/regulation. This aspect of vascular dysfunction is further aggravated by dialysis therapies when large unphysiological changes in hemodynamic responses occur in response to rapid fluid removal by ultrafiltration; attaining normovolemia in a controlled, gradual manner would clearly be less damaging to the patient [35].

The blood volume monitoring helps regulate the fluid removal during the treatment: ideally, the ultrafiltration rate is adjusted automatically, preferably according to the patient's plasma refilling rate. Clinical studies have demonstrated the usefulness of blood volume monitoring in reducing hypotensive episodes. The blood temperature monitoring permits the regulation and stabilization of intradialytic body temperature thereby achieving cardiovascular stability; application of the blood temperature monitoring has been shown to have a considerable positive effect on the intradialytic vascular stability.

Conclusions

The provision of life-saving dialysis treatment for patients worldwide is highly dependent upon the application of distinct technological devices and instrumentation. The early rudimentary appliances have evolved over the years into highly sophisticated equipment that conduct individual functions in a controlled manner. The fundamental impetus to these developments has been based on scientific, medical and technical advancement derived from several decades of experience in the treatment of HD patients.

Together with the practicing nephrologist, industry has played its part in establishing routine dialysis therapies into increasingly reliable and safer procedures. Like the clinician attending to the needs of the individual patient, industry (as one of member of the multidisciplinary patient-care team) strives to contribute towards improved patient care, particularly long-term outcomes and quality of life of patients.

Treatment-modifiable factors have obvious potential to change clinical dialysis practices to impact outcomes and quality of life. In this paper we have sought, firstly, to analyze and structure the factors that contribute towards a dilemma facing the dialysis fraternity today, namely, the unacceptably high CVD-related mortality and morbidity of dialysis patients. Secondly, we have attempted to link the different technological components of the dialysis procedure with recognized cardiovascular risk factors specific to the uremic patient.

It would therefore be reasonable to suggest that the judicious selection and application of the most efficient and beneficial treatment modality (HDF), coupled with ultrapure dialysate, highly biocompatible dialyzers that efficiently remove MM and advanced treatment-surveillance equipment would provide cardioprotection for the dialysis patient. Such an integrated therapy approach, together with the management of URRF and TRF, need to be considered collectively to improve patient outcomes.

References

1   Go AS, Chertow GM, Fan D, McCulloch CE, Hsu CY: Chronic kidney disease and the risks of death, cardiovascular events, and hospitalisation. N Engl J Med 2004;351:1296–1305.
2   Culleton BF, Larson MG, Wilson PW, Evans JC, Parfrey PS, Levy D: Cardiovascular disease and mortality in a community-based cohort with mild renal insufficiency. Kidney Int 1999;56: 2214–2219.
3   Goodkin DA, Bragg-Gresham JL, Koenig KG, Wolfe RA, Akiba T, Andreucci VE, Saito A, Rayner HC, Kurokawa K, Port EK, Held PJ, Young EW: Association of comorbid conditions and mortality in hemodialysis patients in Europe, Japan, and the United States; The Dialysis Outcomes and Practice Patterns Study (DOPPS). J Am Soc Nephrol 2003;14:3270–3277.
4   Foley RN: Cardiac disease in chronic uremia: Can it explain the reverse epidemiology of hypertension and survival in dialysis patients? Semin Dial 2004;17:275–278.
5   Barret BJ, Culleton B: Reducing the burden of cardiovascular disease in patients on dialysis. Dial Transplant 2002;31:155–163.
6   Culleton BF, Hemmelgram B: Inadequate treatment of cardiovascular disease and cardiovascular disease risk factors in dialysis patients. Semin Dial 2004;17:342–345.
7   Port FK, Eknoyan G: The Dialysis Outcomes and Practice Patterns Study (DOPPS) and the Kidney Disease Outcomes Quality Initiative (K/DOQI): A cooperative initiative to improve outcomes for hemodialysis patients worldwide. Am J Kidney Dis 2004;44(suppl 3):1–6.
8   Sarnak MJ, Coronado BE, Greene T, Wang SR, Kusek JW, Beck GJ, Levey AS: Cardiovascular disease risk factors in chronic renal insufficiency. Clin Nephrol 2002;57:327–335.
9   McMahon LP: Hemodynamic cardiovascular risk factors in chronic kidney disease: What are the effects of intervention? Semin Dial 2003;16:128–139.
10   Kalantar-Zadeh K, Block G, Humphreys MH, Kopple JD: Reverse epidemiology of cardiovascular risk factors in maintenance dialysis patients. Kidney Int 2003;63:793–808.
11   Busch M, Franke S, Müller A, Wolf M, Gerth J, Otto LL, Niwa T, Stein G: Potential cardiovascular risk factors in chronic kidney disease: AGEs, total homocysteine and metabolites, and the C-reactive protein. Kidney Int 2004;66:338–347.
12   Uhlig K, Levey AS, Sarnak MJ: Traditional cardiac risk factors in individuals with chronic kidney disease. Semin Dial 2003;16:118–127.
13   Zoccali C, Mallamaci F, Tripepi G: Traditional and emerging cardiovascular risk factors in end-stage renal disease. Kidney Int 2003;63(suppl 85):S105–S110.
14   Levin A: Cardiac disease in chronic kidney disease: Current understandings and opportunities for change. Blood Purif 2004;22:21–27.

15   Sarnak MJ: Cardiovascular complications in chronic kidney disease. Am J Kidney Dis 2003;41 (suppl 5):S11–S17.
16   Krane V, Wanner C: Cardiovascular disease and predisposing factors in chronic renal failure. J Clin Basic Cardiol 2001;4:97–100.
17   Coresh J, Astor B, Sarnak MJ: Evidence for increased cardiovascular disease risk in patients with chronic kidney disease. Curr Opin Nephrol Hypertens 2004;13:73–81.
18   Ritz E: Atherosclerosis in dialyzed patients. Blood Purif 2004;22:28–37.
19   Bammens B, Evenepoel P, Verbeke K, et al: Protein bound solutes: Explaining the dissociation between urea reduction ratio and patient outcome? J Am Soc Nephrol 2003;14:SU-PO911.
20   Jolivot A, Combarnous F, Geleen G, et al: Leptin kinetics and clearance during maintenance hemodialysis: Effect of high flux membranes. Nephrol Dial Transplant 2003;18(suppl 4):M617.
21   Wanner C, Bahner U, Mattern R, Lang D, Passlick-Deetjen J: Effect of dialysis flux and membrane material on dyslipidemia and inflammation in haemodialysis patients. Nephrol Dial Transplant 2004;19:2570–2575.
22   Bowry SK: Dialysis membranes today. Int J Artif Organs 2002;25:447–460.
23   Ayli D, Ayli M, Azak A, Yuksel C, Kosmaz GP, Atilgan G, Dede F, Abayli E, Camlibel M: The effect of high-flux hemodialysis on renal anemia. J Nephrol 2004;17:701–706.
24   Capelli G, Tetta C, Canaud B: Is biofilm a cause of silent chronic inflammation in haemodialysis patient? A fascinating working hypothesis. Nephrol Dial Transplant 2005;20:266–270.
25   Weber V, Linsberger I, Rossmanith E, Weber C, Falkenhagen D: Pyrogen transfer across high- and low-flux hemodialysis membranes. Artif Organs 2004;28:210–217.
26   Schindler R, Christ-Kohlrausch F, Frei U, Shaldon S: Differences in the permeability of high-flux dialyser membranes for bacterial pyrogens. Clin Nephrol 2003;59:447–454.
27   Reznikov LL, Waksman J, Azam T, Kim SH, Bufler P, Niwa T, Werman A, Zhang X, Pischetsrieder M, Shaldon S, Dinarello CA: Effect of advanced glycation end products on endotoxin-induced TNF-alpha, IL-1beta and IL-8 in human peripheral blood mononuclear cells. Clin Nephrol 2004;61:324–336.
28   Hsu PY, Lin CL, Yu CC, Chien CC, Hsiau TG, Sun TH, Hunag LM, Yang CW: Ultrapure dialysate improves iron utilisation and erythropoietin response in chronic hemodialysis patients – A prospective cross-over study. J Nephrol 2004;17.693–700.
29   European Best Practice Guidelines for Haemodialysis (Part 1). EBPG Expert Group on Haemodialysis. Nephrol Dial Transplant 2002;17(suppl 7).
30   Bammens B, Evenepoel P, Verbeke K, Vanrenterghem Y: Removal of protein-bound solute p-cresol by convective transport: A randomized crossover study. Am J Kidney Dis 2004;44:278–285.
31   Ahrenholz PG, Winkler RE, Michelsen A, Lang DA, Bowry SK: Dialysis membrane-dependent removal of middle molecules during hemodiafiltration: The β_2-microglobulin/albumin relationship. Clin Nephrol 2004;62:21–28.
32   Bonomini M, Ballone E, Di Stante S, Bucciarelli T, Dottori S, Arduini A, Urbani A, Sirolli V: Removal of uraemic plasma factor(s) using different dialysis modalities reduces phosphatidylserine exposure in red blood cells. Nephrol Dial Transplant 2004;19:68–74.
33   Locatelli F, Buoncristiani U, Canaud B, Kohler H, Petitclerc T, Zucchelli P: Haemodialysis with on-line monitoring equipment: Tools or toys? Nephrol Dial Transplant 2005;20:22–33.
34   Di Filippo S, Manzoni C, Andrulli S, Tentori F, Locatelli F: How to determine ionic dialysance for the online assessment of delivered dialysis dose. Kidney Int 2001;59:774–782.
35   Donauer J, Böhler J: Rational for the use of blood volume and temperature control devices during haemodialysis. Kidney Blood Press Res 2003;26:82–89.

S.K. Bowry, PhD
Fresenius Medical Care
Else-Kroener Strasse1
D–61352 Bad Homburg (Germany)
Tel. 0049 6172 609 2128, Fax 0049 6172 609 2252, E-Mail sudhir.bowry@fmc-ag.com

Ronco C, Brendolan A, Levin NW (eds): Cardiovascular Disorders in Hemodialysis.
Contrib Nephrol. Basel, Karger, 2005, vol 149, pp 240–260

....

Oxidative Stress and Reactive Oxygen Species

*Francesco Galli, Marta Piroddi, Claudia Annetti, Cristina Aisa,
Emanuela Floridi, Ardesio Floridi*

Department of Internal Medicine, Section of Applied Biochemistry and
Nutritional Sciences, University of Perugia, Perugia, Italy

Abstract

This article discusses different aspects concerning classification/nomenclature, biochemical properties and pathophysiological roles of 'reactive oxygen species' (ROS) which
are pivotal to interpret the concept of 'oxidative stress'. In vitro studies in both the prokaryotes and eukaryotes clearly demonstrate that exogenous or constitutive and inducible endogenous sources of ROS together with cofactors such as transition metals can damage virtually
all the biomolecules. This adverse chemistry is at the origin of structural and metabolic
defects that ultimately may lead to cell dysfunction and death as underlying mechanisms in
tissue degeneration processes. The same biomolecular interpretation of aging has been
proposed to embodies an oxidative stress-based process and oxidative stress may virtually
accompany all the inflammatory events. As a consequence, ROS have proposed to play several roles in the pathogenesis of chronic-degenerative conditions, such as athero-thrombotic
events, neurodegeneration, cancer, some forms of anemia, auto-immune diseases, and the
entire comorbidity of uremia and diabetes. Nowadays, the chance to investigate biochemical
and toxicological aspects of ROS with advanced biomolecular tools has, if needed, still more
emphasized the interest on this area of biomedicine. These technological advancements and
the huge information available in literature represent in our time a challenge to further understand the clinical meaning of oxidative stress and to develop specific therapeutic strategies.

Molecular Oxygen and Reactive Oxygen Species: General Definitions and Nomenclature

The chemical characteristics of molecular oxygen, or dioxygen (O_2) are based
on the presence in the external orbital π^* of two unpaired electrons with the same
spin, therefore, O_2 is by definition a 'free radical' (fig. 1). This conformation

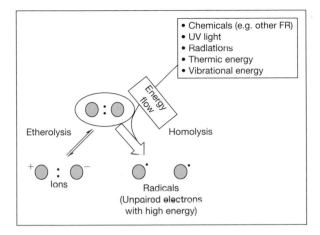

Fig. 1. Mechanism of formation of free radical (FR) species (molecules represented with a superimposed dot). A definition of FR is: any species capable of independent existence (hence the term 'free') that contains one or more unpaired electrons [1].

implies that O_2 attempts to oxidize other molecules, i.e. to capture a pair of electrons to form two electron pairs (this reduced form of O_2 is the peroxide anion, usually reported in its protonated form, i.e. hydrogen peroxide, H_2O_2). The one electron reduction of O_2 forms the superoxide radical ($O_2^{\cdot-}$) while the tetravalent reduction produces intramolecular rearrangements with rupture of the covalent bond of O_2 and the formation of two molecules of O^{2-} that once protonated result into two molecules of water (Reaction [a]).

Reaction [a]

$$O_2 \xrightarrow[\substack{c^-}]{H^+} HO_2^- \xrightarrow[\substack{e^-}]{H^+} H_2O_2 \xrightarrow[\substack{e^-}]{H^+} H_2O + OH^\cdot \xrightarrow[\substack{e^-}]{H^+} 2H_2O$$

$$H^+ + O_2^{\cdot-} \quad H^+ + HO_2^- \quad\quad\quad H^+ + OH^-$$

The pro-oxidant behavior of this atmospheric component is known since ancient times. The rusted iron and the rancid butter are only two of the most popular examples of this chemistry that is intended as the 'bad face' of this Janus molecule, in other words, a molecule that may drive damage to inorganic and organic substrates (i.e. oxidation substrates) representing sources of electrons (or reducing equivalents).

The 'good face' of O_2 is however important to the point that it can give to this molecule the full mark title of 'molecule of life'. Actually, the oxygen-driven

Table 1. Some of the most relevant reactive oxygen species

Radicals	Nonradicals
Superoxide, $O_2^{\cdot-}$	Hydrogen peroxide, H_2O_2
Hydroxyl, $OH^{\cdot}$	Hypochlorous acid, $HOCl$
Peroxyl, $RO_2^{\cdot}$	Ozone, O_3
Alkoxyl, $RO^{\cdot}$	Singlet oxygen, $^1\Delta gO_2$
Hydroperoxyl, $HO_2^{\cdot}$	Peroxynitrite, $ONOO^-$

chemistry plays a central role in the energy metabolism and biosynthetic pathways of the large majority of living organisms. Dioxygen is the final acceptor of electrons during the mitochondrial respiration, i.e. the main source of energy in eukaryotic cells, and the substrate of oxidase and oxygenase enzymes, cytochromes such as the detoxification system represented by the cytochrome P-450 superfamily and the player of direct (nonenzymatic) oxidation reactions important in the processing of metabolic substrates. Moreover, either directly or by the means of its related molecules, O_2 participates in the modulation of cell redox, signaling pathways, and gene expression (discussed more in detail in the next sections). This ultimately may influence either physiologically or pathologically basic cell processes in different tissues such as differentiation, growth and proliferation, and apoptosis.

Aerobic organisms have thus developed strategies to protect their biomolecules from the 'bad face' of O_2 while using its 'good face'. That possibly was one of the former achievements in evolution of present forms of life, i.e. the step toward the colonization of earth by aerobic organisms when more than 2.5×10^9 years ago O_2 appeared in significant amounts in the atmosphere by the evolution of photosynthesis in blue-green algae (*Cyanobacteria*). Then slowly, but inexorably, atmospheric O_2 tension increased to the current levels (21% of dry air; 5×10^6 years ago), a process that was facilitated by the dramatic evolution of plant organisms. That process was again the result of another biologically useful property of O_2, that is, its ability to form ozone (O_3). The formation of the O_3 layer in the stratosphere (Reaction [b]) gave the possibility to screen the harmful solar UV-C radiation, thereby allowing organisms to leave the sea to colonize the land.

Reaction [b]
Solar energy
$$O_2 \longrightarrow 2O + O_2 \longrightarrow 2O_3$$

These introductory concepts suggest the multifaceted nature of oxygen-related molecules, very often described with the name of 'oxygen free radicals'

Table 2. Standard reduction potentials ($E°$) of some biologically relevant redox couples (listed from the highly reducing to the highly oxidizing)

Oxidized form/Reduced form	$E°$ (V)
$H_2O/H^+(e_{aq}^-)^a$	-2.84
$CO_2/CO_2^{\cdot-}$	-1.80
$O_2, H^+/HO_2^\cdot$	-0.46
Fe^{3+}-transferrin/Fe^{2+}-transferrin	-0.40
$O_2/O_2^{\cdot-}$	-0.33
$NAD^+, H^+/NADH$	-0.32
Fe^{3+}-ferritin/ferritin, Fe^{2+}	-0.19
$FAD, 2H^+/FADH_2$	-0.18
Dehydroascorbate/ascorbyl radical (vit. $C^{\cdot-}$)	-0.17
Ubiquinone, $H^+/$ubisemiquinone	-0.04
Fe^{3+}-ADP (or citrate)/Fe^{2+}-ADP (or citrate)	~ 0.10
Fe^{3+}-EDTA/Fe^{2+}-EDTA	0.12
Ubisemiquinone, $H^+/$ubiquinol	0.20
Ferricytochrome c/ferrocytochrome c	0.26
Ascorbyl radical (vit. $C^{\cdot-}$), $H^+/$ascorbate$^-$	0.28
$H_2O_2, H^+/H_2O, OH^\cdot$	0.32
α-tocopheryl radical (α-TO$^\cdot$), $H^+/$ α-tocopherol (α-TOH)	0.50
Urate radical (HU$^{\cdot-}$), $H^+/$urate (UH$_2^-$)	0.59
Cys-S$^\cdot$/Cys-S$^-$	0.92
$O_2^{\cdot-}, 2H^+/H_2O_2$	0.94
$RO_2^\cdot, H^+/ROOH$	~ 0.77–1.44^b
$HO_2^\cdot, H^+/ H_2O_2$	1.06
$RO^\cdot, H^+/ROH$	1.60^c
$OH^\cdot, H^+/ H_2O$	2.31

$E°$ of aqueous suspensions of the redox compounds are determined at a temperature of 25°C and corrected as they are measured at pH 7 (often written as $E°'$).

$^a(e_{aq})$ = Hydrated electron formed by radiolysis of water.

^{b}This value varies depending on the structure of the compound designed with the generic symbol (R).

^{c}Similarly to peroxyl, aliphatic alkoxyl have $E°'$ values that vary according with the nature of (R).

Cys = Cysteine.

Data from reference 1.

or 'reactive oxygen species' (ROS) (table 1). Both these names are obviously not adequate to describe the chemistry of O_2 and the overall chemistry of redox reactions. In fact, not all the oxygen-derived molecules are free radicals (fig. 1, table 2) and the term 'reactive' is undoubtedly relative since some species, such as $O_2^{\cdot-}$ and H_2O_2, are not particularly reactive in aqueous solutions. Another popu-

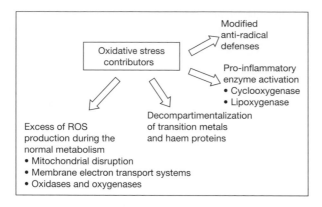

Fig. 2. Some of the most relevant contributors to oxidative stress.

lar collective, but often inappropriate, term is 'oxidants'. Some ROS, in fact, can act both as oxidizing and reducing agents depending on the reaction system in which they are examined. The reactivity and redox chemistry of these compounds is usually described with the thermodynamic parameter 'reducing potential' which expresses the likelihood that a compound can chemically reduce another compound or undergo autoxidation (i.e. the oxygen-mediated oxidation of its reduced form). The rates of these reactions vary substantially as a consequence of the presence in the reaction milieu of factors which can influence the thermodynamics of the reaction. These include catalysts and cofactors such as transition metals and other redox active compounds. Reduction potentials are measured with a standard hydrogen electrode and then referred to as 'standard reduction potentials' ($E°$) which are expressed as negative values (table 2).

Toxicological Aspects of ROS: The Concept of 'Oxidative Stress'

The chemistry of O_2 and ROS has both physiological and toxicological implications that have been extensively investigated in the last century as documented by an impressive number of publications in literature (searching the entry 'oxidative stress' on PubMed without restrictions gave 32,192 results as on February 2005), and now it represents a wide and well-defined area of research in biology and medicine [1]. Toxicological aspects of ROS in aerobes are dependent on a complex series of factors and are commonly described within the unifying concept of 'oxidative stress'. The heterogeneous nature and redundancy of endogenous and exogenous sources of ROS (fig. 2) have represented a major danger for the survival of primordial as well present-time living organisms.

Virtually all the biomolecules are prone to the oxidation chemistry of ROS. When reacting with free radicals such as $OH^·$ or $ROO^·$ or nonradical species such as H_2O_2 and peroxynitrite ($ONOO^-$), biomolecules may form corresponding radicals by hydrogen abstraction, addition, and electron transfer. A classical example is the oxidation of polyunsaturated fatty acids (PUFA) in cell membranes and lipoproteins when exposed to ROS. This process follows a typical chain reaction scheme. PUFA embody the favorite target for oxidation reactions, their double bonds being a source of electrons. The addition of O_2 to PUFA double bonds results in the initiation of the chain reaction in which further oxidation reactions occur by the formation of lipoperoxyl radicals in other PUFA molecules. This self-feeding process can generate critical biochemical lesions in cell membranes that ultimately may result in the rupture of the lipid bilayer and cell death by necrosis.

Several amino acids are modified in their structure by oxidizing reactions thus resulting in functional and structural damage to proteins. Actually, between the former biological consequences of oxidative stress there is the inhibition of some enzyme activities that may lead to metabolic impairment of cells. Damaged proteins may undergo to structural changes that result in an altered life span and turnover by proteases. Again, these changes may lead signaling proteins, membrane receptors, transporters, and cytoskeletal elements to lose their function.

Mutagenic and overall genotoxic properties of ROS also have biological relevance. ROS, in fact, may produce, to different extents, lesions in nucleic acids; these ranging from the chemical modification of a single base to the rupture of DNA filaments. This aspect can be of particular importance in the cytotoxic function that ROS assume in the defense against foreign or anomalous cells (such as bacteria and tumor cells).

Other damaging effects of ROS can derive from the subtraction of electrons from redox reactions of relevance in metabolic pathways and cell homeostasis. Channeling these reactions into sequential enzymes or cytochromes (i.e. in the mitochondria) and the presence of specific systems scavenging unwanted species preserves the redox chemistry of cells.

Within the context of oxidative stress, the damaging function of ROS is influenced by several factors (fig. 2). The catalysis of transition metals such as iron and copper is one of the best known. Of particular relevance in oxidative stress are the complexes that these metals form with chelators and biomolecules such as porphyrins and metal-containing proteins. The best-known example of the latter is the iron-heme complex. These complexes can substantially increase the ability of transition metals to catalyze electron transferring between biomolecules and other generic substrates. However, this property is not always negative since it also represents the catalytic mechanism of some key antioxidant enzymes (see below).

As mentioned in the first section of this paper, important sources of ROS are identified in the cytochromes that allows electron transferring to the final acceptor O_2 during the mitochondrial respiration, but also in the cytochromes present on other organelles such as in the nuclear membrane or microsomes, and the cytochrome P-450 super-family which is involved in detoxification of several substrates. Several oxidases and oxygenases (i.e. xanthine oxidase and the arachidonic acid cascade enzymes, cyclooxygenase and lipoxygenase) are also present in all the cells and their activity and expression can be modulated at different levels. All these are ubiquitary processes and their degree of expression can vary depending on several endogenous and exogenous stimuli influencing the entire metabolic and functional asset of the cell (such as cytokines, mitogens and the same ROS) [1]. In inflammatory cells, that include polymorphonuclear leukocytes, monocytes, and tissue-resident macrophages, these stimuli generate a characteristic activation response with production of considerable amounts of ROS and this is the result of the activity of specific oxidase enzymes (fig. 3) [2, 3]. Representing a chemical weapon for the cytotoxic activity exerted against pathogens during phagocytosis, this cell-mediated immune process, known as metabolic (or respiratory) burst, can produce oxidative stress either at the local or systemic level. The risk of oxidative stress during an inflammatory response obviously depends on the intensity and duration of the response, but is also influenced by the efficacy of the inflammatory feedback due to specific cytokines such as interferon-γ and interleukin-10, the production of glucocorticoids, and from the expression of defense mechanisms in tissues. Actually, chronic activation of inflammatory cells and the consequent secretion of inflammatory mediators (mainly the proinflammatory cytokines IL-1β, tumor necrosis factor- α or TNF-α, and IL-6) and ROS are now universally recognized as pathogenic determinants in different chronic degenerative conditions such as atherosclerosis, neurodegeneration and cancer, and in the overall tissue degeneration associated with aging.

Biomarkers and Molecular Events Associated with
Oxidative Stress

A main undertaking in free radical biology and medicine is the search for reliable biomarkers of oxidative stress [2, 4]. The achievements obtained in the last two decades, thanks to the most advanced analytical and molecular technologies, allowed several of these hallmarks to be identified (fig. 3). These are important to monitor oxidative stress events since they appear at early stages, and then to prove the efficacy of specific intervention measures. Actually, one of the main biases in several clinical trials is the complete absence or weakness of biomarkers that would provide a consistent support to the achievement of such therapeutic or overall clinical outcomes.

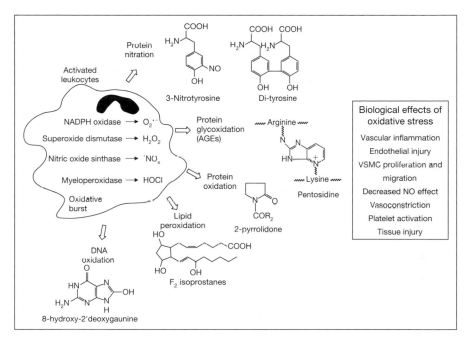

Fig. 3. Main sources of ROS during phagocyte respiratory burst: molecular targets, biomarkers, and biological consequences of oxidative stress. VSMC = vascular smooth muscle cells.

Further biomarkers can be identified at the cellular and molecular level (fig. 4). The activation of immune cells as well as oxidative stress-related responses in other cell types are the consequence of the stimulation of specific signaling pathways such as the protein kinase C (PKC)-dependent signaling, p38-MAPK, JNK and STAT, and the activation of redox-sensitive transcription factors as AP-1 and NFκB [5–9]. Independently or by synergic events, these pathways regulate the expression of different genes that ultimately produce biological responses of fundamental importance in all the steps of the inflammatory response from its onset to termination [10]. These biological responses include the regulation of cell cycle and metabolism, functional activation or negative feedback events, and killing by necrosis or apoptosis. In the latter process, for instance, lipid oxidation products (produced either by enzymatic and nonenzymatic processing of PUFA) trigger specific signals involved in both the mitochondrial-dependent and -independent pathways of programmed cell death of inflammatory cells (fig. 5). Lipid peroxidation can induce an imbalance in the cell redox with a decrease in intracellular thiol (-SH) that together with other events (such as an increase in intracellular calcium) can

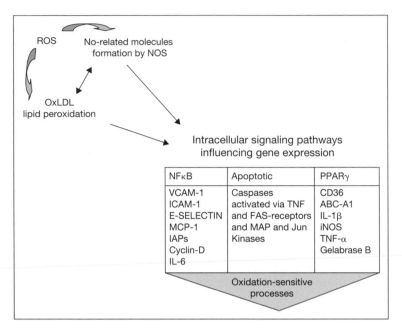

Fig. 4. Some elements of the intracellular signaling pathways influenced by oxidative stress processes. VCAM-1 = Vascular cell adhesion molecule-1; ICAM-1 = intracellular adhesion molecule-1; MCP-1 = monocyte chemoattractant protein-1; IAPs = inhibitor of apoptosis proteins; MAP kinases = mitogen activated protein kinases, ABC-A1 = ATP-binding cassette transporter-1; iNOS = inducible nitric oxide synthase.

activate proapoptotic genes [11–13]. The activation of PKC plays a key role in this process [5, 14, 15]. Members of this kinase family are common checkpoints activated upstream to the aforementioned intracellular events by several inflammatory mediators, mitogens, and stress stimuli.

Some of oxidation-sensitive genes can regulate the expression of several inflammatory or anti-inflammatory cytokines, ROS-generating enzymes, and activated biomolecules, some of which act as paracrine factors influencing inflammatory cell and platelet activation, and vascular cell homeostasis (e.g. nitric oxide and bioactive lipids such as leukotrienes, prostaglandins, and thromboxanes) [5, 16]. Other genes directly or indirectly activated by inflammatory mediators (mainly cytokines and ROS) are responsible of the expression of enzymatic defenses in target tissues (see below), but also induce the expression by liver cells of acute phase proteins (such as C-reactive protein, fibrinogen and many others) and inhibit negative phase proteins (albumin, prealbumin, transferrin, etc.).

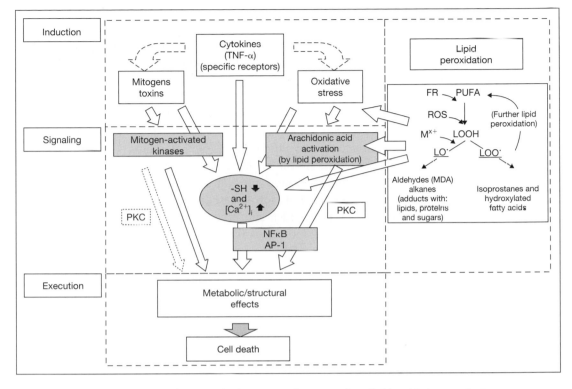

Fig. 5. Apoptotic signaling in immune cells exposed to lipid oxidation products. Reactive species (bioactive lipids) deriving either from enzymatic or nonenzymatic processing of PUFA, can influence both the induction and signaling phases of programmed cell death (apoptosis). In the signaling phase, specific intracellular metabolic events regulate the expression of proapoptotic genes and can trigger mitochondrial-dependent and -independent apoptotic pathways (execution phase). These metabolic events include, among the others, the decrease in cell thiols (possibly by active extrusion of GSH and oxidation of protein -SH groups) and an increase in cytosolic calcium ions. A key player upstream to these events is PKC. FR = Free radical.

Antioxidant Defenses

The adaptation of aerobes to the tension of atmospheric O_2 has lead to the development of multilevel defense systems. Some of these, and possibly the former to appear, are represented by naturally occurring substances such as micronutrients and oligoelements (these include liposoluble compounds such as vitamin E and carotenoids and hydrosoluble elements such as ascorbic acid, polyphenols and bioflavonoids, selenium, copper, zinc and manganese) [1]. Another important class of antioxidants in cells is represented by thiols (-SH). These can be divided in protein and nonprotein -SH, the latter being mainly represented by the cysteine-containing tripeptide glutathione (GSH) and free cysteine. Albumin is the most important protein -SH in serum in which it plays an important antioxidant function. Also some metabolic products are efficient antioxidants (examples are the heme byproduct bilirubin and the catabolic product of pyrimidine uric acid). All these compounds can act individually or synergistically to protect biological targets within the cell and in biological fluids. At the same time, they exert their antioxidant function with different mechanisms such as electron donation or direct scavenging of ROS. During these reactions, similarly to several biomolecules, some of them can be restored to their functional redox state by the exchange of reducing equivalents with coantioxidants and specific metabolic routes (antioxidant networking), others can behave as sacrificial targets thus forming end-products that are excreted. Further antioxidants, such as peptides and proteins in biological fluids may function as chelating or binding molecules for biologically available transition metals or metalloproteins.

More specialized antioxidant defenses are represented by a series of enzymes that have a strategic and balanced distribution with respect to the different cell sources of ROS. Superoxide dismutase, GSH-peroxidases (GPx), GSH-transferase (GST), and catalase (CAT) are the main types of antioxidant enzymes in eukaryotic cells and biological fluids. Superoxide dismutase and GPx use the redox properties of transition metals as mechanism for their catalytic activity. The different isoforms of these enzymes show a characteristic ubiquitary distribution, which mirrors the key role of this line of defense in the control of early and possibly most dangerous events that may result in a condition of oxidative stress. In fact, the different isoenzymes of superoxide dismutase catalyzed the dismutation of $O_2^{\bullet-}$ to H_2O_2 and peroxidases and catalase reduce H_2O_2 to water. GST has a peroxide-like activity and is responsible for the detoxification of many endogenous and exogenous xenobiotics. Other peroxidases are present in biological fluids such as the apolipoprotein paraoxonase and the extracellular GPx.

Some of these are inducible enzymes and their degree of expression depends on the amount and damaging properties of ROS produced. In fact, ROS can directly or indirectly stimulate the expression of several clusters of genes (fig. 4 and followings), some of which are aimed to provide a compensatory/protective response against incipient conditions of oxidative stress. Therefore, some of these enzymes are part of a family of proteins that is often described with the name of stress proteins and include not only antioxidant enzymes, but also components of repairing systems, sacrificial targets, signaling elements, transcription factors, scavenger receptors, detoxification components, hormones and secretory enzymes. All these elements are responsible of the regulation of events such as cell cycle, apoptosis and differentiation, and are also involved in malignant transformation.

Pathogenic Roles of Oxidative Stress

Oxidative stress is universally recognized as a key event in the overall pathophysiology of aging and a unifying pathogenic mechanism for inflammation-based chronic degenerative diseases [2, 4, 17–19]. These include cardiovascular disease, neurodegenerative conditions such as Alzheimer's disease (AD), multiple sclerosis (MD) and Parkinson's disease (PD) autoimmune disorders such as rheumatoid arthritis, cell transformation and progression of several malignancies, uremia and metabolic conditions such as diabetes.

Atherosclerotic Cardiovascular Disease

The most investigated oxidative stress condition was, and still yet remains, atherosclerotic cardiovascular disease (ACVD). Early in the life, endothelial injuries start to occur and, depending on several factors, they may evolve during aging into atherosclerotic lesions. In these lesions, together with the activation and damage of endothelial cells, important aspects are the activation of resident macrophages and other vessel cell components that lead to the production of adhesion and chemotaxis molecules. The latter drive circulating immunoinflammatory cells into the subendothelium. These key steps trigger further molecular events which produce the onset of a chronic inflammatory focus and the formation of the atheromatous plaque. In this context, ROS play a key role in the different steps of plaque formation and progression. The first and possibly more relevant is the damage to LDL particles [20–23]. This can occur already at the level of their transit in the circulation or after the passage throughout the endothelial barrier to reach tissues for lipid and cholesterol delivery (fig. 6). Modified LDL (from minimally modified to fully oxidized) have been proved to trigger key pathogenic events in atherosclerosis such as the formation of 'foam cells' (due to the sustained receptor-mediated scavenging of damaged LDL by phagocytes), the activation of cell components in the vessel wall such as smooth muscle cells (SMC) and fibroblasts and the damage of endothelial cells which are key events in plaque growth and instability. Interestingly, neutrophil adhesion is induced by minimally modified low-density lipoproteins and this effect seems to be mediated by the esterified form of the lipid oxidation products F_2-isoprostanes [24]. Modified LDL have also been reported to interfere with the biological role of NO as vasorelaxing factor, to trigger platelet activation and to have specific immunogenic activity [20, 25].

Some of these effects may be also induced directly by ROS. $ONOO^-$ and H_2O_2 which, in fact, cause SMC and fibroblast activation and proliferation, and endothelial cell apoptosis [22, 26]. Moreover, superoxide can react with NO to form $ONOO^-$ and this reaction was demonstrated to interfere in vivo with the vasorelaxing effect of NO, thus leading to a decreased vessel elasticity and

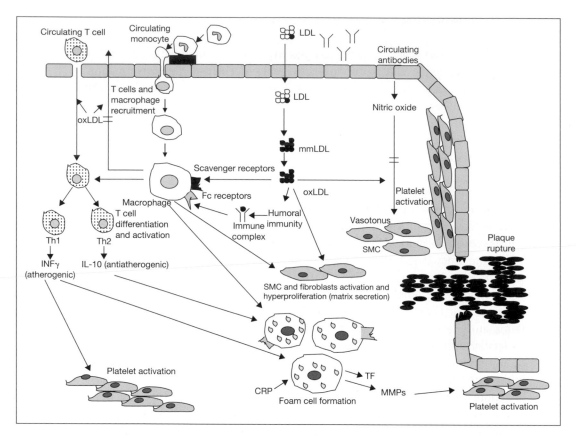

Fig. 6. Molecular and biological factors involved in the formation and progression of an atherosclerotic lesion. A main contributor in the different phases of progression of an atherosclerotic lesion is the oxidative modification of LDL. At different levels of modification, LDL particles can contribute to sustain the generation of the plaque body by foam cell formation, stimulation of SMC and fibroblast activation and proliferation. Moreover, they can play a key role in the events that drive the plaque to instability and rupture. Other details are reported in the text. Modified from [21, 22]. mmLDL = minimally modified LDL; oxLDL = oxidized LDL; TF = tissue factor; MMPs = matrix metalloproteinases; IFN-γ = interferon-γ; IL-10 = interleukin-10; Th1 = T helper type 1 cells; Th2 = T helper type 2 cells.

hypertension [16, 27]. This phenomenology, supported by in vitro and in vivo evidence, has also been confirmed in humans by histological and biochemical examination of atherosclerotic plaques [28]. Biomarkers of oxidative stress in blood of ACVD patients correlate with the presence of lesions (carotideal plaques and aortic calcifications) and vascular dysfunction assessments by clinical and instrumental evaluation [29–31].

Observational and epidemiological studies suggest that diets rich in fruit and vegetables are associated with a lower incidence of fatal and nonfatal events of ACVD [32, 33]. The assumption that these types of diets may exert this beneficial preventive action through a sustained intake of antioxidant nutrients has stimulated several interventional studies based on the administration of supplements containing individual, or mixture, of antioxidants (mostly vitamin E and C, carotenoids, and some oligoelements such as selenium and zinc) [34, 35]. However, conflicting clinical results have been obtained and the absence of a systematic investigation of consistent biomarkers (either in the phases of patient selection or in the monitoring of expected therapeutic effects) has limited the power of many of these studies thereby leaving largely unrevealed the role of antioxidant therapy in ACVD.

Uremic Syndrome

End stage renal disease (ESRD) is associated with an increased incidence of inflammation-based oxidative stress-related diseases. A common trait of these patients is the development of a typical inflammatory syndrome accompanied by an increase in several indices of oxidative stress in blood [36–42]. The uremic intoxication and dialysis-related factors contribute together with several clinical aspects to sustain chronic conditions of inflammation with consequent impact on virtually the entire comorbidity of ESRD, with particular regard to CVD, malnutrition, anemia and immune dysfunction.

In this context, oxidative stress events have detrimental effects on several tissues, with circulating blood cells as the first and more investigated targets [11, 43–46]. In fact, several studies have suggested that a direct proof of the pathogenic role of oxidative stress in uremia is the damage to red cell membranes which can result in a shortened cell life span in the circulation, a well-known contributor of uremic anemia, and defective erythropoietin function (fig. 7). Another consequence of oxidative stress in the uremic blood is the increased death by apoptosis of leukocytes, a recently revealed pathogenic factor in leukopenia and immune dysfunction of ESRD. But of much interest, and intensely investigated in the last years, is the damage to serum proteins that may also occur to a significant extent in diabetic patients [47]. This damage derives from the action of ROS alone (direct damage with formation of protein carbonyls and accumulation of specific modifications on target amino acids such as 3'-N-Tyr and di-Tyr) or in combination with reducing sugars (glycoxidation reactions with formation of advanced glycation end-products or AGEs), free carbonyls, and/or lipoperoxidation products (carbonylation and lipoxidation reactions, respectively; the products of these reactions are often identified as advanced lipoxidation end-produts or ALEs) [48]. In its own complexity, this adverse biochemistry can lead to the onset of a proinflammatory self-feeding

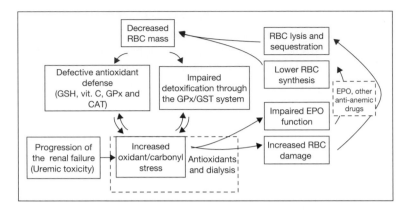

Fig. 7. Factors that may contribute to sustain uremic anemia. The uremic environment and other clinical aspects of ESRD (such as chronic inflammation) generate a condition of oxidative and carbonyl stress that may play a central role in the steady reduction of the RBC mass in the uremic circulation. In fact, this stress can produce the damage and then a shortened life span of circulating RBC. At the same time, it may play a role together with other uremic factors in depressing the erythropoietic response in the bone marrow, which in turn decreases the production of new cells. Chronic and severe conditions of anemia can exacerbate the condition of oxidative stress by weakening the antioxidant and detoxifying network in blood. Actually, the RBC mass represents an important component of this network being rich in antioxidants such as GSH, vitamin C, and enzymes with antioxidant and detoxification activity as GPx and GST. These events can produce a loop of factors that sustain anemia through oxidative stress and vice versa. Dashed boxes indicate intervention strategies that may slacken or eventually interrupt this loop. RBC = Red blood cells; EPO = erythropoietin.

cycle (fig. 8). In this cycle, the accumulation of oxidized and glycoxidized proteins by means of several events (including the proinflammatoy effect of dialysis, the accumulation of pro-oxidant toxins, etc.) sustains the activation of inflammatory cells. In fact, these cells express a highly inducible family of scavenger receptors that recognize damaged proteins (mainly albumin and apolipoproteins) and then produce their elimination by a scavenging process likened to that described in the case of modified LDL by phagocytes within the atherosclerotic plaques (see above). The loop closes when these events further generate serum protein damage. This occurs since the activation of scavenger receptors triggers the respiratory burst of circulating inflammatory (scavenger) cells. Moreover, further protein damage can be due to ROS produced as a consequence of the recruitment of the metabolic response of inflammatory and noninflammatory cells in tissues (due to accumulation of cytokines and other mediators as the same ROS and damaged proteins or other products of biomolecule damage). At the same time, this sustained inflammatory burden and the increased damage to apolipoproteins of circulating LDL may contribute to

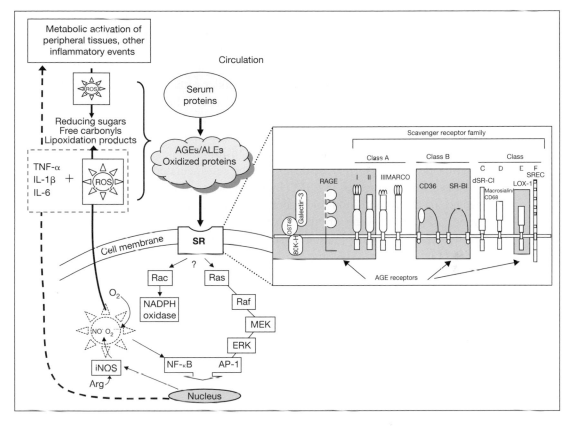

Fig. 8. Basic scheme of the proinflammatory self-feeding cycle generated by the damage to serum proteins in ESRD patients. Inflammatory cells (such as monocytes and polymorphonuclear leukocytes), and also endothelial and SMC cells, express a family of membrane glycoproteins (see the panel on the right) often grouped under the name of scavenger receptors (SR) which recognize and process glycated and oxidized proteins in the context of a phagocyte-like response (ROS and cytokine production). Therefore, a key element in this cycle is the concomitant role of inflammatory leukocyte as scavengers and generators of glycation and oxidation protein products. Other contributors to this adverse loop could be: (1) the metabolic defect of different tissues by inflammatory mediators, oxidative stress, and uremic toxins which might further sustain protein damage by further production of ROS and carbonyl stress; (2) a decreased catabolism of modified proteins; (3) the poor de novo synthesis of serum proteins by the inhibitory effect that inflammatory cytokines such as the IL-6 exert on liver cells, and (4) the impaired negative feedback on monocyte activation by T helper lymphocytes. iNOS = Inducible nitric oxide synthase; Arg = arginine.

extend the pathogenic role of this cycle to vessels (fig. 6), ultimately increasing the number of endothelial lesions and the incidence of ACVD in ESRD [23].

Therapeutic approaches addressed to counteract chronic inflammation by interfering with components of this adverse proinflammatory cycle are still preliminary, but the use of protein-leaking dialyzers or sorbent resins (and thus the clearance of some unidentified, possibly large size, solutes) have suggested a possible way to a breakage [48]. Other strategies may be addressed to the pharmacological control of inflammatory processes, but that seems to be a less practical way. In fact, the scarce selectivity of anti-inflammatory drugs could give, as main side effect, a further weakening of the cell-mediated immune response, which is already depressed in ESRD by the presence of leukopenia, anergy and changes in the composition and functional state of leukocyte subpopulations [12, 46, 49]. These events are due to several factors including uremic toxicity, repeated activation of the immune system by the extracorporeal circulation and recurrent or incidental infections.

The cholesterol-lowering drugs statins and some ACE inhibitors have been suggested to protect from oxidative stress possibly by antioxidant-like activities [50, 51].

Approaches to the antioxidant therapy in ESRD have provided some encouraging results, particularly with regards to some cardiovascular endpoints and anemia [34, 45, 52–55]. However, the majority of the studies available in literature show important limits in relation to the size and homogeneity of the populations studied, the type of design adopted and several deficiencies in laboratory investigations. Therefore, further studies are needed to confirm the beneficial role of the antioxidant therapy in these patients.

Other Conditions

In the last years, the study of a cause-effect association between oxidative stress and mechanisms of neurodegeneration has produced interesting findings and, therefore, a large consensus on its existence. This seems to be particularly true in the case of AD, but a similar coexistence was observed in other forms of dementia and brain aging. In AD, the deposition of the peptide amyloid-β within the senescent plaques of the AD brain has been proposed to cause oxidative stress as demonstrated by the analysis of specific biomarkers (such as F_2-isoprostanes, 5-nitro-γ-tocopherol and 3-nitrotyrosine) either in in vitro or in vivo experimental models [18, 56]. In this human disease, however, encouraging results on the possibility to slow down the progression and detrimental effects of oxidative stress was demonstrated by some pioneering studies based on the administration of antioxidants such as vitamin E and N-acetylcysteine [57–59].

Another model of brain pathology associated with oxidative stress is that of stroke [21, 60]. In this case, and with the same phenomenology of other atherothrombotic conditions [61, 62], ischemia reperfusion of a specific brain area can lead to an important and persistent damage by a sustained production of ROS.

Between the conditions that have a characteristic association with oxidative stress, autoimmune diseases have been largely investigated [1, 33]. Rheumatoid arthritis is a typical example in which biochemical parameters confirm the accumulation of oxidative stress markers, acute phase proteins and the characteristic imbalance between the levels of proinflammatory and anti-inflammatory cytokines that reflects the inflammatory burden of arthritic joints.

Oxidative stress is a pathogenic determinant in many other conditions [1]. These range from the classical retinopathy of prematurity that in the 1940s was quite widespread due to the use of high oxygen concentrations in incubators, to the clinical (neurological and respiratory) profile of subjects exposed to rises in the oxygen tensions in hyperbaric modules, severe hemolytic anemia and porphyrias, sepsis, hepatic degeneration by viral infections or chemical intoxication and many others.

Conclusions

The investigation of oxidative stress mechanisms has represented one of the most intriguing aspects of contemporary science and the understanding of its pathophysiology will remain a major challenge in medicine. The growing support of information and the technological advancements in laboratory techniques have provided robust and easily accessible means to investigate oxidative stress in all its aspects: from basic science to the examination of its pathogenic relevance in several clinical conditions, also including the design and evaluation of new therapeutic strategies.

References

1   Halliwell B, Gutteridge J: Free Radicals in Biology and Medicine, ed 3 New York, Oxford University Press, 1999.
2   Himmelfarb J, Stenvinkel P, Ikizler TA, Hakim RM: The elephant in uremia: Oxidant stress as a unifying concept of cardiovascular disease in uremia. Kidney Int 2002;62:1524–1538.
3   Galle J, Seibold S, Wanner C: Inflammation in uremic patients: What is the link? Kidney Blood Press Res 2003;26:65–75.
4   Shishehbor MH, Hazen SL: Inflammatory and oxidative markers in atherosclerosis: Relationship to outcome. Curr Atheroscler Rep 2004;6:243–250.
5   Kyaw M, Yoshizumi M, Tsuchiya K, Izawa Y, Kanematsu Y, Tamaki T: Atheroprotective effects of antioxidants through inhibition of mitogen-activated protein kinases. Acta Pharmacol Sin 2004;25:977–985.

6 Ceaser EK, Moellering DR, Shiva S, Ramachandran A, Landar A, Venkartraman A, Crawford J, Patel R, Dickinson DA, Ulasova E, Ji S, Darley-Usmar VM: Mechanisms of signal transduction mediated by oxidized lipids: The role of the electrophile-responsive proteome. Biochem Soc Trans 2004;32:151–155.

7 Davignon J, Ganz P: Role of endothelial dysfunction in atherosclerosis. Circulation 2004;109: III27–32.

8 Wautier JL, Schmidt AM: Protein glycation: A firm link to endothelial cell dysfunction. Circ Res 2004;95:233–238.

9 Francois M, Kojda G: Effect of hypercholesterolemia and of oxidative stress on the nitric oxide-cGMP pathway. Neurochem Int 2004;45:955–961.

10 Boullier A, Bird DA, Chang MK, Dennis EA, Friedman P, Gillotre-Taylor K, Horkko S, Palinski W, Quehenberger O, Shaw P, Steinberg D, Terpstra V, Witztum JL: Scavenger receptors, oxidized LDL, and atherosclerosis. Ann NY Acad Sci 2001;947:214–222; discussion 222–223.

11 Galli F, Canestrari F, Buoncristiani U: Biological effects of oxidant stress in haemodialysis: The possible roles of vitamin E. Blood Purif 1999;17:79–94.

12 Galli F, Ghibelli L, Buoncristiani U, Bordoni V, D'Intini V, Benedetti S, Canestrari F, Ronco C, Floridi A: Mononuclear leukocyte apoptosis in haemodialysis patients: The role of cell thiols and vitamin E. Nephrol Dial Transplant 2003;18:1592–1600.

13 Colussi C, Albertini MC, Coppola S, Rovidati S, Galli F, Ghibelli L: H_2O_2-induced block of glycolysis as an active ADP-ribosylation reaction protecting cells from apoptosis. FASEB J 2000;14:2266–2276.

14 Gopalakrishna R, Jaken S: Protein kinase C signaling and oxidative stress. Free Radic Biol Med 2000;28:1349–1361.

15 Curtis TM, Scholfield CN: The role of lipids and protein kinase Cs in the pathogenesis of diabetic retinopathy. Diabetes Metab Res Rev 2004;20:28–43.

16 Chen K, Keaney J: Reactive oxygen species-mediated signal transduction in the endothelium. Endothelium 2004;11:109–121.

17 Zoccali C, Mallamaci F, Tripepi G: Novel cardiovascular risk factors in end-stage renal disease. J Am Soc Nephrol 2004;15(suppl 1):S77–S80.

18 Butterfield DA, Castegna A, Lauderback CM, Drake J: Evidence that amyloid beta-peptide-induced lipid peroxidation and its sequelae in Alzheimer's disease brain contribute to neuronal death. Neurobiol Aging 2002;23:655–664.

19 Gackowski D, Kowalewski J, Siomek A, Olinski R: Oxidative DNA damage and antioxidant vitamin level: Comparison among lung cancer patients, healthy smokers and nonsmokers. Int J Cancer 2004.

20 Stocker R, Keaney JF Jr: Role of oxidative modifications in atherosclerosis. Physiol Rev 2004;84:1381–1478.

21 Napoli C, Palinski W: Neurodegenerative diseases: Insights into pathogenic mechanisms from atherosclerosis. Neurobiol Aging 2005;26:293–302.

22 Napoli C, Quehenberger O, De Nigris F, Abete P, Glass CK, Palinski W: Mildly oxidized low density lipoprotein activates multiple apoptotic signaling pathways in human coronary cells. FASEB J 2000;14:1996–2007.

23 Wratten ML, Tetta C, Ursini F, Sevanian A: Oxidant stress in hemodialysis: Prevention and treatment strategies. Kidney Int Suppl 2000;76:S126–S132.

24 Davi G, Falco A, Patrono C: Determinants of F_2-isoprostane biosynthesis and inhibition in man. Chem Phys Lipids 2004;128:149–163.

25 Jenkins AJ, Rowley KG, Lyons TJ, Best JD, Hill MA, Klein RL: Lipoproteins and diabetic microvascular complications. Curr Pharm Des 2004;10:3395–3418.

26 Quaschning T, Krane V, Metzger T, Wanner C: Abnormalities in uremic lipoprotein metabolism and its impact on cardiovascular disease. Am J Kidney Dis 2001;38:S14–S19.

27 Matsuoka H: Endothelial dysfunction associated with oxidative stress in human. Diabetes Res Clin Pract 2001;54(suppl 2):S65–S72.

28 Torzewski M, Shaw PX, Han KR, Shortal B, Lackner KJ, Witztum JL, Palinski W, Tsimikas S: Reduced in vivo aortic uptake of radiolabeled oxidation-specific antibodies reflects changes in plaque composition consistent with plaque stabilization. Arterioscler Thromb Vasc Biol 2004;24:2307–2312.

29   Nakamura T, Kawagoe Y, Matsuda T, Takahashi Y, Sekizuka K, Ebihara I, Koide H: Effects of LDL apheresis and vitamin E-modified membrane on carotid atherosclerosis in hemodialyzed patients with arteriosclerosis obliterans. Kidney Blood Press Res 2003;26:185–191.

30   Miyazaki H, Matsuoka H, Itabe H, Usui M, Ueda S, Okuda S, Imaizumi T: Hemodialysis impairs endothelial function via oxidative stress: Effects of vitamin E-coated dialyzer. Circulation 2000;101:1002–1006.

31   Palinski W, Napoli C: Unraveling pleiotropic effects of statins on plaque rupture. Arterioscler Thromb Vasc Biol 2002;22:1745–1750.

32   Steptoe A, Perkins-Porras L, Hilton S, Rink E, Cappuccio FP: Quality of life and self-rated health in relation to changes in fruit and vegetable intake and in plasma vitamins C and E in a randomised trial of behavioural and nutritional education counselling. Br J Nutr 2004;92:177–184.

33   Willcox JK, Ash SL, Catignani GL: Antioxidants and prevention of chronic disease. Crit Rev Food Sci Nutr 2004;44:275–295.

34   Boaz M, Smetana S, Weinstein T, Matas Z, Gafter U, Iaina A, Knecht A, Weissgarten Y, Brunner D, Fainaru M, Green MS: Secondary prevention with antioxidants of cardiovascular disease in endstage renal disease (SPACE): Randomised placebo-controlled trial. Lancet 2000;356: 1213–1218.

35   Voko Z, Hollander M, Hofman A, Koudstaal PJ, Breteler MM: Dietary antioxidants and the risk of ischemic stroke: The Rotterdam Study. Neurology 2003;61:1273–1275.

36   Lipinski B: Markers of oxidative stress in uremia. Kidney Int 2004;65:339–340; author reply 340.

37   Tarng DC, Wen Chen T, Huang TP, Chen CL, Liu TY, Wei YH: Increased oxidative damage to peripheral blood leukocyte DNA in chronic peritoneal dialysis patients. J Am Soc Nephrol 2002;13:1321–1330.

38   Kalousova M, Zima T, Tesar V, Sulkova S, Fialova L: Relationship between advanced glycoxidation end products, inflammatory markers/acute-phase reactants, and some autoantibodies in chronic hemodialysis patients. Kidney Int Suppl 2003;S62–S64.

39   Panichi V, Migliori M, De Pietro S, Taccola D, Bianchi AM, Norpoth M, Giovannini L, Palla R, Tetta C: C-reactive protein as a marker of chronic inflammation in uremic patients. Blood Purif 2000;18:183–190.

40   Descamps-Latscha B, Witko-Sarsat V: Importance of oxidatively modified proteins in chronic renal failure. Kidney Int Suppl 2001;78:S108–S113.

41   Ikizler TA, Morrow JD, Roberts LJ, Evanson JA, Becker B, Hakim RM, Shyr Y, Himmelfarb J: Plasma F_2-isoprostane levels are elevated in chronic hemodialysis patients. Clin Nephrol 2002;58: 190–197.

42   Hörl WH: Hemodialysis membranes: Interleukins, biocompatibility, and middle molecules. J Am Soc Nephrol 2002;13(suppl 1):S62–S71.

43   D'Intini V, Bordoni V, Fortunato A, Galloni E, Carta M, Galli F, Bolgan I, Inguaggiato P, Poulin S, Bonello M, Tetta C, Levin N, Ronco C: Longitudinal study of apoptosis in chronic uremic patients. Semin Dial 2003;16:467–473.

44   Canestrari F, Galli F, Giorgini A, Albertini MC, Galiotta P, Pascucci M, Bossu M: Erythrocyte redox state in uremic anemia: Effects of hemodialysis and relevance of glutathione metabolism. Acta Haematol 1994;91:187–193.

45   Galli F, Canestrari F, Bellomo G: Pathophysiology of the oxidative stress and its implication in uremia and dialysis. Contrib Nephrol 1999;127:1–31.

46   Libetta C, Zucchi M, Gori E, Sepe V, Galli F, Meloni F, Milanesi F, Canton AD: Vitamin E-loaded dialyzer resets PBMC-operated cytokine network in dialysis patients. Kidney Int 2004;65: 1473–1481.

47   Galli F: Special issue: Amino acid and protein modifications by oxygen and nitrogen species. Amino Acids 2003;25:205.

48   Galli F, Benedetti S, Floridi A, Canestrari F, Piroddi M, Buoncristiani E, Buoncristiani U: Glycoxidation and inflammatory markers in patients on treatment with PMMA-based protein-leaking dialyzers. Kidney Int 2005;67:750–759.

49   Maccarrone M, Taccone-Gallucci M, Finazzi-Agro A: 5-Lipoxygenase-mediated mitochondrial damage and apoptosis of mononuclear cells in ESRD patients. Kidney Int Suppl 2003;S33–S36.

50 Rosensen RS: Statins in atherosclerosis: Lipid-lowering agents with antioxidant capabilities. Atherosclerosis 2004;173:1–12.

51 Bertrand ME: Provision of cardiovascular protection by ACE inhibitors: A review of recent trials. Curr Med Res Opin 2004;20:1559–1569.

52 Usberti M, Gerardi G, Bufano G, Tira P, Micheli A, Albertini A, Floridi A, Di Lorenzo D, Galli F: Effects of erythropoietin and vitamin E-modified membrane on plasma oxidative stress markers and anemia of hemodialyzed patients. Am J Kidney Dis 2002;40:590–599.

53 Grune T, Sommerburg O, Siems WG: Oxidative stress in anemia. Clin Nephrol 2000;53:S18–S22.

54 Sato M, Matsumoto Y, Morita H, Takemura H, Shimoi K, Amano I: Effects of vitamin supplementation on microcirculatory disturbance in hemodialysis patients without peripheral arterial disease. Clin Nephrol 2003;60:28–34.

55 Morena M, Cristol JP, Canaud B: Why hemodialysis patients are in a prooxidant state? What could be done to correct the pro/antioxidant imbalance. Blood Purif 2000;18:191–199.

56 Williamson KS, Gabbita SP, Mou S, West M, Pye QN, Markesbery WR, Cooney RV, Grammas P, Reimann-Philipp U, Floyd RA, Hensley K: The nitration product 5-nitro-gamma-tocopherol is increased in the Alzheimer brain. Nitric Oxide 2002;6:221–227.

57 Behl C: Vitamin E and other antioxidants in neuroprotection. Int J Vitam Nutr Res 1999;69: 213–219.

58 Fang YZ, Yang S, Wu G: Free radicals, antioxidants, and nutrition. Nutrition 2002;18:872–879.

59 Calabrese V, Scapagnini G, Colombrita C, Ravagna A, Pennisi G, Giuffrida Stella AM, Galli F, Butterfield DA: Redox regulation of heat shock protein expression in aging and neurodegenerative disorders associated with oxidative stress: A nutritional approach. Amino Acids 2003;25:437–444.

60 Zhang WR, Hayashi T, Sasaki C, Sato K, Nagano I, Manabe Y, Abe K: Attenuation of oxidative DNA damage with a novel antioxidant EPC-K1 in rat brain neuronal cells after transient middle cerebral artery occlusion. Neurol Res 2001;23:676–680.

61 Becker LB: New concepts in reactive oxygen species and cardiovascular reperfusion physiology. Cardiovasc Res 2004;61:461–470.

62 Wainwright CL: Matrix metalloproteinases, oxidative stress and the acute response to acute myocardial ischaemia and reperfusion. Curr Opin Pharmacol 2004;4:132–138.

Francesco Galli, PhD
University of Perugia, Department of Internal Medicine
Section of Applied Biochemistry and Nutritional Sciences
Via del Giochetto, IT–06126 Perugia (Italy)
Tel. +39 075 585 7445, Fax +39 075 585 7441, E-Mail f.galli@unipg.it

Ronco C, Brendolan A, Levin NW (eds): Cardiovascular Disorders in Hemodialysis.
Contrib Nephrol. Basel, Karger, 2005, vol 149, pp 261–271

....

Calcium, Phosphorus and Vitamin D Disorders in Uremia

Eduardo Slatopolsky, Alex Brown, Adriana Dusso

Washington University School of Medicine, Department of Medicine,
Renal Division, St. Louis, Mo., USA

Abstract

Background: Alterations in calcium, phosphate (P) and vitamin D metabolism play a
critical role in the development of secondary hyperparathyroidism (SH), parathyroid hyperplasia and soft tissue and vascular calcification. **Methodology:** Studies were performed in
uremic dogs and rats fed a low and high P diet over a period of 1–4 months. In addition, in vitro
studies were performed in normal parathyroid glands incubated in culture media containing
0.2 mM P (low) or 2.0 mM P (high). **Results:** Uremic rats maintained on a low P diet did not
develop SH or parathyroid hyperplasia. There was an enhancement of p21, the suppressor of
the cell cycle, in these parathyroid glands. Opposite results were obtained using a high P diet.
There was an enhancement of transforming growth factor-α and epidermal growth factor
receptor, known enhancers of cell proliferation. In vitro studies demonstrated the direct effect
of P on parathyroid hormone secretion. **Conclusions:** Early dietary P restriction prevents
the development of SH and parathyroid hyperplasia. If dietary P restriction is applied to
rats with established SH, there is a significant amelioration of SH and parathyroid hyperplasia.
In addition, control of serum P in uremic patients is crucial in the prevention of vascular
calcification.

Introduction

Secondary hyperparathyroidism (SH) and hyperplasia of the parathyroid
glands are universal complications in patients with chronic kidney disease (CKD).
Abnormal calcium, phosphorus and vitamin D metabolism in chronic renal
failure play a key role in the development of SH and renal osteodystrophy. In
addition, hyperphosphatemia is a major factor in the pathogenesis of vascular
calcification.

The Role of Phosphate Retention

Dietary phosphate content influences parathyroid hormone (PTH) synthesis and secretion. In renal failure, high dietary phosphate worsens SH, whereas phosphate restriction totally prevents any enhancement of parathyroid function [1, 2]. Although hyperphosphatemia promotes calcification in soft tissues and the vascular tree, the mechanisms by which phosphate retention affects serum ionized calcium (ICa) are not completely understood. In renal failure, hyperphosphatemia decreases the release of calcium from bone and further inhibits the already reduced $1,25(OH)_2D_3$ [1,25D] synthesis, worsening 1,25D deficiency. Low serum 1,25D reduces both intestinal calcium absorption and bone calcium mobilization, thus, decreasing serum ICa concentrations. In contrast, hypophosphatemia enhances renal 1α-hydroxylase activity, increasing serum 1,25D and consequently serum calcium. Numerous studies in rat and dog models of renal failure have shown that phosphate restriction prevents the development of SH [1–3]. In studies in patients with moderate renal insufficiency, phosphate restriction increased plasma levels of 1,25D with a concomitant normalization of PTH levels [4]. In severe renal insufficiency in patients and experimental animals, however, the marked reduction in renal mass precludes low phosphate induction of hydroxylase activity. Nevertheless, low phosphate per se, independent of changes in either 1,25D or ICa, can ameliorate SH [5–6]. Direct control of PTH release by phosphate can be shown in vitro when intact parathyroid glands or bovine parathyroid tissue slices are exposed to different concentrations of phosphate in the medium [3, 7, 8]. Different from the inhibitory effect of 1,25D on PTH gene transcription, phosphate control of PTH synthesis appears to involve post-transcriptional mechanisms. Moallem et al. [9] showed the existence of proteins in the parathyroid gland that bind the 3′ untranslated region of PTH mRNA preventing transcript degradation. The binding of these proteins to PTH mRNA is increased by hypocalcemia and decreased by hypophosphatemia, suggesting that phosphate restriction may reduce serum PTH by decreasing the stability of PTH mRNA. Conversely, high serum phosphate, likely through induction of hypocalcemia, may prolong the half-life of PTH mRNA and therefore PTH synthesis.

Phosphate Regulation of Parathyroid Hyperplasia

In the last decade, advances in the understanding of the mechanisms responsible for parathyroid hyperplasia and renal failure, as well as those involved in the regulation of uremia-induced parathyroid growth by calcium, phosphate or 1,25D, has been clarified by numerous investigators. Under normal circumstances, parathyroid cells rarely divide but preserve the potential to proliferate in response to mitogenic stimuli. In renal failure, high phosphate, low calcium and vitamin D deficiency markedly increase the number of parathyroid cells which stain positively for proliferating cell nuclear antigen, a marker of mitotic activity [10, 11]. In contrast, dietary phosphate restriction totally prevents, whereas vitamin D administration reduces, uremia-induced parathyroid cell growth. In studying the effect of phosphate on parathyroid growth in 5/6 nephrectomized rats, our laboratory showed that 60–80% of parathyroid growth induced by a high phosphate diet in the month following nephrectomy occurred between days 3–5 [1, 2, 12]. Parallel increases in parathyroid gland weight, protein and DNA content showed hyperplasia as the main contributor to parathyroid gland enlargement [3]. After mitogenic stimulus, such as high dietary phosphate, the commitment of a parathyroid cell to either remain quiescent or divide depends on the net balance between two opposing forces. The mitogenic force responsible for the progression through the cell cycle and completion of mitosis is determined by the levels and/or activity of specific complexes of cyclin and cyclin-dependent kinases (CDK). The antimitogenic force to arrest growth depends on the levels of CDK inhibitors, which regulate the activity of the cyclin-CDK complexes. Studies in vivo showed potential roles for the promoter of growth, transforming growth factor-α (TGF-α) and the CDK inhibitor p21 in dietary phosphate regulation of parathyroid cell growth in early renal failure [13]. High phosphate induces parathyroid TGF-α expression which functions as an autocrine signal to further stimulate uremia-induced parathyroid cell growth. Conversely, the antimitogenic effect of phosphate restriction involves a specific induction of p21 expression in the parathyroid glands. This occurs independently and changes in serum 1,25D levels and counteracts early mitogenic signals triggered by uremia, thus preventing parathyroid gland growth. On the other hand, in intestine and liver, dietary phosphate content has no effect on either growth or p21 and TGF-α expression [13]. It is still unclear how the parathyroid gland senses inorganic phosphorus. Although the phosphate cotransporter PiT-1 has been cloned from parathyroid glands [14], at the present time there is no link between expression of the PiT-1 cotransporter and regulation of parathyroid function. A phosphate receptor or phosphate-sensing molecule in the parathyroid gland is another hypothetical candidate, but at this time there is only indirect evidence of its existence. Further investigations are necessary to prove the existence of such a phosphate receptor in the parathyroid glands.

Control of Serum Phosphate

Preventing hyperphosphatemia is mandatory for controlling parathyroid hyperplasia, SH and bone disease in patients with advanced renal failure.

In addition, it is well known that phosphate plays a critical role in the development of vascular calcification. Furthermore, hyperphosphatemia-mediated calcification may not be a simple passive process of deposition of calcium-phosphate crystal in vascular walls. The work of several investigators suggest that calcification may be an active process under specific genetic and molecular control [15]. Recent in vitro work has demonstrated an active process in which phosphate enters the cells via a sodium-dependent phosphate-cotransporter-mediated mechanism, PiT-1, [16, 17] inducing the expression of the 'master gene', Cfa-1, and setting into place the active deposition of calcium into vascular walls. The end result is a change from a vascular cell type to an osteoblastic-like cell. Calcification is an inherent part of atherosclerosis and is the most frequent cause of cardiovascular disease in patients with CKD. Furthermore, recent studies using electron beam computer tomography showed a progressive increase in the amount of calcium in the coronary arteries in mitral and aortic valves of patients with chronic renal failure. Clearly, vascular calcification, metastatic calcifications, and calciphylaxis also contribute to increased morbidity and mortality in these patients. Dietary phosphate and an excess calcium load in the diet play a key role in these pathological manifestations. Although 95% of dialysis patients use phosphate binders to reduce dietary phosphate absorption, more than half of these patients do not achieve good control of phosphate or calcium-phosphate (Ca × P) product. The most commonly used phosphate binders contain aluminum salts, calcium carbonate or calcium acetate. Calcium contained in phosphate binders can lead to hypercalcemia, thus worsening soft tissue calcification and vascular disease, especially in patients on vitamin D therapy. It is important to emphasize that patients receiving large amounts of calcium as a phosphate binder may not develop hypercalcemia; however, the increase in the 'calcium load' per se plays an important pathogenetic mechanism in the development of vascular calcification. Aluminum salts were used in the 60 s and 70 s. However, due to aluminum accumulation resulting in significant morbidity and mortality in numerous patients, their use has been reduced significantly at present. To avoid the deleterious side effects of aluminum or calcium carbonate or acetate, new phosphate binders have been developed. One of them, an aluminum- and calcium-free phosphate binder, sevelamer, is a hydrogel of cross-linked polyallylamin that is resistant to digestive degradation and not absorbed in the gastrointestinal tract. Studies have shown that this agent can effectively and safely lower serum phosphate without changing serum calcium. Long-term studies have shown a decrease in low-density lipoprotein and in some patients, also an increase in high-density lipoprotein cholesterol [18]. The mechanisms may be similar to those of cholesteramine, which bind bile salts. Chertow et al. [19] in a multicenter study in the United States and Europe compared the fate of sevelamer and calcium carbonate or calcium acetate on the cardiovascular system. One group of patients received sevelamer and the other received calcium salts (calcium carbonate or calcium acetate) for a period of 52 weeks. The calcium content of the coronary artery and aorta was assessed by electron beam computed tomography. Both calcium salts and sevelamer controlled the Ca × P product but patients receiving calcium salts became hypercalcemic more frequently (16 vs. 5% in the sevelamer group). More importantly, at the completion of the study, electron beam computed tomography demonstrated that the increase in the mean calcium score in the coronary artery and aorta was greater in the subjects treated with calcium than in those treated with sevelamer [19]. In addition, C-reactive protein decreased in the patients ingesting sevelamer and increased in the patients ingesting calcium salts. Since cardiovascular mortality in dialysis patients is approximately 50–60%, alterations in mineral metabolism are critical as are inflammatory processes, hypertension and alterations in lipid metabolism. The control of phosphorus is crucial since an increased Ca × P product not only increases soft tissue calcification, but phosphorus per se increases the expression of the transcriptional factor, Cbfa-1 [16, 17].

Lanthanum carbonate is a trivalent cation that binds phosphate at all pHs to form lanthanum phosphate, which is insoluble. Hutchinson [20] demonstrated that the phosphate-binding capacity of lanthanum is similar to that of aluminum in vitro. Studies in patients maintained on hemodialysis or continuous ambulatory peritoneal dialysis demonstrated that lanthanum carbonate can reduce serum phosphate to approximately 5.0 mg/dl. Experimental work in uremic rats, however, demonstrated a significant accumulation of lanthanum in the liver [21]. Thus, further long-term studies are necessary in dialysis patients to determine the potential toxic effect of lanthanum accumulation, since a small amount of lanthanum is absorbed in the gastrointestinal tract.

The Role of Calcium

The major factor involved in the regulation of PTH secretion is the concentration of ICa in the extracellular fluid. A variety of experimental studies in vivo have demonstrated an inverse correlation between extracellular calcium ion concentrations and hormone secretion. It is important to note, however, that hormone secretion is not completely suppressed during hypercalcemia, and a basal rate of hormone secretion persists. The initial secretory response to hypocalcemia occurs within seconds, suggesting that calcium acts directly on the plasma membrane. The mechanism may relate to an effect of calcium on the parathyroid cell membrane calcium receptor.

Brown et al. [22] cloned a calcium receptor localized in bovine parathyroid cell membranes. Parathyroid cells possess an extracellular calcium-sensing mechanism that also recognizes trivalent and polyvalent cations (such as neomycin) and regulates PTH secretion by changes in phosphoinositide turnover and cytosolic calcium. This receptor features a large, extracellular domain containing clusters of acidic amino acids that are possibly involved in calcium binding. The extracellular domain is coupled to a cellular membrane-spanning domain similar to the G-protein coupled receptor superfamily. Pollack et al. [23] demonstrated that mutation X in the human calcium-sensing receptor causes familial hypocalciuric hypercalcemia and severe neonatal hyperparathyroidism.

Regulation of the expression of the calcium sensor could have major physiological and pathological implications. Studies in vivo in rat parathyroid glands and kidney and in vitro in cultured bovine parathyroid cells have indicated that this receptor is not regulated by extracellular calcium [24]. Vitamin D deficiency has been shown to decrease and 1,25D to increase calcium sensor mRNA in the rat, suggesting an additional level of control by 1,25D [25].

Calcium has also been demonstrated to control PTH mRNA levels in vivo and in vitro. In the rat, Naveh-Many et al. [25] showed that hypocalcemia increases PTH mRNA, whereas hypercalcemia has no effect. Okazaki et al. [26], using baby hamster kidney cells, identified a negative regulatory element, located 3.5 kb upstream from the transcriptional start site of the PTH gene that modulates transcriptional suppression by extracellular calcium when fused to a chloramphenicol acetyltransferase reporter. This negative regulatory element consists of a putative palindromic structure formed from two 6-base-pair (bp) elements separated by 3 bp and is similar to the negative calcium-responsive elements in the atrial natriuretic peptide and renin genes. The relevance of this calcium response element in the PTH gene will require further confirmation in a parathyroid cell line. While considerable information has been obtained regarding the control of PTH secretion from normal parathyroid tissue, the control of PTH secretion in abnormal parathyroid glands is not well defined. The observation that an elevated calcium concentration does not completely suppress PTH secretion in normal parathyroid tissue and that hypercalcemia can be induced by the transplantation of multiple normal parathyroid glands suggests that the persistent basal secretion of PTH may be of physiological significance when parathyroid mass is increased. This phenomenon may contribute to the apparent nonsuppressibiltiy of PTH secretion in hyperparathyroidism. Alternatively, it is also possible that there is an intrinsic defect in the response of abnormal parathyroid tissue to calcium as has been demonstrated in parathyroid tissue from patients with primary hyperparathyroidism. Since parathyroid hyperplasia is a universal phenomenon in chronic renal failure, it is important to consider the possibility that the response of parathyroid tissue to calcium in patients with chronic renal failure might also be abnormal. Bellorin-Font et al. [27] demonstrated that parathyroid adenylate cyclase kinetics were abnormal in hyperplasic parathyroid glands from patients with chronic renal failure, but similar to that seen in parathyroid adenomas. The adenylate cyclase of pathological parathyroid tissue has an increased affinity for magnesium and a reduced sensitivity to inhibition by calcium. The altered affinity for magnesium could be corrected by guanosine triphosphate in vitro, suggesting the possibility that the abnormal regulation of adenylate cyclase was at or closely related to the guanyl nucleotide regulator site of adenylate cyclase.

An additional effect of calcium on PTH secretion results from calcium-dependent regulation of hormone degradation within the parathyroid gland, which can alter the amount of hormone available for secretion. The second response occurs when newly synthesized PTH is available for secretion after a prolonged decrease in extracellular ICa (several hours). The third response is the increase in parathyroid cell growth and number in response to hypocalcemia, resulting in elevation in PTH secretion in several days.

Brown et al. [28] demonstrated altered calcium-regulated PTH release in isolated parathyroid cells from uremic patients. These hyperplasic glands displayed an increased 'set-point' for calcium suppression of PTH release.

Abnormal Regulation of PTH by Calcium in Renal Failure

Hyperplasic parathyroid glands display less sensitivity to calcium than does normal tissue [28]. This observation suggests that a mechanism for the increased PTH levels seen in chronic renal failure may be a shift in the set-point for calcium-regulated PTH secretion as well as an increase in parathyroid tissue mass. The shift in the set-point for calcium-regulated PTH secretion is also manifested by an increase in the calcium concentration required for inhibition of adenylate cyclase activity in membranes prepared from hyperplasic parathyroid glands obtained from patients with chronic renal insufficiency. Not only is the set-point for calcium-regulated hormone secretion elevated in cells from hyperplasic parathyroid tissue, but also the degree of responsiveness across the calcium-sensitive range is altered.

Several factors can thus lead to elevated serum PTH levels in man. These include: (i) an increase in tissue mass either from parathyroid cell hypertrophy and/or hyperplasia; (ii) an increase in the set-point for calcium needed to inhibit hormone secretion; and (iii) a change in the degree of suppression by calcium throughout the calcium-sensitivity range (e.g. slope of the suppression line). Kifor et al. [29], using immunohistochemistry, demonstrated a 59% decrease in the calcium-sensing receptor in the parathyroid glands of uremic patients. Similar results were demonstrated in parathyroid adenomas from patients with primary hyperparathyroidism. These investigators concluded that the degree of calcium-sensing receptor reduction would be sufficient to account for at least in part, the altered sensitivity to calcium that is observed in SH.

The Role of Vitamin D

In stage 3 of CKD, impaired renal production of 1,25D, the hormonal form of vitamin D, is a major contributor to the generation and maintenance of SH and parathyroid hyperplasia. 1,25D directly represses both parathyroid cell proliferation and PTH synthesis. Thus, its deficiency in renal failure causes high serum PTH levels and parathyroid gland enlargement. 1,25D deficiency also indirectly leads to SH by decreasing intestinal calcium absorption. The resulting hypocalcemia is the most potential stimulus for the rapid increase in PTH synthesis and secretion directed toward normalizing serum calcium. The efficacy of 1,25D in inhibiting parathyroid cell growth and PTH synthesis renders calcitriol replacement therapy a valuable tool in the treatment of hyperparathyroidism. In CKD, serum 1,25D decreases as renal function deteriorates, starting stage 3. Since 1,25D increases vitamin D receptor (VDR) messenger RNA levels and protein stability, the latter by preventing VDR degradation, a low serum 1,25D level in itself is partially responsible for the reduced parathyroid VDR content. Our laboratory has shown a strong direct correlation between serum 1,25D and parathyroid VDR content in 5/6 nephrectomized rats [30]. More importantly, prophylactic administration of either calcitriol or its analog (22-oxacalcitriol) prevents the marked reduction in parathyroid VDR content induced by renal failure [30]. Thus, in kidney disease, the 1,25D/VDR complex, the first step in 1,25D action, results from the combination of a decrease in both renal 1,25D synthesis and parathyroid VDR content. Therefore, early therapeutic intervention with oral 1,25D or one of its analogs in stage 3 of CKD would delay the onset of vitamin D resistance by preventing 1,25D deficiency and its critical consequence, the reduction in VDR content. Several laboratories have identified factors contributing to impaired heterodemineralization and VDR/retinoid X receptor binding to the vitamin D responsive element including (1) reduced retinoid X receptor, (2) accumulation of uremic toxins, (3) increased parathyroid calreticulin and (4) activation of a VDR/unrelated pathway, which might interfere with 1,25D signaling.

Little is known at present as to how renal failure affects transactivation/transrepression, the most critical step in the 1,25D-VDR-mediated regulation of gene expression. The numerous protein DNA interactions that occur upon the binding of the VDR/retinoid X receptor with the vitamin D responsive element and nuclear coregulator molecules suggest that in uremia, vitamin D resistance could also result from the increased expression of essential coactivator or corepressor molecules or from the defective recruitment of these molecules by the VDR. Metabolic acidosis, during the course of renal insufficiency, could also decrease the levels of 1,25D and prevent the appropriate increase in 1,25D levels seen with higher levels of PTH. An additional mechanism that could limit the production of 1,25D in CKD could be decreased delivery of the 1,25D precursor, 25-hydroxy vitamin D, which is bound to a circulating vitamin D-binding protein, to proximal tubule uptake mechanisms. This has been shown to involve megalin, a molecule required for uptake of 25-hydroxy vitamin D-bound vitamin D-binding protein into the cells that facilitate the delivery of the precursor to the 1-hydroxylase.

Integrated Management of Renal Osteodystrophy

It would seem that an increase in parathyroid activity begins in the early stage of CKD when the GFR is only slightly decreased. When PTH levels are elevated, it is reasonable to evaluate vitamin D status by measuring 25-hydroxy vitamin D levels. If the serum levels are less than 30 ng/ml, vitamin D_2 supplementation should be provided at a dose of roughly 2,000 units per day or 50,000 units twice per month. When the levels of 25-hydroxy vitamin D are adequate, dietary phosphate restriction should be initiated and the resultant effect of this restriction on PTH levels monitored. This is done by prescribing phosphate binders to be taken with meals. Initially, calcium salts, e.g. 1.5 g of elemental calcium per day, can be used since this calcium load can be handled by the kidney. If larger doses are required, however, consideration should be given to non-calcium-containing phosphate binders. Currently sevelamer is an excellent calcium-free phosphate binder. If acidosis is present, the patient should be treated with sodium bicarbonate. Also in ESRD patients (CKD, stage 5) intact serum PTH should be maintained between 150 and 300 pg/ml. Values below 150 pg/ml may predispose the patient to the development of adynamic bone disease. On the other hand, values above 300 pg/ml may promote the development of osteitis fibrosa. If calcitriol is required, intermittent administration of intravenous calcitriol at a dose of 1–4 μg/dialysis should be instituted. The higher the level of PTH, the greater the amount of calcitriol that should be given to the patients. If an analog like paricalcitol (Zemplar) is utilized, the initial dose should be calculated by dividing the PTH level by 100, e.g. if the PTH is 1,000 pg/ml the initial dose of paricalcitol should be 10 μg/dialysis. The amount can then be adjusted up or down to achieve a PTH level between 150 and 300 pg/ml. Once this is achieved, the dose should be substantially reduced since further suppression will only induce adynamic bone disease and severe hypercalcemia in these patients. If patients are taking aluminum-based phosphate binders, serum aluminum levels should be monitored. Aluminum binders, if prescribed, should be reduced to a minimum and given for as short a period as possible. Calcium citrate should be avoided since citrate greatly increased the absorption of aluminum. If aluminum is found in bone, chelation should be induced by administration of desferrioxamine. To prevent side effects, patients should not receive more than 500 mg/week of desferrioxamine. After 4–6 months of treatment, serum aluminum should be measured before and after the administration of desferrioxamine (5 mg/kg). The combination of an increment in serum aluminum of greater than 50 μg/l and an intact PTH level of less than 150 pg/ml implies the greatest risk of aluminum bone disease.

References

1   Slatopolsky E, Caglar S, Gradowska L, Canterbury JM, Reiss E, Bricker N: On the prevention of secondary hyperparathyroidism in experimental chronic renal disease using a 'proportional reduction' of dietary phosphorus intake. Kidney Int 1972;2:147–151.
2   Rutherford WE, Bordier P, Marie P, Hruska K, Harter H, Greenwalt A, Blondin J, Haddad J, Bricker N, Slatopolsky E: Phosphate control and 25-hydroxy-cholecalciferol administration in preventing experimental renal osteodystrophy in the dog. J Clin Invest 1977;60:332–341.
3   Slatopolsky E, Finch J, Denda M, Ritter C, Zhong M, Dusso A, Macdonald P, Brown AJ: Phosphorus restriction prevents parathyroid cell growth. High phosphorus directly stimulates PTH secretion in vitro. J Clin Invest 1996;97:2534–2540.
4   Portale AA, Booth BE, Halloran BP, Morris RC Jr: Effects of dietary phosphorus on circulating concentrations of 1,25 dihydroxyvitamin D and immunoreactive parathyroid hormone in children with moderate renal insufficiency. J Clin Invest 1984;73:1580–1589.
5   Lucas PA, Brown RC, Woodhead JS, Coles GA: 1,25 Dihydroxycholecalciferol and parathyroid hormone in advanced chronic renal failure: Effect of simultaneous protein and phosphorus restriction. Clin Nephrol 1986;25:7–10.
6   Lopez-Hilker S, Dusso A, Rapp N, Martin K, Slatopolsky E: Phosphorus restriction reverses hyperparathyroidism in uremia independent of changes in calcium and calcitriol. Am J Physiol 1990;259:F432–F437.
7   Almaden Y, Canalejo A, Hernandez A, Ballesteros E, Garcia-Navarro S, Torres A, Rodriguez M: Direct effect of phosphorus on PTH secretion from whole rat parathyroid glands in vitro. J Bone Miner Res 1996;11:970–976.
8   Nielsen PK, Feldt-Rasmussen U, Olgaard K: A direct effect in vitro of phosphate on PTH release from bovine parathyroid tissue slices but not from dispersed parathyroid cells. Nephrol Dial Transplant 1996;11:1762–1768.
9   Moallem E, Kilav R, Silver J, Naveh-Many T: RNA-protein binding and post-transcriptional regulation of parathyroid hormone gene expression by calcium and phosphate. J Biol Chem 1998;273: 5253–5259.
10 Naveh-Many T, Rahamimov R, Livni N, Silver J: Parathyroid cell proliferation in normal and chronic renal failure rats. The effects of calcium, phosphate and vitamin D. J Clin Invest 1995;96: 1786–1793.
11 Yi H, Fukagawa M, Yamato H, Kumagai M, Watanabe T, Kurokawa K: Prevention of enhanced parathyroid hormone secretion, synthesis and hyperplasia by mild dietary phosphorus restriction in early chronic renal failure in rats: Possible direct role of phosphorus. Nephron 1995;70:242–248.
12 Denda M, Finch J, Slatopolsky E: Phosphorus accelerates the development of parathyroid hyperplasia and secondary hyperparathyroidism in rats with renal failure. Am J Kidney Dis 1996;28: 596–602.

13   Dusso A, Lu Y, Pavlopoulos T, Naumovich L, Lu Y, Finch J, Brown AJ, Morrissey J. Slatopolsky E: p21 (WAF1) and transforming growth factor-alpha mediate dietary phosphate regulation of parathyroid cell growth. Kidney Int 2001;59:855–865.

14   Tatsumi S, Segawa H, Morita K, Haga H, Kouda T, Yamamoto H, Inoue Y, Nii T, Katai K, Taketani Y, Miyamoto KI, Takeda E: Molecular cloning and hormonal regulation of PiT-1, a sodium-dependent phosphate cotransporter from rat parathyroid glands. Endocrinology 1998;139:1692–1699.

15   Cozzolino M, Dusso AS, Slatopolsky E: Role of calcium-phosphate product and bone-associated proteins on vascular calcification in renal failure. J Am Soc Nephrol 2001;12:2511–2516.

16   Jono S, McKee MD, Murry CE, Shioi A, Nishizawa Y, Mori K, Mori H, Giachelli CM: Phosphate regulation of vascular smooth muscle cell calcification. Circ Res 2000;87:E10–E17.

17   Giachelli CM: Vascular calcification: In vitro evidence for the role of inorganic phosphate. J Am Soc Nephrol 2003;14:S300–S304.

18   Chertow GM, Burke SK, Dillon MA, Slatopolsky E: Long-term effects of sevelamer hydrochloride on the calcium × phosphate product and lipid profile of hemodialysis patients. Nephrol Dial Transplant 1999;14:2907–2914.

19   Chertow GM, Burke SK, Raggi P: Sevelamer attenuates the progression of coronary and aortic calcification in hemodialysis patients. Kidney Int 2002;62:245–252.

20   Hutchinson AJ: Calcitriol, lanthanum carbonate and other phosphate binders in the management of renal osteodystrophy. Perit Dial Int 1999;14:408–412.

21   Lacour B, Lucas A, Auchere D, Ruellan N, de Serre Patey NM, Drueke TB: Chronic renal failure is associated with increased tissue deposition of lanthanum after 28-day oral administration. Kidney Int 2005;67:1062–1069.

22   Brown EM, Gamba G, Riccardi D, Lombardi M, Butters R, Kifor O, Sun A, Hediger MA, Lytton J, Herbert SC: Cloning and characterization of an extracellular Ca^{2+}-sensing receptor from bovine parathyroid. Nature 1993;366:575–580.

23   Pollak MR, Brown EM, Wu-Chou Y-H, Herbert SC, Marx SJ, Steinmann B, Levi T, Seidman CE, Seidman JG: Mutations in the human Ca^{2+}-sensing receptor gene cause familial hypocalciuric hypercalcemia and neonatal severe hyperparathyroidism. Cell 1993;75:1297–1303.

24   Brown AJ, Zhong M, Finch J, Ritter C, McCracken R, Morrissey J, Slatopolsky E: Rat calcium-sensing receptor is regulated by vitamin D but not calcium. Am J Physiol 1996;270:F454–F460.

25   Naveh-Many T, Friedlander MM, Mayer H, Silver J: Calcium regulates parathyroid hormone messenger ribonucleic acid (mRNA), but not calcitonin mRNA in vivo in the rat. Dominant role of 1,25-dihydroxyvitamin D_3. Endocrinology 1989;125:275–280.

26   Okazaki T, Zajac JD, Igarashi T, Ogata E, Kronenberg HM: Negative regulatory elements in the human parathyroid hormone gene. J Biol Chem 1991;266:21903–21910.

27   Bellorin-Font E, Martin KJ, Freitag JJ, Slatopolsky E: Altered adenylate cyclase kinetics in hyper-functioning human parathyroid glands. J Clin Endocrinol Metab 1981;52:499–507.

28   Brown EM, Wilson RE, Thatcher JG, Marynick SP: Abnormal calcium regulated PTH release in normal parathyroid tissue from patients with adenoma. Am J Med 1981;71:565–571.

29   Kifor O, Moore FD, Wang P, Goldstein M, Vassilev P, Kifor I, Hebert SC, Brown EM: Reduced immunostaining for the extracellular Ca sensing receptor in primary and uremic secondary hyper-parathyroidism. J Clin Endocrinol Metab 1996;81:1598–1606.

30   Denda M, Finch J, Brown AJ, Nishii Y, Kubodera N, Slatopolsky E: 1,25 dihydroxyvitamin D_3 and 22-oxacalcitriol prevent the decrease in vitamin D receptor content in the parathyroid glands of uremic rats. Kidney Int 1996;50:34–39.

Eduardo Slatopolsky, MD, Professor of Medicine
Washington University School of Medicine
Renal Division, 660 S. Euclid Avenue, Box 8126
St. Louis, MO–63110 (USA)
Tel. +1 314/362 7208, Fax +1 314/362 7875, E-Mail eslatopo@im.wustl.edu

Ronco C, Brendolan A, Levin NW (eds): Cardiovascular Disorders in Hemodialysis.
Contrib Nephrol. Basel, Karger, 2005, vol 149, pp 272–278

....

Cardiovascular Calcification in End Stage Renal Disease

Paolo Raggi

New Orleans, La., USA

Abstract

Extensive atherosclerosis and heavy vascular and valvular calcifications are common complications of end stage chronic kidney disease (CKD-V) and are related to a high incidence of cardiovascular events. In CKD-V vascular calcifications occur both in the subintimal space and in the runica media. Intimal calcification is associated with atherosclerosis and is therefore a universal finding. On the contrary, medial calcification is characteristically associated with advanced CKD and diabetes mellitus. Numerous metabolic and endocrine abnormalities, primarily involving calcium and phosphorus metabolism, are found in CKD. Furthermore, CKD-V is believed to be a state of heightened inflammation and oxidative stress. All of these dysfunctions occur early in the course of the renal failure and likely contribute to the development and progression of vascular calcification and atherosclerosis. This review is centred on the pathobiology of vascular calcification in CKD-V, its detection with modern imaging modalities and the therapeutic approaches currently available to slow its progression.

Introduction

Cardiovascular disease is the most common cause of morbidity and mortality in patients suffering from end stage renal disease (ESRD) and it accounts for 40–50% of deaths in this population [1]. The annual cardiovascular mortality by age, race, and gender in dialysis patients is much higher compared to the general population [2, 3]. Arterial calcification represents a significant risk factor for cardiovascular morbidity and mortality in subjects with normal renal function, and ESRD patients present extensive calcification of the cardiovascular system [4]. Calcification is typically located in the subintimal space where atherosclerosis usually begins and expands [4], but also in the media layer. Several mediators of bone mineralization are believed to contribute to these processes, though the initiating stimuli remain partly unknown. It is probable that some of the arterial pathological changes may be secondary to uremia-specific factors such as deregulation of calcium, phosphorus, and parathyroid hormone balance, and treatments such as calcium-based phosphate binders and Vitamin D. Indeed, it is emerging that cardiovascular disease is strongly related to bone disease in renal failure.

Pathology and Etiopathogenesis of Vascular Calcification

Extraskeletal calcification is common in ESRD and for a long time it was believed to be a mere consequence of secondary hyperparathyroidism. However, increasing experimental evidence demonstrates that it is an active process involving enzymes and cells similar to those found in healthy bone [5]. Moe et al. [5] showed moderate calcification in human epigastric arteries collected during renal transplantation. Calcification was mainly localized in the arterial media layer and colocalized with bone matrix protein [5]. A large number of in vitro studies have highlighted the involvement in the process of vascular calcification of osteoblast-like cell [6], cytokines [7], transcription factors [8], and bone morphogenic proteins [9–11] found in normal bone.

Vascular calcification can be found in the intimal layer and in the tunica media of the arterial wall. Intima layer calcification is closely associated with atherosclerotic changes in the vessel wall. It occurs in a patchy or disseminated distribution, it may lead to ischemic events and it is often associated with them. Medial layer calcification, also known as Monckeberg's sclerosis, is character-istic of ESRD patients; it is more linearly and diffusely distributed than subinti-mal calcification. It is responsible for arterial stiffening and decrease in arterial compliance, leading to systolic hypertension, left ventricular hypertrophy and ultimately left ventricular failure [12]. Though rare and poorly understood, calciphylaxis is a serious uremia-related disorder characterized by systemic calcification of small arterioles with necrosis of the subcutaneous tissues [13]. Finally, valvular calcification and dysfunction is also common in patients on maintenance dialysis and its pathogenesis appears to be very similar to that of intimal atherosclerosis and medial calcification.

The incipient stimuli for the deposition of hydroxyapatite crystals in the arterial wall and cardiac valves are partly atherogenic and partly secondary to uremic factors. Indeed, hypercalcemia and hyperphosphatemia have both been implicated in the process of calcification. This process also seems dependent upon the presence and activity of factors favoring and others antagonizing the precipitation of calcium crystals. Matrix GLA protein, a vitamin-K dependent factor, and fetuin-A (α_2-Heremans-Schmid glycoprotein), a negative acute-phase reactant, reduce calcium crystal formation in-vitro. Furthermore, knock-out mice for both Matrix GLA protein or fetuin-A develop early and severe vascular calcifications leading to cardiovascular death [14]. In an observational study, a low level of fetuin-A was identified as an independent predictor of all-cause and cardiovascular mortality in ESRD [15].

Diagnostic Tools for Cardiovascular Calcification Assessment

A number of noninvasive imaging techniques are available to screen for the presence of cardiovascular calcification. Plain X-rays of abdomen and extremities can reveal the presence of macroscopic calcifications in the arteries of these territories. Echocardiography is the primary noninvasive tool used for assessment of morphological and functional changes of myocardium and cardiac valves. Similarly, 2D-ultrasound can be used to identify presence of vascular calcification in various arterial territories such as carotid arteries, femoral arteries, and aorta. However, none of these radiological and ultrasound techniques permits quantification of the extent of calcification. On the contrary, radiological imaging modalities such as electron beam computed tomography and multislice computed tomography, play a relevant role in the detection and quantification of cardiovascular calcifications. In both techniques image acquisition and data reconstruction are synchronized to the patient's cardiac cycle and generally performed during diastole, when coronary artery displacement is minimal. To prevent artifacts due to respiratory motion, the scans are typically performed during a single inspiratory breath hold. The Agatston score is utilized to quantify coronary artery calcification [16]. The score is calculated as the product of a calcified area by its density and it is therefore a reflection of the amount of calcium deposited in the plaque.

Clinical Impact of Cardiovascular Calcification

Several lines of evidence support the association of cardiovascular calcification and morbidity and mortality in patients undergoing maintenance dialysis. Wang et al. [17], found a positive correlation between valve calcification and cardiovascular as well as all-cause mortality in a cohort of 192 patients undergoing peritoneal dialysis. All subjects were screened at baseline by means of echocardiography for calcification of the aortic and mitral valve and follow-up lasted on average 18 months. Cardiovascular mortality was 7- fold higher in the group when compared to the group without valvular calcifications (22 vs. 3%;

p < 0.0001), and mortality was greater when both valves were calcified compared to single valve calcification. Valve calcification was a predictor of cardiovascular risk independent of age, gender, dialysis duration, C-reactive protein level, diabetes mellitus, and prior atherosclerotic vascular disease. In another cross-sectional and follow-up study [18], the investigators performed baseline soft-tissue radiograms of the pelvis and ultrasound scanning of the extracranial arteries to detect calcification of the femoral and carotid arteries, respectively. The study cohort included 202 patients on maintenance hemodialysis. At the end of follow-up (~7 years), multivariable Cox regression analysis showed that arterial calcification was a predictor of all-cause as well as cardiovascular mortality independent of classical risk factors for atherosclerosis. Blacher et al. [19] studied prospectively (average follow-up 53 ± 21 months) 110 patients on maintenance hemodialysis stratified according to the severity of arterial calcification. At baseline the patients were divided into four groups according to the number of arterial sites (common carotid arteries, abdominal aorta, iliofemoral axes and legs) showing calcification assessed either ultrasonographically or radiographically. The primary end-points were all-cause as well as cardiovascular mortality. This study showed that the presence and extent of cardiovascular calcifications were strongly associated with an adverse outcome. In fact, the adjusted hazard ratios for all-cause and cardiovascular mortality for each point increase in calcification severity increased by 1.9 (95% confident interval: 1.4–2.6) and 2.6 (95% confident interval. 1.2–2.4), respectively (p < 0.01 for both).

Matsuoka et al. [20] followed 104 chronic hemodialysis patients for an average of 43 months after a screening electron beam computed tomography. Patients were divided into two groups according to a baseline coronary calcium score falling below or above the median (measured at a value of 200). The 5-year cumulative survival was significantly lower for patients with a score ≥200 than for those with a score <200 (67.9 vs. 84.2%; p = 0.0003).

Therapeutic Approaches

Given the high prevalence of risk factors for atherosclerosis, it is imperative to pay careful attention to treatment of such factors in ESRD. However, as mentioned above, secondary hyperparathyroidism and abnormalities of mineral metabolism are probably partly responsible for the high cardiovascular morbidity and mortality seen in ESRD patients. The recent Kidney Disease Outcomes Quality Initiative (K/DOQI) guidelines recommended lower serum phosphorus (3.5–5.5 mg/dl), serum calcium (8.4–9.5 mg/dl), and serum calcium-phosphorus product (<55 mg^2/dl^2) targets than ever before [21]. Current medical treatments for secondary hyperparathyroidism include dietary phosphate restriction, oral phosphate binders, calcimimetics, vitamin D and its analogs.

Calcium-Based Phosphate Binders

Calcium-based salts have been the preferred phosphate binders for the past two decades. They replaced aluminum salts that were found to cause long-term toxicities such as encephalopathy, adynamic bone disease, and microcytic anemia [22]. Still widely used, calcium-containing binders provide a substantial calcium load often leading to hypercalcemia and an increased risk of metastatic calcification. In ESRD patients with evidence of metastatic calcification, the use of calcium-based phosphate binders or active vitamin D should be carefully monitored, and new calcium and aluminum-free phosphate binders are preferred [21].

Sevelamer

Sevelamer hydrochloride is a calcium-free nonabsorbable phosphate-binding polymer [23]. As shown in a randomized clinical trial, Sevelamer is capable of controlling serum phosphate to a degree similar to calcium-based binders without causing hypercalcemia [24]. Furthermore, it has been shown to arrest progressive coronary, aortic, and valvular calcification while promoting greater bone mineral density in maintenance hemodialysis patients [24, 25]. Beyond the effects on calcium and phosphorus metabolism, Sevelamer has other favorable effects such as reduction in serum uric acid [26], improvement in lipid profile [23], and an anti-inflammatory effect [27].

Lanthanum Carbonate

Lanthanum carbonate is a calcium- and aluminum-free phosphate binder. Lanthanum is a rare earth metal (atomic number 57) that has been shown to be an effective phosphate binder in preclinical and clinical studies, with reported low systemic absorption and a good short-term safety profile [28, 29]. Studies are in progress to establish the long-term safety and efficacy of lanthanum carbonate though laboratory experiments which suggest a significant accumulation of this metal in the liver and bone of animals.

Calcimimetic Agents

These new agents simulate the effect of calcium on calcium-sensing receptors, expressed on the surface of parathyroid cells, to regulate the secretion of parathyroid hormone. Recent reports showed that calcimimetic agents have a favorable impact on serum calcium, phosphorus, and calcium-phosphorous product [30, 31]. However, contextual administration of active vitamin D and calcium may occasionally be necessary to increase intestinal absorption of calcium and maintain normocalcemia.

Conclusions

Cardiovascular morbidity and mortality are highly prevalent in patients with ESRD and several factors are contributors. Among others: prevalent risk factors for atherosclerosis, a heightened inflammatory state, alterations in calcium and phosphorus metabolism and use of medical remedies with potentially damaging effects on the cardiovascular and bone tissues. The modern treatment approaches outlined above should be tailored to the individual patient and should be aimed at improving the overall metabolic state as well as reducing the risk of cardiovascular disease and soft tissue calcification related to it. Sevelamer is a compound with several ideal pharmacodynamic characteristics for the control of serum phosphate in ESRD. It is implicit that aggressive management of all traditional risk factors for atherosclerosis should also be implemented to reduce the long-term cardiovascular complication rate in this condition.

References

1   U.S. Renal Data System: USRDS annual report. Am J Kidney Dis 1998;32:S81–S88.
2   Parfrey PS, Foley RN: The clinical epidemiology of cardiac disease in chronic renal failure. J Am Soc Nephrol 1999;10:1606–1615.
3   Parfrey PS, Foley RN, Harnett JD, Kent GM, Murray DC, Barre PE: Outcome and risk factors for left ventricular disorders in chronic uraemia. Nephrol Dial Transplant 1996;11:1277–1285.
4   Schwarz U, Buzello M, Ritz E, Stein G, Raabe G, Wiest G, Mall G, Armann K: Morphology of coronary atherosclerotic lesions in patients with end stage renal failure. Nephrol Dial Transplant 2000;15:218–223.
5   Moe SM, O'Neill KD, Duan D, Ahmed S, Chen NX, Leapman SB, Fineberg N, Kopecky K: Medial artery calcification in ESRD patients is associated with deposition of bone matrix proteins. Kidney Int 2002;61:638–647.
6   Cheng SL, Shao JS, Charlton-Kachigian N, Loewy AP, Towler DA: MSX2 promotes osteogenesis and suppresses adipogenic differentiation of multipotent mesenchymal progenitors. J Biol Chem 2003;14:45969–45977.
7   Demer LL: Vascular calcification and osteoporosis: Inflammatory responses to oxidized lipids. Int J Epidemiol 2002;31:737–741.
8   Ducy P: Cbfa1: A molecular switch in osteoblast biology. Dev Dyn 2000;219:461–471.
9   Ducy P, Karsenty G: The family of bone morphogenetic proteins. Kidney Int 2000;57:2207–2214.
10 Deckers MM, van Bezooijen RL, van der Horst G, Hoogendam J, van Der Bent C, Papapoulos SE, Lowik CW: Bone morphogenetic proteins stimulate angiogenesis through osteoblast-derived vascular endothelial growth factor A. Endocrinology 2002;143:1545–1553.
11 Zebboudj AF, Imura M, Bostrom K: Matrix GLA protein, a regulatory protein for bone morphogenetic protein-2. J Biol Chem 2002;277:4388–4394.
12 London GM, Marchais SJ, Guerin AP, Metivier F, Adda H: Arterial structure and function in end-stage renal disease. Nephrol Dial Transplant 2002;17:1713–1724.
13 Androgue HJ, Frazier MR, Zeluff B, Suki WN: Systemic calciphylaxis revisited. Am J Nephrol 1981;1:177–183.
14 Schafer C, Heiss A, Schwarz A, Westenfeld R, Ketteler M, Floege J, Muller-Esterl W, Schinke T, Jahnen-Dechent W: The serum protein alpha 2-Heremans-Schmid glycoprotein/fetuin-A is a systemically acting inhibitor of ectopic calcification. J Clin Invest 2003;112:357–366.

15 Ketteler M, Bongartz P, Westenfeld R, Wildberger JE, Mahnken AH, Bohm R, Metzger T, Wanner C, Jahnen-Dechent W, Floege J: Association of low fetuin-A (AHSG) concentrations in serum with cardiovascular mortality in patients on dialysis: A cross-sectional study. Lancet 2003;361: 827–833.

16 Agatston AS, Janowitz WR, Hildner FJ, Zusmer NR, Viamonte M Jr, Detrano R: Quantification of coronary artery calcium using ultrafast computed tomography. J Am Coll Cardiol 1990;15: 827–832.

17 Wang AY, Wang M, Woo J, Lam CW, Li PK, Lui SF, Sanderson JE: Cardiac valve calcification as an important predictor for all-cause mortality and cardiovascular mortality in long-term peritoneal dialysis patients: A prospective study. J Am Soc Nephrol 2003;14:159–168.

18 London GM, Guerin AP, Marchais SJ, Metivier F, Pannier B, Adda H: Arterial media calcification in end-stage renal disease: Impact on all-cause and cardiovascular mortality. Nephrol Dial Transplant 2003;18:1731–1740.

19 Blacher J, Guerin AP, Pannier B, Marchais SJ, London GM: Arterial calcifications, arterial stiffness, and cardiovascular risk in end-stage renal disease. Hypertension 2001;38:938–942.

20 Matsuoka M, Iseki K, Tamashiro M, Fujimoto N, Higa N, Touma T, Takishita S: Impact of high coronary artery calcification score (CACS) on survival in patients on chronic hemodialysis. Clin Exp Nephrol 2004;8:54–58.

21 National Kidney Foundation (NKF) K/DOQI guidelines. K/DOQI clinical practice guidelines for bone metabolism and disease in chronic kidney disease. Am J Kidney Dis 2003;42:S7–S201.

22 Parkinson IS, Ward MK, Feest RWP, Kerr DNS: Fracturing dialysis osteodystrophy and dialysis encephalopathy. Lancet 1979;1:406–409.

23 Chertow GM, Burke SK, Dillon MA, Slatopolsky E: Long-term effects of sevelamer hydrochloride on the calcium × phosphate product and lipid profile of hemodialysis patients. Nephrol Dial Transplant 1999;14:2907–2914.

24 Chertow GM, Burke SK, Raggi P, Treat to Goal Working Group: Sevelamer attenuates the progression of coronary and aortic calcification in hemodialysis patients. Kidney Int 2002;62:245–252.

25 Raggi P, James G, Burke S, Bommer J, Taber SC, Holzer H, Braun J, Chertow GM: Paradoxical decrease in vertebral bone density with calcium-based phosphate binders in hemodialysis. J Bone Miner Res, in press.

26 Garg JP, Chasan-Taber S, Blair A, Plone M, Bommer J, Raggi P, Chertow GM: Effect of sevelamer and calcium-based phosphate binders on uric acid concentrations in hemodialysis patients. Arthritis Rheum 2005;52:290–295.

27 Ferramosca E, Burke S, Chasan-Taber S, Ratti C, Chertow GM, Raggi P: Potential antiatherogenic and anti-inflammatory properties of sevelamer in maintenance hemodialysis patients. Am Heart J, in press.

28 Joy MS, Finn WF: LAM-302 Study Group: Randomized, double-blind, placebo-controlled, dose-titration, phase III study assessing the efficacy and tolerability of lanthanum carbonate: A new phosphate binder for the treatment of hyperphosphatemia. Am J Kidney Dis 2003;42:96–107.

29 Hutchison AJ, Speake M, Al-Baaj F: Reducing high phosphate levels in patients with chronic renal failure undergoing dialysis: A 4-week, dose-finding, open-label study with lanthanum carbonate. Nephrol Dial Transplant 2004;19:1902–1906.

30 Lindberg JS, Moe SM, Goodman WG, Coburn JW, Sprague SM, Liu W, Blaisdell PW, Brenner RM, Turner SA, Martin KJ: The calcimimetic AMG 073 reduces parathyroid hormone and calcium × phosphorus in secondary hyperparathyroidism. Kidney Int 2003;63:248–254.

31 Block GA, Martin KJ, de Francisco AL, Turner SA, Avram MM, Suranyi MG, Hercz G, Cunningham J, Abu-Alfa AK, Messa P, Coyne DW, Locatelli F, Cohen RM, Evenepoel P, Moe SM, Fournier A, Braun J, McCary LC, Zani VJ, Olson KA, Drueke TB, Goodman WG: Cinacalcet for secondary hyperparathyroidism in patients receiving hemodialysis. N Engl J Med 2004;350:1516–1525.

Paolo Raggi, MD
1430 Tulane Avenue, SL-48
New Orleans, LA–70112 (USA)
Tel. +1 504 988 6139, Fax +1 504 988 4237, E-Mail praggi@tulane.edu

Ronco C, Brendolan A, Levin NW (eds): Cardiovascular Disorders in Hemodialysis.
Contrib Nephrol. Basel, Karger, 2005, vol 149, pp 279–286

·······················

The Mechanism of Calcium Deposition in Soft Tissues

Diego Brancaccio, Mario Cozzolino

Renal Unit, San Paolo Hospital, Milan, Italy

Abstract

The current understanding of the mechanisms of calcium deposition in soft tissues in chronic kidney disease (CKD) patients has been deeply investigated in the last ten years. Because of higher morbidity and mortality due to cardiovascular disease in dialysis patients compared to general population, several studies showed that extraskeletal calcification may play a major role in the pathogenesis of cardiovascular events in CKD patients. Traditionally, the pathogenesis of vascular and soft tissue calcification has been associated with a passive calcium-phosphate deposition. Actually, it is well known that extraskeletal calcification is also related to an active process. In this review, we analyzed some of the factors potentially involved in the pathogenesis of vascular calcification in CKD patients.

Introduction

Severe cardiovascular calcification is very common in chronic kidney disease (CKD) [1]. Patients develop extensive medial calcification, which causes increased arterial stiffness and high morbidity and mortality due to cardiovascular events [2]. A large number of risk factors are associated with vascular calcification in dialysis patients (time on dialysis, uremic toxins, history of diabetes, inflammation), but abnormalities in bone mineral metabolism may play a critical role [3]. In fact, elevated serum phosphate, calcium-phosphate product, and parathormone levels contribute to vascular calcification, although their roles are incompletely understood [4].

In addition to these classic alterations in bone and mineral metabolism, an active process has been documented [5]. In the last decade, several studies defined calcification of atherosclerotic lesions as an active process similar to bone formation. Several gene products may induce or inhibit the process of soft tissue calcification. However, the potential role of 'protective' proteins associated with reduced vascular calcification in CKD needs to be better elucidated.

Cardiovascular Calcifications in CKD

Dialysis patients develop atherosclerotic vascular disease earlier than the general population [2]. In CKD-induced cardiovascular disease, two groups of vascular calcification risk factors need to be considered: (1) 'classic' risk factors are age, gender, family history, smoking, obesity, hypertension, diabetes, and dyslipidemia; (2) 'uremia-associated' risk factors are time on dialysis, uremic toxins, inflammation (advanced-glycation end products, oxidative stress and nitric oxide, asymmetric dimethylarginine, homocysteine), and increased serum levels of phosphate, calcium-phosphate product, and parathormone.

Elevations of serum phosphate and calcium-phosphate product levels may worsen cardiovascular events in uremic population, by causing a progressive increase in calcium deposition in the coronary arteries and cardiac valves. Furthermore, calcium-containing phosphate binders can increase calcium load [5]. Recently, an elegant study by Goodman et al. [6] in hemodialysis children and young adults showed a correlation between coronary artery calcification detected by electron beam computed tomography and age of dialysis, serum phosphorus, calcium-phosphate product levels, and daily intake of calcium. Moreover, experimental observations in uremic rats [7, 8] and in dialysis patients [9] showed that a new calcium-free and aluminum-free phosphate binder (sevelamer HCl) attenuated the progression of cardiovascular calcification better than calcium-based phosphate binders.

In the last decade, several authors analyzed the pathogenic mechanisms of phosphate-induced vascular calcification, pointing out not only a calcium-phosphate 'passive' deposition in the vessels, but also an 'active' role of genes associated with osteoblastic functions in vascular smooth muscle cells (VSMCs) [3]. Recently, very enlightening in vitro studies by Giachelli and coworkers [10] in Seattle, demonstrated that high phosphorus levels in the culture media (2 mmol/l) stimulated calcification in VSMCs. Moreover, additional in vitro studies indicated that high phosphate increases vascular calcification by a direct induction of the osteoblast-specific gene Cbfa-1, which regulates the expression of several bone morphogenic proteins (BMPs).

In CKD the expression of these BMPs was also induced by uremic serum independently of phosphate concentrations, in an in vitro model of bovine

VSMCs. Importantly, Moe et al. [11] showed an increased expression of Cbfa1 and osteopontin in the media and intima of calcified epigastric arteries from uremic patients.

Clearly, these data suggest that vascular calcification is an active process. The cellular formation of a bone-like structure in calcified vessel walls indicates that the uremic environment and elevations in serum phosphate levels may be regulatory keys for the pathogenesis of vascular calcification.

Considering serum calcium and phosphate levels, physiochemically crystallization should immediately occur, in the absence of active inhibitors. However, serum biological macromolecules have an inhibitory effect on calcium-phosphate precipitation. Loss of these inhibitory proteins determines mineralization in the arteries.

Bone Matrix Protein 7

Recently, exciting new data from the group of Hruska [12, 13] in Saint Louis suggested that a new player, bone morphogenetic protein-7 (BMP-7), is involved in the process of vascular calcification of chronic renal failure, as well as in the pathogenesis of renal osteodystrophy.

BMPs are members of the transforming growth factor-β superfamily of cytokines and consist of a group of at least 15 morphogens involved in intracellular messaging.

BMP-7 is a crucial element for the development of kidneys, eyes, and bones. In the skeleton, BMP-7 deficiency produces a patterning defect causing extra digits and rib to be formed and leads to a deficiency of precursor cell commitment to the osteoblastic differentiation and mineralization program. In the adult, BMP-7 maintains a role in osteoblast function, suggesting a hormonal role in bone metabolism. Interestingly, BMP-7 expression decreases early in the course of renal failure [14].

This state of BMP-7 deficiency has important consequences in the pathogenesis and treatment of chronic renal insufficiency, but is also very interesting for the pathogenesis and treatment of vascular calcifications. Indeed, BMP-7 maintains VSMCS differentiation and prevents their transformation into an osteoblastic phenotype [12]. Thus, the state of BMP-7 deficiency, characteristic of chronic renal failure, could favor vascular calcification, especially in the context of atherosclerotic lesions.

In renal osteodystrophy, BMP-7 affects osteoblast morphology and number, eliminates peritrabecular fibrosis, decreases bone resorption, and increases bone formation in secondary hyperparathyroidism. Moreover, it restores normal rates of bone formation in the adynamic bone disorder, as demonstrated in a renal ablation animal model. In this study, reversal of adynamic bone disease with a physiologic bone anabolic factor was demonstrated: the results of BMP-7 treatment differs from that of elevating PTH levels because bone resorption is not stimulated and increased bone formation rates are thereby truly anabolic. Therefore, BMP-7 is broadly efficacious in renal osteodystrophy, and importantly increases the skeletal deposition of ingested phosphorus and calcium, improving ion homeostasis in CKD.

Given the known role of BMP-7 in osteoblast biology it may seem paradoxical that it would reduce vascular calcification. However, the effects of individual BMP are highly dependent on the target cell type, the receptors expressed, and the state of differentiation and maturity. Transdifferentiation of VSMC in osteoblastic phenotype could be the critical first step in the etiology of vascular calcification and it is clear that BMP-7 has a positive influence in maintaining VSMC differentiation.

Suggesting that BMP-7 may be viewed as a fourth renal hormone, along with renin, erythropoietin, and calcitriol, Davies et al. [13] demonstrated that BMP-7 is an effective treatment of vascular calcification in the context of a murine model of atherosclerosis and chronic renal failure, a finding that may have important implications for the development, in humans, of future therapies for this condition, which is currently without treatment and with strong negative influences on cardiovascular morbidity and mortality.

α_2-Heremans-Schmid Glycoprotein (Human Fetuin)

α_2-Heremans-Schmid glycoprotein (AHSG) also known as human fetuin, is another important inhibitor of extraskeletal calcification. Serum concentration of AHSG fall during the cellular immunity phase of inflammation. In vitro, fetuin inhibits the de novo formation and precipitation of calcium-phosphate, with no effects on hydroxyapatite once it is formed. AHSG antagonizes the antiproliferative action of TGF-β_1 and blocks osteogenesis and deposition of calcium-containing matrix in dexamethasone-treated rat bone marrow cells.

Fetuin-deficient mice develop extensive ectopic calcifications in myocardium, kidney, lung, tongue, and skin [15]. In addition, hemodialysis patients with lower serum fetuin levels have increased mortality due to cardiovascular events [16]. This observation by Ketteler et al. [16] suggests that AHSG is involved in preventing the accelerated extraskeletal calcification observed in CKD.

Price et al. [17] have deeply investigated the mechanisms by which proteins inhibit vascular calcification. They showed a high-molecular-weight complex of a calcium-phosphate mineral phase and the two inhibitory proteins AHSG and matrix GLA protein (MGP) in the serum of rats treated with the bone-active bisphosphonate etidronate. Furthermore, in a recent in vivo study, Price found that vitamin D-induced vascular calcification associates with a 70% reduction of serum fetuin, probably due to the clearance of the fetuin-mineral complex from serum.

The potential role of fetuin in the pathogenesis of extraskeletal calcification in CKD patients is still poorly understood. Clearly, AHSG binds calcium and hydroxyapatite, fetuin knock-out mice have soft tissue calcifications, and serum AHSG levels are reduced in both uremic and inflammatory conditions. Therefore, loss of serum and local fetuin could directly promote vascular calcification in uremic patients.

MGP

MGP is a member of the vitamin K-dependent protein family with unique structural and physical properties. During the first 2 months of life, MGP-deficient mice develop diffuse arterial calcification, osteoporosis, and pathological fractures [18]. For its properties as extracellular matrix protein with high affinity for calcium and phosphate, MGP plays an important role in the prevention of vascular calcification and in the pathogenesis of osteoporosis, although its effects in CKD patients are still unclear [19].

By binding BMP-2, MGP elicits an inhibitory mechanism on mineralization. A recent study by Wajih et al. [20] demonstrated an uptake mechanism for serum fetuin by cultured human VSMCs. Fetuin uptake and secretion by these cells could represent an additional protective mechanism against arterial calcification.

The localization of MGP and other bone matrix proteins, such as osteopontin, has been investigated by Canfield et al. [21] in calcified atherosclerotic arteries and in normal vessel walls. While MGP was not detected in normal blood vessels, its expression was enhanced at loci of arterial calcification, such as atherosclerotic and calciphylactic lesions. Therefore, the MGP localization in calcified arteries suggests an etiopathogenic role for this inhibitory protein on the development of vascular calcification.

An association between polymorphisms of the MGP gene and myocardial infarction in low-risk individuals has been described [22]. Potentially, definition of polymorphisms of the MGP gene represents a critical step in understanding pathogenic mechanisms of vascular calcification in CKD and dialysis patients. Recently, Jono et al. [23] reported an association between serum MGP levels and coronary artery calcification, detected by electron beam computed

Table 1. MGP, fetuin, and BMP-7

	MGP	Fetuin	BMP-7
Molecular weight	10 kDa	62 kDa	46.75 kDa
Serum levels	60–130 U/l	0.5–1.0 g/l	0.1–0.5 ng/ml
Synthesis	VSMCs, chondrocytes	Hepatocytes	Osteoblasts
Phenotype of knock-out mice	Aortic media calcification, cartilage calcification, lethal due to aortic rupture	Diffuse metastatic soft-tissue and intra-arterial calcification, impaired survival	Hypoplastic-dysplastic kidneys, bone mineralization deficiency, developmental defects of eyes
Properties	Vitamin K-dependent γ-carboxylation necessary for activation	Negative acute phase reactant, TGF-β antagonist, inhibitor of insulin receptor tyrosine kinase activity	TGF-β superfamily member, regulator of osteoblast differentiation, VSMCs differentiation

tomography in 115 subjects with suspected coronary artery disease and normal renal function. Patients with coronary artery calcification had lower serum MGP levels compared to those with no calcium in the coronary tree, suggesting the potential role of MGP on prevention of vascular calcification [23].

Conclusions

Vascular calcification is a complex phenomenon in which basic mechanisms have been identified. Lack of inhibitory proteins (BMP-7, fetuin, MGP) is a first possible principle of calcium-phosphate deposition in soft tissues (table 1). In addition, an organized calcification process similar to bone formation has been clearly identified in calcified atherosclerotic lesions. In the setting of CKD, different stimuli such as hyperphosphatemia, hyperparathyroidism, diabetes, dyslipidemia, hypertension, and iatrogenic vitamin D administration have a common target in increase of calcium load. In particular, alteration of bone mineral metabolism is a 'typical' risk factor for higher incidence of vascular calcification and cardiovascular events in CKD patients. Therefore, it becomes mandatory to prevent vascular calcification by decreasing serum phosphate and calcium-phosphate product levels, and reducing the active process through regulation of specific genes. In fact, modulation of the production and activity of BMP-7, AHSG, and MGP could offer a novel approach to the treatment of several vascular diseases.

The advent of these new pieces of information has opened important pathways for better understanding the puzzle of the pathogenesis of increased risk of extraskeletal calcification and cardiovascular events in CKD patients.

References

1   Christian RC, Fitzpatrick LA: Vascular calcification. Curr Opin Nephrol Hypertens 1999;8:443–448.
2   London GM: Cardiovascular calcifications in uremic patients: Clinical impact on cardiovascular function. J Am Soc Nephrol 2003;14:S305–S309.
3   Cozzolino M, Dusso AS, Slatopolsky E: Role of calcium-phosphate product and bone-associated proteins on vascular calcification in renal failure. J Am Soc Nephrol 2001;12:2511–2516.
4   Ganesh SK, Stack AG, Levin NW, Hulbert-Shearon T, Port FK: Association of elevated serum PO(4), Ca × PO(4) product, and parathyroid hormone with cardiac mortality risk in chronic hemodialysis patients. J Am Soc Nephrol 2001;12:2131–2138.
5   Cozzolino M, Brancaccio D, Gallieni M, Slatopolsky E: Pathogenesis of vascular calcification in chronic kidney disease. Kidney Int 2005; in press.
6   Goodman WG, Goldin J, Kuizon BD, Yoon C, Gales B, Sider D, Wang Y, Chung J, Emerick A, Greaser L, Elashoff RM, Salusky IB: Coronary-artery calcification in young adults with end-stage renal disease who are undergoing dialysis. N Engl J Med 2000;342:1478–1483.
7   Cozzolino M, Dusso AS, Liapis H, Finch J, Lu Y, Burke SK, Slatopolsky E: The effects of sevelamer hydrochloride and calcium carbonate on kidney calcification in uremic rats. J Am Soc Nephrol 2002;13:2299–2308.
8   Cozzolino M, Staniforth ME, Liapis H, Finch J, Burke SK, Dusso AS, Slatopolsky E: Sevelamer hydrochloride attenuates kidney and cardiovascular calcifications in long-term experimental uremia. Kidney Int 2003;64:1653–1661.
9   Chertow GM, Burke SK, Raggi P, Treat to Goal Working Group: Sevelamer attenuates the progression of coronary and aortic calcification in hemodialysis patients. Kidney Int 2002;62:245–252.
10 Jono S, McKee MD, Murry CE, Shioi A, Nishizawa Y, Mori K, Morii H, Giachelli CM: Phosphate regulation of vascular smooth muscle cell calcification. Circ Res 2000;87:E10–E17.
11 Moe SM, Duan D, Doehle BP, O'Neill KD, Chen NX: Uremia induces the osteoblast differentiation factor Cbfa1 in human blood vessels. Kidney Int 2003;63:1003–1011.
12 Lund RJ, Davies MR, Hruska K: Bone morphogenetic protein-7: An anti-fibrotic morphogenetic protein with therapeutic importance in renal disease. Curr Opin Nephrol Hypertens 2002;11:31–36.
13 Davies MR, Lund RJ, Hruska KA: BMP-7 is an efficacious treatment of vascular calcification in a murine model of atherosclerosis and chronic renal failure. J Am Soc Nephrol 2003;14:1559–1567.
14 Wang S, Chen Q, Simon TC, Strebeck F, Chaudhary L, Morrissey J, Liapis H, Klahr S, Hruska KA: Bone morphogenic protein-7 (BMP-7), a novel therapy for diabetic nephropathy. Kidney Int 2003;63:2037–2049.
15 Schafer C, Heiss A, Schwarz A, Westenfeld R, Ketteler M, Floege J, Muller-Esterl W, Schinke T, Jahnen-Dechent W: The serum protein alpha 2-Heremans-Schmid glycoprotein/fetuin-A is a systemically acting inhibitor of ectopic calcification. J Clin Invest 2003;112:357–366.
16 Ketteler M, Bongartz P, Westenfeld R, Wildberger JE, Mahnken AH, Bohm R, Metzger T, Wanner C, Jahnen-Dechent W: Association of low fetuin-A (AHSG) concentrations in serum with cardiovascular mortality in patients on dialysis: A cross-sectional study. Lancet 2003;361:827–833.
17 Price PA, Williamson MK, Nguyen TM, Than TN: Serum levels of the fetuin mineral complex correlate with artery calcification in the rat. J Biol Chem 2004;279:1594–1600.
18 Luo G, Ducy P, McKee MD, Pincro GJ, Loyer E, Behringer RR, Karsenty G: Spontaneous calcification of arteries and cartilage in mice lacking matrix GLA protein. Nature 1997;386:78–81.

19   Shearer MJ: Role of vitamin K and Gla proteins in the pathophysiology of osteoporosis and vascular calcification. Curr Opin Clin Nutr Metab Care 2000;3:433–438.
20   Wajih N, Borras T, Xue W, Hutson SM, Wallin R: Processing and transport of matrix Gla protein (MGP) and bone morphogenetic protein-2 (BMP-2) in cultured human vascular smooth muscle cells: Evidence for an uptake mechanism for serum fetuin. J Biol Chem 2004; in press.
21   Canfield AE, Farrington C, Dziobon MD, Boot-Handford RP, Heagerty AM, Kumar SN, Roberts IS: The involvement of matrix glycoproteins in vascular calcification and fibrosis: An immuno-histochemical study. J Pathol 2002;196:228–234.
22   Hermann SM, Whatling C, Brand E, Nicaud V, Gariepy J, Simon A, Evans A, Ruidavets JB, Arveiler D, Luc G, Tiret L, Henney A, Cambien F: Polymorphisms of the human matrix Gla protein (MGP) gene, vascular calcification, and myocardial infarction. Arterioscler Thromb Vasc Biol 2000;20:2386–2393.
23   Jono S, Ikari Y, Vermeer C, Dissel P, Hosegawa K, Shioi A, Taniwaki H, Kizu A, Nishizawa Y, Saito S: Matrix Gla protein is associated with coronary artery calcification as assessed by electron-beam computed tomography. Thromb Haemost 2004;91:790–794.

Diego Brancaccio, MD
Renal Unit, Azienda Ospedale San Paolo
Via A. di Rudinì, 8, IT–20142 Milano (Italy)
Tel. +39 02/81844371, Fax +39 02/89129989
E-Mail diego.brancaccio@tiscalinet.it

Ronco C, Brendolan A, Levin NW (eds): Cardiovascular Disorders in Hemodialysis.
Contrib Nephrol. Basel, Karger, 2005, vol 149, pp 287–294

....

Biosimilars, Generic Versions of the First Generation of Therapeutic Proteins: Do They Exist?

Daan Crommelin[a], Theresa Bermejo[b], Marco Bissig[c], Jaak Damiaans[d], Irene Krämer[e], Patrick Rambourg[f], Giovanna Scroccaro[g], Borut Štrukelj[h], Roger Tredree[i], Claudio Ronco[j]

[a]Faculty of Pharmaceutical Sciences, Utrecht University, Utrecht, The Netherlands; [b]Servicio de Farmacia, Hospital Ramón y Cajal, Madrid, Spain; [c]Servzio di Farmacia, Ospedale Civico Lugano, Lugano, Switzerland; [d]Pharmacy Department, Virga Jesse Hospital, Hasselt, Belgium; [e]Pharmacy Department, Johannes Gutenberg-University Hospital, Mainz, Germany; [f]Pharmacie Saint-Eloi, CHU de Montpellier, Montpellier, France; [g]Servizio di Farmacia, A.O., Verona, Italy; [h]Faculty of Pharmacy, University of Ljubljana, Ljubljana, Slovenia; [i]Pharmacy Department, St George's Hospital, London, UK; [j]Dipartimento Interaziendale di Nefrologia Dialisi e Trapianto Renale, Ospedale San Bortolo, Vicenza, Italy

Abstract

This contribution describes the present regulatory status in the EU of biosimilars, the generic versions of the first generation of therapeutic proteins. It points out why and where recombinant protein molecules and low-molecular-weight drugs differ in their behaviour and why biosimilars should be handled differently than generic low-molecular-weight drugs. This information is important for practitioners (pharmacists and physicians) while selecting the best supplier of a therapeutic protein.

Copyright © 2005 S. Karger AG, Basel

Introduction

Recombinant therapeutic proteins (biopharmaceuticals), e.g. recombinant insulin and human growth hormone, were first introduced in the early 1980s.

Since then more than 150 products with biopharmaceuticals were marketed in Europe, North America, Japan and Australia. USA sales reached almost 40 billion USD in 2003. Among these biopharmaceuticals are erythropoietin and the family of the interferons. The contingent of biopharmaceuticals is growing fast with over 370 products in development for a wide range of serious diseases.

The patents protecting the first generation of biopharmaceuticals expire or will do so in the near future. In principle, this opens the possibility to launch generic versions as we have seen with low-molecular-weight products.

For nephrologists, who are used to prescribe biopharmaceuticals such as erythropoietin to their patients, the issue of emerging biosimilars is of particular importance. This, since biopharmaceuticals can elicit an immune response, which can lead to severe reactions, e.g. pure red cell aplasia. Nephrologists have, therefore, the responsibility to review the safety profile of the biopharmaceutical products they prescribe and to evaluate the safety profile of emerging biosimilar products.

What is required for low-molecular-weight drugs to have a generic version approved? Basically, the requirement is to prove 'essential similarity' between the innovator's product and the generic version. That means that the generic product has the same qualitative and quantitative composition in terms of its active substance, the active substance is in the same physicochemical form and bioequivalence is proven in normal, healthy volunteers. Why is essential similarity difficult to achieve with biopharmaceuticals? In this contribution the specific characteristics that make biopharmaceuticals stand out as a special group of pharmaceuticals will be discussed. The authors were part of a working group discussing the biosimilar issue from different professional and scientific angles (full report submitted to the European Journal of Hospital Pharmacy, 2005).

Before discussing differences between pharmaceutical proteins and low-molecular-weight drugs, a brief definition of terms is necessary. Table 1 gives definitions of the different terms used in this contribution. The terms 'biogeneric', 'second entry biological', 'subsequent entry biological', 'nonpatented biological product' and 'multisource product' will not be used as the European Agency for the Evaluation of Medicinal Products (EMEA) prefers the term 'biosimilar' (the FDA coined the term 'follow-on biologic').

Complexity of Their Structure

Biopharmaceuticals differ from low-molecular-weight drugs in a number of aspects listed in table 2 [1]. Biopharmaceuticals are proteins composed of a

Table 1. Definitions

Low-molecular-weight drug	Classical medicinal product
Generic drug	Chemical and therapeutic equivalent of low-molecular-weight drug whose patent has expired
Biopharmaceutical	A medical product developed by means of one or more of the following biotechnology practices: rDNA, controlled gene expression, antibody methods[1]
Biosimilar, or similar biological medicinal product	A biological medicinal product referring to an existing one and submitted to regulatory authorities for marketing authorization by an independent applicant after the time of protection of the data has expired for the original product[1]

[1]EMEA definition.

Table 2. How biopharmaceuticals differ from low-molecular-weight drugs

Molecular weight
Complexity of structure
Production and characterization
Structure and physicochemical properties
Biopharmaceutical purity
Biological activity
Stability
Immunogenicity

chain of amino acids. In glycoproteins these amino acids are glycosylated, i.e. they carry sugar groups. Their molecular weight can vary between 5 kDa (such as insulin) up to over 500 kDa for molecules such as IgM. These large molecules are folded and form complex structures. α-helices and β-sheets are local conformational entities and are called secondary structures. When these local secondary structures interact with each other a tertiary structure is formed. The exact nature of this conformation is critical for the therapeutic action of the biopharmaceutical and its adverse effects. Conformational stability is brought about mainly by rather weak physical interactions that are readily disturbed by changes in the milieu, such as temperature variation, or even by shaking. Thus, these complex (glyco)proteins are prone to degradation reactions. Their secondary and tertiary structure can be readily changed by misfolding or their primary structure can be oxidized or degraded by hydrolytic reactions (e.g. deamidation or amino acid chain breakage).

Production and Characterization

Biopharmaceuticals are produced by genetically modified and carefully selected, stable production cells taken from a company-owned host cell bank. These production cells are either prokaryotes, such as *E. coli*, or eukaryotes such a yeast cells or Chinese hamster ovary cells. Eukaryotes are used for glycosylated proteins as prokaryotes miss the capacity to produce glycoproteins. The glycosylation patterns (so called isoform patterns) of proteins depend on the production cell and the conditions under which the organisms are growing. As the structure of the isoforms determines the pharmacological effect, a strict definition of isoform pattern is essential. After finishing the fermentation process the biopharmaceutical is isolated from the fermentation media and cell contents by an elaborate process of purification steps, called downstream processing to yield the proper purity.

For low-molecular-weight drugs, an arsenal of physicochemical techniques, such as different chromatographic and spectroscopic techniques, is available to fully characterize the active molecule including its impurity profiles and contaminants. With our present toolbox of analytical techniques including the bioassays, it is very difficult to fully characterize proteins at the end of the downstream processing procedure. For the smaller proteins such as insulin, it is already hard but not fully impossible; for the larger proteins, it is beyond our present analytical capacity.

This leads to the conclusion that for biopharmaceuticals 'the process is the product'. One has to control all different steps in the production process and run-in process checks. Changes in the production process may lead to changes in the biopharmaceutical conformation and therefore therapeutic value.

An example from 'real life' is the publication of the results of the analysis of 12 erythropoietin batches in Brazil taken from five manufacturers. Potencies were between 68 and 119% of the potency on the label, isoform patterns varied between manufacturers and unacceptably high bacterial endotoxin levels were found in 3 samples [2]. The Brazilian regulatory agency suspended the importation of two epoetin α biosimilars from one manufacturer for failing to meet the required standards [3].

Stability

Biopharmaceuticals are rather unstable structures. The manufacturers have tested the formulations for optimum product integrity circumstances. In many cases the protein is distributed and dispensed in the freeze-dried form. The patient or the hospital staff has to prepare the final dosage form by reconstituting the dried powder according to the manufacturer's guidelines. When the proper excipients and freeze-drying protocol are chosen, a product will result that is indeed stable for the required 2 years (shelf-life). But, nowadays more and more ready-to-use, aqueous systems (in vials or in prefilled syringes) are marketed. Here, maintaining 'cold chain' conditions is the rule. Biopharmaceuticals in their final dosage form should never be shaken or heated (or accidentally frozen) as aggregation of the protein is facilitated. Aggregates are generally seen as major sources of immunogenicity. A complete set of recommendations for storage and handling of biopharmaceuticals has been published before by members of this group [4].

Immunogenicity

In general, low-molecular-weight drug molecules do not show immune reactions. However, biopharmaceuticals may (and often do) elicit an immune response. Factors influencing the immune response are listed in table 3. An immune response per se may not have clinical consequences. But, in a number of cases the immune response may lead to clinical inefficacy and generalized immune effects (allergy, anaphylactic reactions and serum sickness) [5]. If the biopharmaceutical is similar to an endogenous protein and causes neutralizing antibodies, then severe side-effects may occur if the action of the endogenous protein cannot be taken over by other bioactives [no redundant pathway(s)]. With erythropoietin such a situation exists. The sudden increase in the occurrence of pure red cell aplasia upon administration of human recombinant erythropoietin via the subcutaneous route which occurred between 1998–2003 was ascribed to changes in the formulation of the erythropoietin-injections in combination with the use of pre-filled syringes with uncoated rubber stoppers [6].

Predicting immunogenicity of biopharmaceuticals is difficult. The presence (and detection) of aggregates is often seen as an immune stimulating

Table 3. Factors influencing immunity

Product-related factors
Sequence variation
Glycosylation
Host cells
Contaminants and process-related impurities
Formulation
Handling and storage
Patient-related factors
Route of administration
Dose and treatment duration
Concomitant diseases and/or medication
Congenital deficiencies

signal and breaking of tolerance. But, in general, one has to admit that our analytical 'toolbox', including bioassays, is less sensitive in detecting subtle changes in the protein structure than the immune system in patients. It is not surprising that developing better strategies to predict immunogenicity attracts a lot of interest from academic and industrial groups. Some success with transgenic animals has been reported.

The Regulatory Situation

The regulatory situation can be described as an evolving process. In Europe legislation concerning biosimilars was published in a directive 2004/27/EC in April 2004 [7, 8]. Two key messages from this directive are:
'biosimilar products are distinguished from low-molecular-weight generics and will be assessed on a case-to-case basis, with additional data being mandatory and including at least some preclinical and clinical data.

The Committee for Human Medicinal Products (CPMP) of the EMEA will decide what kind of tests will be required to show that biosimilars are safe and efficacious, which implies a need for EMEA guidelines. The FDA will bring out a draft guidance on biosimilars in 2005, probably taking a risk-based approach for the assessment of immunogenicity for biosimilars in development [9].

Concluding Remarks

Biosimilars, generic versions of the first generation of therapeutic proteins: do they exist? This is the title of this contribution. From the above, it should be clear that on the basis of EU-legislation biosimilars should undergo a more rigorous test protocol in the development process than low-molecular-weight generics. There is a substantial amount of scientific data supporting the EMEA policy of requesting preapproval clinical testing of efficacy and safety, including collecting the first set of data on immunogenicity, and to continue monitoring the performance of the product after approval by a robust postmarketing surveillance plan.

Practitioners, pharmacists and/or physicians, have to take decisions regarding the purchase of a biopharmaceutical or its biosimilar. The authors of this contribution see the availability of data on the similarity in performance and safety of the originator's product and a biosimilar published in peer-reviewed journals and/or from EMEA evaluation reports as important material to base a purchasing decision on. There are more considerations and those can be found in a full-length article submitted to the European Journal of Hospital Pharmacy, 2005.

Thus, biosimilars are structurally more complex and inherently less stable than 'classic', low-molecular-weight generics, and their efficacy and safety can only be assessed thoroughly on the basis of clinical data, including postmarketing data.

Acknowledgement

This work was supported by Johnson & Johnson Pharmaceutical Services, LLC.

References

1   Crommelin DJ, Storm G, Verrijk R, Leede L de, Jiskoot W, Hennink WE: Shifting paradigms: Biopharmaceuticals versus low molecular weight drugs. Int J Pharm 2003;266:3–16.
2   Schmidt CA, Ramos AS, da Silva JEP, Fronza M, Dalmora SL: Activity evaluation and characterization of recombinant human erythropoietin in pharmaceutical products. Arq Bras Endocrinol Metabol 2003;47:183–189.
3   Agencia Nacional de Vigilancia Sanitaria (ANVISA) resolutions 1.174 (22 July 2003) & 1.250 (1 August 2003).
4   Crommelin DJA, Bissig M, Gouveia W, Tredree R: Storage and handling of biopharmaceuticals: Problems and solutions – A workshop discussion. European Journal of Hospital Pharmacy 2003;8:89–93.
5   Schellekens H: Bioequivalence and the immunogenicity of biopharmaceuticals. Nat Rev Drug Discov 2002;1:457–462.
6   Sharma B, Lisi P, Ryan M, Bader F: Technical investigations into the potential cause of the increased incidence of epoetin-associated pure red cell aplasia. European Journal of Hospital Pharmacy 2004;5:86–91.
7   European Agency for the Evaluation of Medicinal Products, Committee for Proprietary Medicinal Products. EMEA/CPMP/BWP/3207/00/Rev 1. Guideline on comparability of medicinal products containing biotechnology-derived proteins as active substance: Quality issues. London, 11 December 2003.

8   European Agency for the Evaluation of Medicinal Products, Committee for Proprietary Medicinal Products. EMEA/CPMP/3097/02/Final. Guideline on comparability of medicinal products containing biotechnology-derived proteins as active substance: Non-clinical and clinical issues. London, 17 December 2003.

9   Crommelin DJA, Bermejo TH, Bissig M, Damiaans J, Krämer I, Rambourg P, Scroccaro G, Štrukelj B, Tredree R: Pharmaceutical evaluation of biosimilars: Important differences from generic low molecular weight pharmaceuticals. Submitted to the European Journal of Hospital Pharmacy, 2005.

Prof. Dr. Daan J.A. Crommelin
Utrecht University, Sorbonnelaan 16
NL–3584 CA Utrecht (The Netherlands)
Tel. +31 30 2536973/7306, Fax +31 30 2517839, E-Mail d.j.a.crommelin@pharm.uu.nl

New Technologies

Ronco C, Brendolan A, Levin NW (eds): Cardiovascular Disorders in Hemodialysis.
Contrib Nephrol. Basel, Karger, 2005, vol 149, pp 295–305

....

Electrophysiological Response to Dialysis: The Role of Dialysate Potassium Content and Profiling

A. Santoro[a], *E. Mancini*[a], *R. Gaggi*[a], *S. Cavalcanti*[b],
S. Severi[b], *L. Cagnoli*[c], *F. Badiali*[c], *B. Perrone*[d],
G. London[e], *H. Fessy*[f], *L. Mercadal*[g], *F. Grandi*[h]

[a]Malpighi Nephrology Division, Policlinico S. Orsola-Malpighi, Bologna, Italy,
[b]Biomedical Engineering Laboratory, DEIS, University of Bologna, Bologna,
[c]Nephrology Department, Ospedale Degli Infermi, Rimini, Italy, [d]Centre Hospitalier
F.H. Manhes, Fleury-Mérogis, [e]Unité de Nèphrologie et Dialyse, Hôpital René Dubos,
Pontoise, [f]B Hôpital Tenon, Paris, [g]Service de Néphrologie, Hopital La Pitié, Paris,
France, and [h]Hospal S.p.A., Bologna, Italy

Abstract

The task of dialysis therapy is, amongst other things, to remove excess potassium (K^+)
from the body. The need to achieve an adequate K^+ removal with the risk of cardiac arrhythmias due to sudden intra–extracellular K^+ gradient advises the distribution of the removal
throughout the dialysis session instead of just in the first half. The aim of the study was to
investigate the electrical behavior of two different K^+ removal rates on myocardial cells (risk
of arrhythmia and ECG alterations). Constant acetate-free biofiltration (AFB) and profiled
K^+ (decreasing during the treatment) AFB (AFBK) were used in a patient sample to understand, first of all, the effect on premature ventricular contraction (PVC) and on repolarization
indices [QT dispersion (QTd) and principal component analysis (PCA)]. The study was
divided into two phases: phase 1 was a pilot study to evaluate K^+ kinetics and to test the
effect on the electrophysiological response of the two procedures. The second phase was set
up as an extended cross-over multicenter trial in patient subsets prone to arrhythmias during
dialysis. Phase 1: PVC increased during both AFB and AFBK but less in the latter in the middle of dialysis (298 in AFB vs. 200 in AFBK). The PVC/h in a subset of arrhythmic patients
was 404 ± 145 in AFB and 309 ± 116 in AFBK (p = 0.0028). QT interval (QTc) prolongation was less pronounced in AFBK than in AFB. Phase 2: The PVC again increased in both
AFB and AFBK but less in the latter mid-way through dialysis (79 ± 19 AFB vs. 53 ± 13
AFBK). Moreover, in the most arrhythmic patients the benefit accruing from the smooth K^+
removal rate was more pronounced (103 ± 19 in AFB vs. 78 ± 13 in AFBK). **Conclusion:** It
is not the K^+ dialysis removal alone that can be destabilizing from an electrophysiological

standpoint, but rather its removal dynamics. This is all the more evident in patients with arrhythmias who benefit from the K^+ profiling during their dialysis treatment.

Introduction

Patients with end-stage renal disease show a prevalence of heart diseases nearly twice that of the general population. Data from United States Renal Data System clearly demonstrate that the yearly death rates per 1,000 patients at risk are mainly related to cardiac dysfunctions (from cardiac arrest or sudden death due to valvular heart disease) and range from 1.2 to 37% in the age-interval between 45 and 65 years [1]. Moreover, this incidence worsens even more for patients aged over 65 years. The main causes can be classified as mechanical heart dysfunctions, such as cardiomyopathy, atherosclerosis, ischemia, valvular heart disease; functional dysfunctions as acute myocardial infarction and coronary heart disease; and electrical dysfunctions, mainly ECG abnormalities and cardiac arrhythmias. Very often, more than one comorbid condition is present contemporaneously, leading to a negative vicious circle. In fact, in the presence of cardiomyopathy, left ventricular hypertrophy or valvular heart disease, one may observe the onset of risky cardiac arrhythmias contributing to the impairment of already critical hemodynamic conditions, which in turn lead to further mechanical heart stress and a worsening in the disease [2]. Many of these conditions are related to the primary causes of end-stage renal disease, such as hypertension or diabetes, but are also complicated and even worsened by the renal replacement therapy [3]. In fact, despite the continuous improvements in dialysis technology, replacement therapy is indubitably unphysiological due to its intermittent regimen nature, which leads to saw-toothed equilibrium states for fluids and electrolytes.

Electrolytes disorders are, therefore, one of the main causes of cardiac arrhythmias especially during the dialysis session because of their involvement in the genesis, duration, morphology and propagation of the cellular action potential.

One of the most important electrolyte disorders of uremia is the increase in the serum potassium (K^+) levels [4]. Clinical consequences of hyperkalemia are the hyperpolarization blockage of neuromuscular cells, starting with asthenia, muscular pain and constipation. The most worrying clinical outcomes are at cardiac level, inducing severe hypokinetic arrhythmias leading up to a total atrioventricular block [5]. Severe hyperkalemia leads to K^+ transfer, Na^+/K^+-ATPase pump-mediated, towards the intracellular space increasing the concentration at this site. The task of dialysis treatments is to remove the extra amount of K^+ of the interdialysis period due to the exogenous intake induced by vegetables, fruits and their juices. The amount of K^+ removed by dialysis is related to the concentration gradient between the serum K^+ levels and the K^+ content of dialysis bath [4, 6, 7]. Very low concentrations in the dialysate can remove large amounts of K^+ from the body but with an altered concentration gradient between the intra- and extracellular space, with the appearance of electrical instability of the cellular membrane, in particular in the cardiac cells. Hyperkinetic arrhythmias, which can be harmful for the cardiac function [8, 9], have been observed.

The need to compensate an adequate K^+ removal with the risk of cardiac failure has advised the use of a smooth K^+ removal during dialysis [10–12]. Markers of altered electrical cardiomyocyte behavior have been investigated for their power to predict cardiac arrhythmias [13]. Nevertheless, there is a lack of general consensus around these techniques. The most investigated index is the QT length and its dispersion [14]. Despite some criticisms leveled against the measurement method and the reproducibility of the QT interval (QTc) and QT dispersion (QTd), these indices are assumed to be predictors of ventricular arrhythmias [15]. In fact, QT length is considered to be a surrogate of cellular action potential duration, but it yields a limited view of the complex electrogenesis of the ventricular repolarization. Emerging techniques, such as the principal component analysis (PCA) of a spatial T wave [16], seems to be more promising for understanding the heterogeneity of ventricular repolarization. PCA allows the characterization of repolarization, regardless of the precise determination of the T-wave offset. This is of great importance during hemodialysis, where the electrolyte imbalances can lead to the inversion of the T wave or to the onset of the U wave, which could measure the actual T-wave offset.

The aim of the present study was to investigate the effect of two different K^+ removal rates on the onset of cardiac arrhythmias and their relationship with some ECG parameters, including global indices of ventricular repolarization homogeneity.

Study Design, Materials and Methods

The Study Was Designed in Two Phases
Phase 1. Phase 1 was designed as a single center pilot study on a small pool of patients to investigate the K^+ balance and the effects on cardiac arrhythmias (premature ventricular contractions, PVC) and on the ventricular repolarization indices of the two K^+ removal rates. The experimental set-up was a two-arm cross-over with AB and BA sequences, where A is a standard acetate-free biofiltration (AFB) and B is a profiled AFB.

Ten hemodialysis patients (mean age 66 ± 8 years) were selected from the dialysis population with no particular inclusion or exclusion criteria, except for the pacemaker holders, age below 18 or above 80 years, and the use of a central venous catheter as vascular access.

None of the patients received medications that could affect the cardiac rhythm (digoxin, antiarrhythmic, antihypertensive), at least the day before the dialysis session under study.

Phase 2. On the basis of the results obtained in the previous phase of the study, the second part was mainly addressed to investigating the real clinical effect of the two K^+ removal rates in an enlarged sample of patients. Phase 2 was designed as a multicenter cross-over experimental design with two arms and the sequences AAB and BBA. Patients were considered eligible if they had been on renal replacement therapy for at least 6 months with three times per week regimen and were treated with bicarbonate-buffered dialysate. Moreover, patients were excluded if they were pacemaker holders, or they were on home therapies with digitalis or β-blockers, or had a residual diuresis greater than 500 ml/day, or were affected by neoplastic disease. Patient selection was also based on the presence of cardiac arrhythmias in standard treatment evaluated by an ECG Holter recording during the session. Patients were classified as arrhythmic if grade 2 at least according to Lown's classification.

After assessing the selection criteria the patients were centrally enrolled and randomized to one of the two arms by the balanced block randomization technique.

The results found in Phase 1 of the study were used to design the second phase. The sample size was calculated assuming the count of ectopic beats as the main response variable and was made on the basis of the results recorded in the previous study. An average difference of 107 premature ventricular complexes was found between AFB and K^+ profiled AFB (AFBK) with a standard deviation of 120 in the non-selected sample. A two-tailed paired t test was used for the computation assuming a significance level (an error) of 0.05, and a power of 0.8. In this case, the total number of patients was 34.

Treatments

Both the treatments were AFB, a hemodiafiltration technique which is a buffer-free dialysate [17, 18]. The correction of the acid-base balance is achieved by intravenously infusing a sterile solution of sodium bicarbonate (145 mEq/l) in post-dilution mode. The infusion flow rate (usually in the range 2–2.4 l/h) is set to overcompensate the bicarbonate loss on the filter side and to gain as much bicarbonate as to achieve a plasma bicarbonate of around 28 mEq/l by the end of the treatment. The dialysate composition was the same for all the components except for K^+ [sodium (Na^+) 139 mEq/l, calcium (Ca^{2+}) 1.25 mEq/l, magnesium (Mg^{2+}) 0.37 mEq/l, chloride (Cl^-) 145 mEq/l, glucose 1 g/dl]. Standard AFB has a K^+ dialysate content equal to 2 mEq/l while in AFBK the dialysate K^+ decreases along with the treatment from an initial value of 1–1.5 mEq/l lower than the serum K^+ and a final value of 1.5 mEq/l [19].

The patient weight loss was set according to the interdialytic weight gain. Filters, blood and dialysate flow rates were set equally in both the treatments, as well as the extracorporeal circuit heparinization.

Data Acquisition

Serum Na^+ and K^+ were recorded at the beginning and at the end of each treatment, and hourly. The outlet dialysate K^+ was also recorded hourly in order to assess the K^+ removal during the treatments.

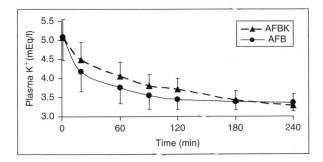

Fig. 1. Serum potassium time course during AFB and AFBK. The time course was statistically time dependent in both the procedures.

A 12-lead ECG Holter recording (H-12 Holter; Mortara Instrument Inc., Milwaukee, Wisc., USA) was obtained in all the patients. ECG recording started 15 min before the start of dialysis and lasted for 24 h. The data were then downloaded into a PC for the subsequent analysis of the cardiac arrhythmias and ECG indices.

Statistical Analysis

The descriptive analyses performed on the main variable were mean, standard deviation and standard error of the mean. All the analyzed variables were tested for their distribution normality and whenever they were found far from normal values they were log-transformed. The ANOVA for repeated measures were used for the analysis assuming time and treatments as factors. The Wilcoxon test was also used whenever appropriate. The significance level was assumed to be equal to 5%.

Results

Phase 1 Results

The main electrolytes (Na^+, Ca^{2+}, HCO_3^-) behaved similarly except for K^+. In particular, the ionized Ca^{2+} increased similarly in both the treatments starting from an initial value of 1.20 ± 0.1 in AFB versus 1.17 ± 0.13 in AFBK and rising to 1.26 ± 0.1 in AFB versus 1.22 ± 0.1 in AFBK. Serum K^+ showed a marked decrease in conventional AFB during the first half of dialysis, more than in AFBK (fig. 1). The highest difference in serum K^+ was achieved at 60 min after the start of dialysis (4.1 ± 0.4 mEq/l in AFBK vs. 3.8 ± 0.4 mEq/l in AFB), while the final values were equivalent in both the treatments. Despite this difference in serum K^+, the K^+ removed by the end of AFB and the end of AFBK were comparable (88 ± 15 mEq/l in AFBK vs. 92 ± 19 mEq/l in AFB).

Hemodynamic parameters did not change significantly between the treatments showing a decrease in mean arterial blood pressure (from 92 ± 15 to

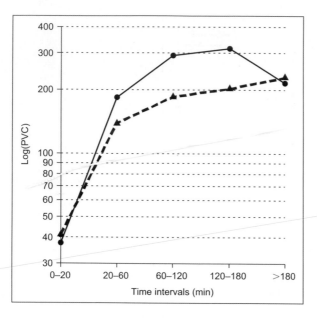

Fig. 2. Log plot of premature ventricular contractions (PVCs) in the two treatments. The figure shows the data for AFB treatment (solid line) and AFBK (dashed line).

87 ± 13 mm Hg in AFB and from 88 ± 15 to 86 ± 13 mm Hg in AFBK) offset by an increase in the heart rate from 71 ± 10 to 82 ± 17 bpm in AFB and from 72 ± 9 to 78 ± 16 bpm in AFBK.

The number of PVC increased in both treatments during the sessions but less in AFBK than in AFB. The highest difference was achieved before halfway through dialysis, while the arrhythmogenic effect of the treatments seems to be similar at the end of dialysis. As the distribution of these variables were very asymmetric (asymmetry 1.4 and kurtosis 0.5) and far from normal values, the variables were first logarithmically transformed prior to analyzing any statistical differences (see fig. 2). The ANOVA for repeated measures applied to the transformed variables did not show any statistical difference ($p = 0.428$). Since the number of PVCs is widely variable from patient to patient, we selected those patients with remarkable PVCs per hour. In particular, we chose those patients who could be classified at least within grade II of Lown's grading (number of PVC/h >30).

The average PVCs per hour are reported in table 1. Despite the small number of cases, we analyzed the statistical difference by means of the Wilcoxon test and found a significance level of 0.02.

Figure 3 shows the results of the ventricular repolarization measurements in this subgroup of patients. The QTc intervals tend to increase during dialysis

Table 1. PVCs per hour in patients selected according to Lown's grade

Patient	AFBK	AFB
1	35	45
2	626	716
3	72	100
4	685	954
5	129	290
6	302	437
Mean	309	424
SEM	116	145

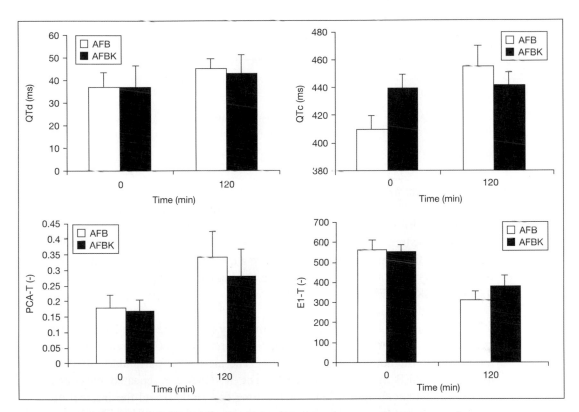

Fig. 3. Simple (QTd and QTc) and global (PCA-T and E1-T) repolarization indexes at 0 and 120 min after AFB and AFBK.

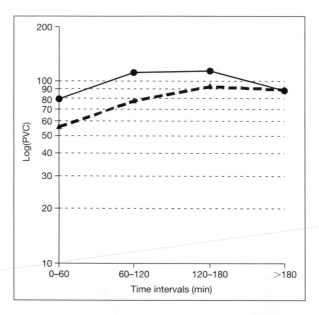

Fig. 4. Analysis during dialysis of premature ventricular contractions (PVCs) in the Phase 2 of the study. Figure reports the data for AFB (solid line) and AFBK (dashed line).

in both the treatments even though this rise is much more evident for AFB than for the AFBK. The QTd tends to increase more markedly in AFB than in AFBK although the difference did not achieve a statistical significance.

The same trend was also seen in the PCA analysis (fig. 3). Again it is worth pointing out that dialysis seems to introduce a marked ventricular repolarization heterogeneity with an effect twice that at the beginning of dialysis. Moreover, during AFBK the variations in PCA-T and E1-T were much greater than in conventional AFB.

Finally, figure 4 only reports the preliminary data of the Phase 2 part of the study. It can be noted that in arrhythmic patients, the PVCs increase in both treatments throughout the dialysis session. The increase is particularly marked in standard AFB and lower in AFBK. The ANOVA for repeated measures showed a statistical significance for the time (p = 0.004) and the interaction time by treatment (p = 0.015), but not for the treatment (p = 0.11). The difference between the two treatments becomes higher when calculating the hourly PVC (fig. 5). The scatter plot reports the PVC per hour in AFBK (thick line) against the PVC per hour in standard AFB. The scatter plot also shows the identity (thin line) and the regression line for comparison. It can be seen that the frequency of ectopic beats in AFBK lies below the identity line particularly for

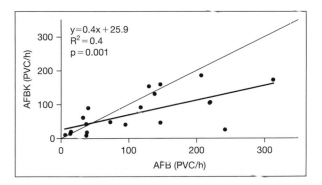

Fig. 5. Scatter plot of PVC per hour in AFBK against standard AFB. The thin line is the identity line, the thick one is the regression line.

high values in standard AFB. The regression analysis gave a value of Pearson's coefficient of 0.44. Moreover, the confidence interval at 95% of β coefficient is 0.19–0.67, meaning that it is statistically below 1.

Discussion

The Phase 1 study has shown that the K^+ removal rate can affect ECG alterations in patients who receive hemodialysis, and higher the removal rate, the more evident the arrhythmogenic effect. The stimulus provided by dialysis treatment to arrhythmias is a well-known phenomenon. In the first large survey on the dialysis population in 1988 by the *Gruppo Emodialisi e Patologie Cardiovascolari*, the incidence of ventricular and supraventricular arrhythmias was reported in 76% and 69% of the studied patients, amongst whom 21% were grade 4a and 4b arrhythmic according to the Lown's classification [20]. Dialysis therapy was reported as the first cause of arrhythmias not only during the dialysis session but also several hours after the end of the treatment. Moreover, Abe et al. [21] have shown, in a large sample of hemodialysis patients, the dependence of ECG abnormalities on several factors, such as age, time on dialysis, the presence of hypertension and diabetes. Previously, Morrison et al. [5] have highlighted an unexpected frequency of ventricular arrhythmias in patients using digoxin and with left ventricular hypertrophy. They also found that a dialysate K^+ of 2 mEq/l could lead to arrhythmias more than a dialysate K^+ content of 3.5 mEq/l could.

The present study confirms that the ventricular repolarization homogeneity can be preserved not so much by a different K^+ removal (which resulted to be equal both in AFB and AFBK) as by a smooth K^+ removal during the treatment. The indices of ventricular repolarization homogeneity, expressed by QTd, seem to provide qualitative information similar to that of more complex analysis as provided by the PCA. However, PCA analysis seems to be more sensitive to dialysis-induced electrophysiological imbalances. The Phase 2 study confirms that a smooth reduction in serum K^+ results to be less arrhythmogenic.

In 1990 Ebel et al. [22] introduced a dialysate K^+ profiling by changing the concentrates during dialysis, thus obtaining a stepwise time course. The arrhythmia frequency decreased from nearly 60% in constant dialysate K^+ to nearly 25% in dialysate K^+ profiling. Similar results were obtained by Redaelli et al. [12] using an exponential dialysate K^+ profiling versus a constant one. The reduction in ventricular ectopic beats was equal to 32%. In these studies, however, the K^+ removed was not equalized and so the real benefits of K^+ profiling could have also been affected by differences in the amount of K^+ removed. Indeed, in our study, the reduction in the arrhythmogenic effect was obtained at the same amount of K^+ removed, thus indicating that the removal rates predispose to ventricular arrhythmias.

In the second phase of the study it emerged that in the arrhythmic patients, also indeed more clearly in them, the effect of a more physiological electrophysiological response is obtained, thanks to a K^+ profile that conditions the gradual decrease in the serum K^+. Very relevant are the data on the adequate repolarization in AFBK as testified by the behavior of the simple and the complex repolarization indices (less altered in AFBK as compared with the AFB), because the arrhythmogenic stimulus originates from the altered repolarization.

References

1   United States Renal Data System: USRDS 2003 Annual Data Report. Bethesda, National Institute of Diabetes and Digestive and Kidney Diseases, NIH 2003.
2   Zipes DP, Wellens HJJ: Sudden cardiac death. Circulation 1998;98:2334–2351.
3   Leevin A: Clinical epidemiology of cardiovascular disease in chronic kidney disease prior to dialysis. Semin Dial 2003;16:101–105.
4   Ketchersid TL, Van Stone JC: Dialysate potassium. Semin Dial 1991;4:46–51.
5   Morrison G, Michelson EL, Brown S, Morganroth J: Mechanism and prevention of cardiac arrhythmias in chronic hemodialysis patients. Kidney Int 1980;17:811–819.
6   Sanders HN, Tyson IB, Bittle PA, Ramirez G: Effect of potassium concentration in dialysate total body potassium. J Ren Nutr 1998;8:64–68.
7   Sherman RA, Hawang ER, Bernholc AS, Eisinger RP: Variability in potassium removal by Hemodialysis. Am J Nephrol 1986;6:284–288.
8   Redaelli B, Bonoldi G, Di Filippo G, Viganò MR, Malnati A: Behaviour of potassium removal in different dialytic schedules. Nephrol Dial Transplant 1998;13(suppl 6):35–38.
9   Rombolà G, Colussi G, De Ferrari ME, Frontini A, Minetti L: Cardiac arrhythmias and eletctrolyte changes during haemodialysis. Nephrol Dial Transplant 1992;7:318–322.
10   Palmer BF: The effect of dialysate composition on systemic hemodynamics. Semin Dial 1992;5: 54–60.

11 Redaelli B, Limido D, Beretta P, Viganò MR: Hemodialysis using a constant potassium gradient: Rationale of a multicenter study. Int J Artif Organs 1995;8:731–734.

12 Redaelli B, Locatelli F, Limido D, Andrulli S, Signorini MG, Sforzini S, Bonoldi L, Vincenti A, Cerutti S, Orlandini G: Effect of a new model of haemodialysis potassium removal on the control of ventricular arrhythmias. Kidney Int 1996;5:609–617.

13 Suzuki R, Tsumura K, Inoue T, Kishimoto H, Morij H: QT interval prolongation in the patients receiving maintenance hemodialysis. Clin Nephrol 1998;49:240–244.

14 Day CP, McComb JM, Campbell RW: QT dispersion: An indication of arrhythmia risk in patients with long QT intervals. Br Heart J 1990;63:342–344.

15 Coumel P, Maison-Blanche P, Badalini F: Dispersion of ventricular repolarization. Reality? Illusion? Significance? Circulation 1998;97:2491–2493.

16 Priori SG, Mortara DW, Napolitano, C, Diehl L, Paganini V, Cantu F, Cantu G, Schwartz PJ: Evaluation of the spatial aspects of T-wave complexity in the long-QT syndrome. Circulation 1997;96:3006–3012.

17 Santoro A, Ferrari G, Spongano M, Badiali F, Zucchelli P: Acetate-free biofiltration: A viable alternative to bicarbonate dialysis. Int J Artif Organs 1989;13:476–485.

18 Santoro A, Spongano M, Ferrari G, Bolzani R, Augella F, Borghi M, Briganti M, Cagnoli L, Docci D, Feletti C, Fusaroli M, Gattiani A, Sanna G, Stallone C, Zucchelli P: Analysis of the factors influencing bicarbonate balance during acetate-free biofiltration. Kidney Int 1993;43(suppl 41): S184–S187.

19 Paolini F, Santoro A, Mancini E, Bosetto A: Acetate-Free Biofiltration (AFB) with constant and profiled potassium (K) concentration in the dialysate. J Am Soc Nephrol 1999;10:194.

20 Gruppo Emodialisi e Patologie Cardiovascolari: Multicentre, cross-sectional study of ventricular arrhythmias in chronically hemodialysed patients. Lancet 1988;6:305–308.

21 Abe S, Yoshizawa M, Nakanishi N, Yazawa T, Yokota K, Honda M, Slogan G: Eletcrocardiographic abnormalities in patients receiving hemodialysis. Am Heart J 1996;131:1137–1144.

22 Ebel H, Saure B, Laage C, Dittmar A, Keuchel M, Stellwagg M, Lange H: Influence of computer-modulated profile hemodialysis on cardiac arrhythmias. Nephrol Dial Transplant 1990;5 (suppl 1):165–166.

Antonio Santoro, MD
U.O. Nefrologia e Dialisi Malpighi
Policlinico S.Orsola-Malpighi, Va P.Palagi 9
IT–40138 Bologna (Italy)
Tel. +39 (0) 51 6362430, Fax +39 (0) 51 6362511, E-Mail santoro@aosp.bo.it

Ronco C, Brendolan A, Levin NW (eds): Cardiovascular Disorders in Hemodialysis.
Contrib Nephrol. Basel, Karger, 2005, vol 149, pp 306–314

....

Noninvasive Assessment of Vascular Function

Daniel Schneditz, Thomas Kenner

Institute of Physiology, Medical University Graz, Graz, Austria

Abstract

Impaired arterial compliance contributing to increased blood pressure and cardiac workload is well accepted as a major factor in cardiovascular disease. Information on local arterial compliance is obtained when analyzing the deformation of selected arterial segments under stress. A more global measure of arterial compliance is obtained by analyzing the arterial pulse by so-called pulse wave analysis. The arterial pulse, even when measured locally, carries characteristic information from the whole arterial system because of reflection of waves at distinct sites of the arterial system. Pulse wave velocity and the transfer function for pulse transmission is obtained from the combined measurement of arterial pulses at proximal and distal measuring points. Both pulse wave velocity and transfer function importantly, but not exclusively, depend on arterial compliance. The reconstruction of the aortic pulse from peripheral pulse measurements using a population-based transfer function finally provides information on central effects of reduced arterial compliance and increased peripheral resistance which may help in the diagnosis and treatment of vascular disease.

Introduction

Cardiovascular events remain the prime cause of morbidity and mortality in end-stage renal disease patients treated by hemodialysis. This involves direct cardiac as well as vascular factors. The scope of this contribution, however, is to focus on noninvasive determination of vascular factors with direct and indirect effects on cardiac function. Direct effects, for example, relate to endothelial function adjusting the lumen of the arterial conduit, whereas indirect effects, for example, relate to the local augmentation of aortic pressure, thereby contributing to left ventricular hypertrophy. Furthermore, it is the aim to present a short overview and the physiological background of established and of new techniques to measure vascular function by noninvasive means suitable for testing asymptomatic subjects where invasive techniques are not warranted and which are also suitable for repeated bedside testing to monitor the effect and success of therapy.

Arterial Compliance

Decreased arterial compliance is well recognized as a strong indicator of cardiovascular disease, peripheral vascular occlusive disease, diabetes, and renal failure. Compliance (C, in ml/mm Hg) measures the change in vessel volume per unit length (ΔV) caused by a change in pressure (Δp) and is defined as

$$C = \frac{\Delta V}{\Delta p} \tag{1}$$

Arterial compliance can be determined noninvasively from the pulsatile deformation of the arterial wall measured by imaging techniques and simultaneous pressure measurements. One limiting factor with this method is that arteries distended by arterial pressure produce only small deformations because of their nonlinear elastic behavior (fig. 1). Notice that with regard to the operating range compliance is rather constant in the aorta but nonlinear especially in more distal and muscular arteries [1, 2]. Thus, especially for the investigation of peripheral arteries, by lowering transmural pressure using external compression, the operating point can be moved to regions of increased compliance where the same pulse pressure (Δp) produces much larger deformations (ΔV) which are therefore more accurately measured by noninvasive techniques [3]. Thus, it should be easier to differentiate the diseased from the normal arterial wall. In a small group of 5 end-stage renal disease patients relative brachial artery deformation was significantly smaller (4.9 $\pm$ 1.8%) when compared to controls (32.9 $\pm$ 10.2%) measured in the undistended state when applying external compression [3]. The technique provides local information and is limited to measurements in the limbs.

Pulse Wave Velocity

Pulse waves are transmitted very rapidly through the arterial tree. The arterial pressure wave starting at the aortic root reaches the radial artery in about

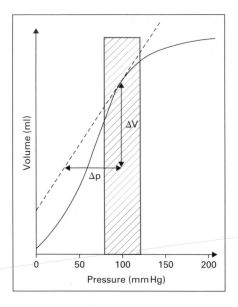

Fig. 1. Schematic representation of thoracic aortic compliance. Arterial compliance is defined as the slope of the volume versus pressure curve (full line). In central arteries compliance is more or less constant up to 100 mm Hg and rapidly decreases for higher pressures. The shaded area shows the normal operating range of the thoracic aorta. The broken line shows the slope ($\Delta V/\Delta p$) at a distending pressure of 100 mm Hg.

0.1 s, whereas a drop of blood takes at least 8 s to make the same journey. Pulse wave velocity (c, in m/s) is given by the formula:

$$c = \sqrt{\frac{V}{C\rho}} \qquad (2)$$

where C and V refer to the compliance (Eq. 1) and to the volume of the vessel per unit length, respectively, and where ρ is the density of blood. The less compliant the vessel, the greater the speed. An equivalent relationship exists for the velocity of sound in blood which is used to measure hemoconcentration during hemodialysis [4, 5]. Pulse wave velocities in the circulation range from 1 m/s in the very compliant pulmonary artery to more than 15 m/s in small systemic arteries. Pulse wave velocity was long regarded as a matter of more theoretic than practical interest, but has been rediscovered as a potential measure of vascular compliance. To determine pulse wave velocity (c = d/t), carotid and radial, femoral or tibial pulses are recorded at proximal and distal measuring sites using the R wave of a simultaneously recorded ECG as reference frame to measure the transit time (t). The distance (d) traveled by the pulse wave is approximated from anthropometric measurements taken over the body surface. In a recent study done in 10 end-stage renal disease patients carotid pulse wave velocity (9.9 ± 3.1 m/s) was high before hemodialysis when mean arterial pressure was high (108 ± 18 mm Hg) and fell to control levels (8.0 ± 1.4 m/s) following hemodialysis and angiotensin-converting enzyme inhibition when mean arterial pressure was low (87 ± 10 mm Hg) [6]. One important caveat in the interpretation of the measured decrease in pulse wave velocity, however, is the effect of blood pressure. A decrease in blood pressure by 10 mm Hg alone will lead to a reduction in pulse wave velocity by 0.5–1 m/s because arterial compliance increases as transmural pressure decreases (fig. 1) [7]. On the other hand, when the mechanical properties of the system are known and stable, the measurement of pulse wave velocity may be used to determine blood pressure [8].

Flow-Mediated Dilatation

Endothelial dysfunction is considered the first functionally significant stage of atherosclerosis and primarily refers to impaired endothelium-dependent vasodilatation even though it probably includes other aspects of impaired endothelial function [9]. The most widely used noninvasive test for endothelial dysfunction involves the measurement of brachial artery diameter in response to increased shear stress and NO release which causes endothelium-dependent dilatation. Arterial diameter is usually measured by ultrasound before and during a hyperemic phase following a short period of arterial occlusion of the forearm by means of a conventional cuff sphygmomanometer inflated to a pressure above systolic pressure. Arterial diameters are measured at the end of diastole using the easily visualized lumen-intima interface. The change in arterial diameter in the hyperemic phase is expressed as a percentage of the original value. In a retrospective analysis of flow-mediated dilatation measured in patients with suspected coronary artery disease, endothelial dysfunction has been defined as a dilatation less than 4.5% [10]. The main limitation to this technique is its inherent biological variability and its dependency on the examiner [11]. The vasodilator response varies largely within an individual in response to many different variables such as baseline artery diameter and the degree of ischemia which can be increased by ischemic handgrip exercise. As there is no ideal measurement to assess the accuracy of the technique, rigorous attention to protocol standardization and training is critical for generating valid and reproducible data. It also should be noted that the ultrasound technique is intended to measure vasodilatation in large conduit arteries which may react differently than small peripheral arteries.

Pulse Wave Analysis

The arterial pulse is one of the most fundamental signs in clinical medicine and its interpretation has been a cornerstone of medical practice since antiquity. The arterial pulse which is composed of pressure and flow components is the result of a complex interaction between the heart and the vasculature and carries global information on functional aspects of the high pressure part of the cardiovascular system [12]. This overview deals with the pressure component of the arterial pulse. While the arterial pulse is characteristic when measured at different locations of the vascular tree, it is apparently influenced by all parts of the arterial system. The identification of a specific contribution of selected components, however, remains difficult. Recently, improved and more reliable techniques have become available to measure arterial pulses noninvasively which has led to a revival of pulse wave analysis.

Physiological Considerations

With each heartbeat the pressure in the ascending aorta rises and falls in a characteristic, asymmetrical way causing a pulsatile distension of the ascending aorta. The magnitude of the pressure oscillations between diastole and systole, the pulse pressure, essentially depends on stroke volume and arterial compliance. Arterial compliance is influenced by mean blood pressure because compliance decreases as pressure rises (fig. 1), the rate of ventricular ejection because the arterial wall is viscoelastic, and mechanical properties of the arterial wall which change with age and as a consequence of vascular disease. As pulsations move on through the vascular system, the properties of the arterial tree transform the shape and timing of the waves mainly because of wave reflection in the periphery, changes in mechanical properties of the arterial conduit, and the tapering of the vessels [13]. The magnitude of peripheral wave reflection importantly depends on peripheral resistance and hence on the degree of vasoconstriction or vasodilatation [14].

The actual shape of the arterial pulse at a given location can thus be thought as a superposition of the pulse caused by the expulsion of stroke volume into the viscoelastic arterial tree and multiple (forward and backward) reflections at hemodynamic discontinuities in the artery (fig. 2). Reflection at an obstacle such as peripheral resistance or the closed aortic valve is positive and causes pressure augmentation while reflection at a hemodynamic discontinuity such as the opening of the brachial artery into the aorta is negative and causes pressure diminution. The superposition of incident and reflected waves may occur at different phases of the pulse cycle and depends on pulse wave velocity as well as on the location of the sampling site. Thus, the reflected wave may be added to systolic or diastolic parts of the local arterial pulse.

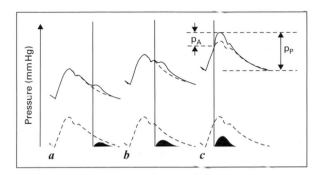

Fig. 2. Effect of wave reflection on aortic pulse. The aortic pulse (full line) results from a superposition of incident and reflected wave components originating from different sites of the arterial tree. The effect of one particular reflected wave (black area) on the pulse contour lacking this reflection (broken line) is shown. Since transmission times of incident and reflected wave components decrease with increased pressure and/or decreased compliance, the superposition of wave components may occur at different phases of the aortic pulse. The superposition may occur late in the diastolic phase of the aortic pulse such as in hypotension (*a*), early in the diastolic phase such as under normal conditions (*b*), or during the systolic part of the aortic pulse such as with reduced arterial compliance and/or hypertension (*c*). With systolic superposition (*c*), the aortic pulse often shows a systolic inflection which can be used to determine aortic pulse augmentation (p_A). The augmentation index is defined as ratio of p_A to pulse pressure (p_P).

Transfer Function

The transformation of the arterial pulse measured at two locations of the arterial tree can be described by its transfer function. This function describes the characteristics of the transmission of pulses between measuring sites determined by physical properties such as vessel compliance, length, and diameter. The transfer function of a linear system is obtained by computing the amplitudes and phase angles for the frequencies obtained by Fourier analysis for both pulses and further computing ratios of amplitudes and differences of phase angles as described elsewhere [13]. In the so-called locus plot the transfer function presents as a characteristic loop circling the origin. Once the transfer function for a given transmission line is known, the input to the transmission line can be reconstructed from the measured output. This approach assumes linear behavior of the transmission line, i.e. doubling the input is expected to double the output. However, arterial compliance is nonlinear, which should be kept in mind when dealing with transfer functions.

Aortic Pulse Analysis

Aortic pressure appears to be of special clinical interest as it determines cardiac afterload. Noninvasive aortic pulse analysis first and foremost depends on accurate recording of the radial pulse wave by tonometric or plethysmographic techniques.

Tonometry

The principle of applanation tonometry is based on flattening the artery by pressure applied from the pressure sensor against a fixed support behind the artery. In this state the vessel wall is relaxed and the pressure is entirely transmitted to the piezoelectric transducer located in the sensor head. Reasonable confidence is gained if the measured amplitude is the greatest that can be reached. Prime locations for tonometric measurements are radial, tibial, and carotid arteries. The technique is realized in the CBM-7000 (Colin Corp., Japan) and in the Sphygmocor (AtCor Medical, Australia).

Finger Plethysmography

A continuous record of digital arterial pressure is obtained by finger plethysmography using the so-called vascular unloading technique. In this technique the finger is wrapped by a small inflatable cuff and finger blood volume which oscillates with each arterial pulse is measured by optical means. The volume signal is fed to a pump to change the cuff pressure and to compensate for the oscillations in finger blood volume [15]. If this condition is fulfilled the cuff pressure follows digital arterial pressure. Comparable to tonometry, the oscillations in counter-pressure reach a maximum when the arterial wall is relaxed, i.e. when the transmural pressure between the vessel and the tissue disappears. In this case, the cuff pressure, which is easily measured, matches the pressure in the artery. The technique is realized in the Finometer (Finapres Medical Systems, The Netherlands) and in the TaskForce Monitor (CNS Systems, Austria).

Aortic Pressure Reconstruction

If direct measurements are not available aortic pressure pulses can be reconstructed from the radial pulse measured by noninvasive techniques using the transfer function approach described above. However, without additional measurements the specific transfer function in the individual to be studied is in fact unknown. This problem has been addressed by assuming a population-based transfer function for the upper limb segment of the arterial tree [16, 17]. To support the assumption of a unique and stable transfer function it is claimed that upper arm length is not very different between adults, that the brachial artery is rarely involved in atherosclerotic lesions, and that brachial pulse wave velocity shows little changes with age.

Augmentation Index

Reflection of the arterial pulse wave into the systolic phase of the aortic pressure pulse leads to an increase in aortic systolic pressure (fig. 2). The augmentation index, which is considered a measure of systemic arterial stiffness, is calculated as the difference between primary and augmented systolic peaks expressed as a percentage of the pulse pressure [18]. In a recent study it was seen that both administration of the ß$_2$-agonist albuterol activating the L-arginine-NO pathway as well as of nitroglycerine significantly and repeatedly reduced the augmentation index through an endothelium-dependent effect [19].

There are a few questions regarding aortic pulse analysis technique. First, since aortic pressure is reconstructed from the radial pulse using a constant transfer function, all the information must be contained in the radial pulse curve. The question therefore arises why the radial pulse curve is not directly analyzed? Second, there is no need to assume a constant transfer function when the actual transfer function can be determined from the measurement of a second arterial pulse. Why is the actual transfer function not determined? And third, it must always be remembered that the use of transfer functions is based on linear system behavior which may not be applicable in a variety of pathological settings.

Conclusion

The techniques discussed in this short overview circulate around the theme of arterial compliance which, for example, is approached by local measurements done at selected points of the vascular tree. On the other hand, arterial compliance has systemic effects which can be observed throughout the arterial system. Different techniques have been developed to identify altered arterial compliance from the noninvasive measurement of pulse wave velocity or from noninvasive pulse wave analysis. However, changes in pulse wave velocity or pulse wave contour are not exclusively and not always related to altered arterial compliance. Therefore, results obtained from noninvasive measurements and based on many inferences need to be interpreted with special caution. On the other hand, noninvasive techniques have the potential of widespread application which will help to test and validate the concepts developed in theoretic and experimental models over many decades and which hopefully will finally be of help in early diagnosis and improved treatment of vascular disease.

References

1   Weizsäcker HW, Kampp TD: Passive elastic properties of the rat aorta. Biomed Technik 1990;35:224–234.
2   Weizsäcker HW, Holzapfel GA, Desch GW, Pascale K: Strain energy density function for arteries from different topographical sites. Biomed Technik 1995;40(suppl 2):139–142.

3   Weitzel WF, Kim K, Rubin JM, Xie H, O'Donnell M: Renal advances in ultrasound elasticity imaging: Measuring the compliance of arteries and kidneys in end-stage renal disease. Blood Purif 2005;23:10–17.

4   Schneditz D, Heimel H, Stabinger H, Kenner T, Leopold H: Determination of velocity of sound in flowing blood and its application in hemodialysis; in Hutten H (ed): Science and Technology for Medicine: Biomedical Engineering in Graz. Lengerich, Pabst Science Publishers, 1995, pp 479–493.

5   Schneditz D: Extracorporeal sensing techgniques. Contrib Nephrol 2005;149:35–41.

6   Tycho Vuurmans JL, Boer WH, Bos WJ, Blankestijn PJ, Koomans HA: Contribution of volume overload and angiotensin II to the increased pulse wave velocity of hemodialysis patients. J Am Soc Nephrol 2002;13:177–183.

7   Schimmler W: Über die Beziehung zwischen der Pulswellengeschwindigkeit in der Aorta-Iliaca und dem Blutdruck bei longitudinaler Beobachtung. Basic Res Cardiol 1975;70:46–57.

8   Kenner T: Indirect measurement of blood pressure without a cuff. Z Kardiol 1996;85(suppl 3):45–50.

9   Lüscher TF: The endothelium and cardiovascular disease – A complex relation. N Engl J Med 1994;330:1081–1083.

10 Schroeder S, Enderle MD, Ossen R, Meisner C, Baumbach A, Pfohl M, Herdeg C, Oberhoff M, Haering HU, Karsch KR: Noninvasive determination of endothelium-mediated vasodilation as a screening test for coronary artery disease: Pilot study to assess the predictive value in comparison with angina pectoris, exercise electrocardiography, and myocardial perfusion imaging. Am Heart J 1999;138:731–739.

11 Šejda T, Pit'ha J, Švandovà E, Poledne R: Limitations of non-invasive endothelial function assessment by brachial artery flow-mediated dilatation. Clin Physiol Funct Imaging 2005;25:58–61.

12 Kenner T: Physical and mathematical modeling in cardiovascular systems; in Hwang NHC, Gross DR, Patel DJ (eds): Quantitative Cardiovascular Studies. Baltimore, University Park Press, 1979, pp 41–109.

13 Wetterer E, Kenner T: Grundlagen der Dynamik des Arterienpulses. Berlin/Heidelberg/New York, Springer Verlag, 1968.

14 Kenner T: Die Beziehung zwischen dem endsystolischen Druck und dem peripheren Widerstand am Schlauchmodell. Z Kreislaufforsch 1959;48:570–574.

15 Peñáz J, Voigt A, Teichmann W: Beitrag zur fortlaufenden indirekten Blutdruckmessung. Z Gesamte Inn Med 1976;31:1030–1033.

16 Karamanoglu M, O'Rourke MF, Avolio AP, Kelly RP: An analysis of the relationship between central aortic and peripheral upper limb pressure waves in man. Eur Heart J 1993;14:160–167.

17 O'Rourke MF, Pauca A, Jiang XJ: Pulse wave analysis. Br J Clin Pharmacol 2001;51:507–522.

18 Safar ME, London GM: Therapeutic studies and arterial stiffness in hypertension: Recommendations of the European Society of Hypertension. The Clinical Committee of Arterial Structure and Function. Working Group on Vascular Structure and Function of the European Society of Hypertension. J Hypertens 2000;18:1527–1535.

19 Wilkinson IB, Hall IR, MacCallum H, Mackenzie IS, McEniery CM, van der Arend BJ, Shu YE, MacKay LS, Webb DJ, Cockcroft JR: Pulse-wave analysis: Clinical evaluation of a noninvasive, widely applicable method for assessing endothelial function. Arterioscler Thromb Vasc Biol 2002;22:147–152.

Daniel Schneditz, PhD
Institute of Physiology, Center for Physiological Medicine
Medical University Graz, Harrachgasse 21/5
AT–8010 Graz (Austria)
Tel. +43 (316) 380 4269, Fax +43 (316) 380 9630, E-Mail daniel.schneditz@meduni-graz.at

Ronco C, Brendolan A, Levin NW (eds): Cardiovascular Disorders in Hemodialysis.
Contrib Nephrol. Basel, Karger, 2005, vol 149, pp 315–324

....

New Insights in Uremic Toxicity

R. Vanholder, G. Glorieux, N. Lameire

Nephrology Section, Department of Internal Medicine,
University Hospital, Ghent, Belgium

Abstract

The uremic syndrome is characterised by the retention of a host of compounds that in healthy subjects are secreted by the kidneys into normal urine. These compounds disturb many physiologic functions, resulting in toxicity. Many of the responsible compounds remain unknown, however, as well as many patho-physiologic actions of the known retention solutes. In this publication, we review recent new information regarding uremic toxicity. Especially difficult to remove compounds, such as protein bound and larger molecules, seem to play a role. New strategies enhancing their removal might be highly useful.

Introduction

The uremic syndrome is attributed to the retention of a myriad of compounds, which under normal conditions are cleared by the healthy kidneys. If those compounds interfere with biological/biochemical functions, they are called uremic toxins. This toxicity emanates in the loss of a substantial number of body functions, and their deterioration progresses gradually as uremic retention becomes more important due to loss of kidney function.

Initial characterization and classification of relevant toxins depends to a large extent on in vitro studies, whereby potential toxins are brought into contact with cell systems as a reproduction of uremic functional changes. Those studies help to identify new potential toxins and/or to characterize new as yet unrecognized functional defects attributable to already known compounds. The present review gives a summary of the most recent information that has been obtained regarding newly identified potential toxins and newly characterized toxic effects of known solutes.

For a summary of known uremic toxins, we refer to a recent publication by the European Uremic Toxin Work Group (EUTox) which offered an encyclopedic review of the known retention solutes together with their normal and uremic concentrations [1]. In the latter analysis, 90 different uremic solutes were identified and subdivided according to their physicochemical characteristics into: (1) small water-soluble compounds (MW <500 Da); (2) protein-bound solutes; (3) larger middle molecules (MW >500 Da). Although most protein-bound solutes also have a MW <500 Da, some of those are substantially larger (e.g. leptin, MW = 16,000). The physicochemical characteristics are important to explain the behavior of compounds during dialysis, the small water-soluble compounds being removed easily by all dialysis strategies, and the larger middle molecules being difficult to remove unless if dialyzers with a large pore size are applied (so-called high-flux membranes). The protein-bound compounds are difficult to remove in spite of low molecular weight because protein binding imposes resistance against transmembrane transfer. In what follows, we will indicate, if possible, the physicochemical characteristics of the compounds under discussion.

Old Known Uremic Toxins with New Toxic Effects

Guanidino Compounds

Guanidino compounds are small water-soluble solutes to which until recently essentially neurotoxic effects had been attributed [2]. Many guanidino compounds, especially guanidine, guanidinosuccinic acid and creatinine as a precursor of methylguanidine, and methylguanidine itself are highly increased in uremic biological fluids and tissues [3].

Until recently, no effects playing a role in cardiovascular damage had been attributed to the guanidines, except for asymmetric dimethylarginine [4], which was shown to be a strong inhibitor of inducible nitric oxide synthase (iNOS) [5], an enzyme with protective impact on vascular endothelium. The concentrations of this compound as they are encountered in uremia [6] however not necessarily correspond to those inhibiting iNOS [5].

In a recent study, Glorieux et al. [7] demonstrated, however, how several other guanidino compounds than asymmetric dimethylarginine also had a potential damaging effect on the vessels, as they appeared to stimulate leukocyte function. Leukocytes have been attributed a role in atherogenesis. Atheromatosis, formerly considered a degenerative disease, is now seen as an inflammatory disorder whereby activated leukocytes adhere to the endothelial vessel wall, and induce the first steps of a process eventually resulting in plaque formation, stenosis and thrombosis [8]. Of note, kidney failure is linked to inflammation, which in its turn is related to atherogenesis, malnutrition and mortality [9].

In the in vitro study by Glorieux et al. [7], several guanidino compounds elicited proinflammatory effects on leukocytes. Methylguanidine and guanidine stimulated the proliferation of undifferentiated HL-60 cells and the antiproliferative effect of calcitriol was neutralized in the presence of methylguanidine and guanidinosuccinic acid. The phorbol myristate-acetate-stimulated chemiluminescence production of the calcitriol differentiated HL-60 cells was enhanced in the presence of guanidine. Methylguanidine and guanidinoacetic acid enhanced the lipopolysaccharide-stimulated intracellular production of TNF-α by normal human monocytes. Until then, most guanidino compounds had been attributed neutral or even leukocyte inhibiting effects [10], be it with less refined techniques.

In another recent study, Perna et al. [11] evaluated the impact of a panel of uremic solutes on the structural modification of albumin by deamidation. Plasma proteins in hemodialysis display an increase in deamidated/isomerized asparagine and aspartic acid content, and also incubation of normal plasma in the presence of various uremic toxins led to an increased deamidated/isomerized content. The same effect was seen in purified human albumin. The two most active toxins in this respect, showing a dose-response effect, were guanidine and guanidinopropionic acid. Deamidated human albumin showed a reduced binding capacity to homocysteine, hence increasing the availability of nonprotein-bound homocysteine, and the toxic potential of this compound.

Guanidino compounds, are small water-soluble substances, and in this way, could be considered to show similar kinetic behavior as urea, which is easily removed by any dialysis strategy, even standard methods such as low-flux dialysis. Until recently, the kinetic behavior of guanidines had, however, not been submitted to evaluation. Eloot et al. [12] subjected several guanidino compounds, such as guanidine, methylguanidine, creatinine, creatine, guanidinosuccinic acid and guanidinoacetic acid to intradialytic kinetic analysis, and compared those kinetics to that of urea. In spite of being water-soluble and small like urea, all guanidino compounds showed a different kinetic behavior than urea, and especially their distribution volume was markedly higher with the exception of guanidinosuccinic acid. The distribution volumes of methylguanidine and guanidinoacetic acid were even more than twice as large as that of urea. This resulted in a dramatical decrease of effective removal which can only be corrected by changing the dialytic concept, essentially by increasing dialysis duration and/or frequency.

In summary, recent data point to a role for guanidino compounds in vascular toxicity, next to their traditionally accepted neurotoxicity. In spite of their physicochemical characteristics which seemingly are comparable to those of urea, their intradialytic behavior is markedly different, so that they can be classified as difficult to remove molecules. In this way, their behavior resembles that of other water-soluble compounds, such as phosphate and the purines xanthine and hypoxanthine [13].

Advanced Glycation End Products

Advanced glycation end products (AGEs) are generated through chemical rearrangements and degradation reactions from stable Amadori products and Schiff-base adducts which are the result of a nonenzymatic reaction of glucose or reducing sugars with free amino groups. Next to their generation during the ageing process, they were also recognized to occur at increased concentrations in diabetes mellitus and renal failure. Oxidative and carbonyl stress were pointed out as responsible actors in the generation of AGEs in uremia, rather than reactions with glucose [14].

Repeated studies demonstrated the proinflammatory effect of AGEs, mediated by leukocyte activation, and pointing to their potential role in uremic vascular damage [15, 16]. All previous studies in this context had, however, been undertaken with artificially generated mixes of AGEs, whereby it remained unclear whether these structures were representative for the AGEs encountered in vivo in uremia.

In a recent study, Glorieux et al. [17] applied albumin preparations, which had been modified chemically at lysine and arginine residues, to contain N-ε-carboxymethyllysine, N-ε-carboxyethyllysine, glyoxal-induced imidazolinones (Arg I) and methylglyoxal-induced imidazolinones (Arg II). These four compounds are also found in vivo in uremia [18]. N-ε-carboxyethyllysine, N-ε-carboxymethyllysine as well as Arg I albumin markedly induced leukocyte response, as characterized by increased chemiluminescence production, an indicator of free radical production and enhanced CD14 expression, a lipopolysaccharide receptor. Arg I albumin also induced leukocyte proliferation.

These data confirm the proinflammatory and hence atherogenic role of the AGEs, this time in the presence of genuine AGEs, as they are accumulated in the body of uremic patients [18].

Protein-Bound Uremic Solutes

The group of protein-bound solutes contains a heterogeneous set of molecules, such as indoxyl sulfate, hippuric acid, homocysteine, and the phenolic compounds which are all characterized by the difficulty to remove them during dialysis [19].

Multiple toxic effects have been attributed to these compounds, but up till recently no data were available regarding a potential vascular impact, except for the presumed effect of homocysteine.

Dou et al. [20] recently investigated the in vitro effect of a large panel of uremic-retention solutes on endothelial proliferation and wound repair on human umbilical vein endothelial cells. Endothelial proliferation was inhibited by two protein-bound uremic-retention solutes: p-cresol and indoxyl sulfate. Inhibition of endothelial proliferation by p-cresol was dose dependent. Moreover, p-cresol and indoxyl sulfate decreased endothelial wound repair. The presence of albumin did not affect the inhibitory effect of these solutes on endothelial proliferation. Hence, these solutes could play a role in endothelial dysfunction observed in uremic patients.

Cerini et al. [21] tested the effect of p-cresol on the barrier function and the permeability of the endothelium. Permeability was markedly increased in the presence of p-cresol, in presence or absence of albumin, and at the same time a reorganization of the cytoskeleton and an alteration of the adherent junctions was observed.

To these data should be added the studies by D'Hooge et al. [22], showing neurotoxic effects for indoxyl sulfate, spermine, phenol and p-cresol, and by Canalejo et al. [23], showing that HPLC-fractions containing protein-bound compounds, and phenol as individual compound, inhibited the impact of calcitriol on the parathyroid gland and induced calcitriol resistance.

Of note, two recent clinical studies pointed to a relationship between p-cresol and the clinical condition in uremia [24, 25].

The potential impact of guanidino compounds on the liberation of homocysteine from its albumin-binding sites was already pointed out above [11]. Homocysteine has further also been linked to altered DNA methylation and altered gene expression patterns, and treatment with folic acid not only decreases plasma homocysteine levels, but it also restores DNA and gene expression back to normal [26].

New Unknown Uremic Toxins with Old Toxic Effects

Phenylacetic Acid

Phenylacetic acid is a protein-bound compound which previously had been identified at increased concentrations in uremic plasma and was shown to inhibit Ca^{2+}-ATPase [27].

More recently, Jankowski et al. [28] tested the hypothesis that uremic toxins are responsible for reduced iNOS expression. Nitric oxide (NO) prevents atherogenesis and inflammation in vessel walls by inhibition of cell proliferation and cytokine-induced endothelial expression of adhesion molecules and proinflammatory cytokines. Reduced NO production due to inhibition of either endothelial NOS or iNOS may therefore reinforce atherosclerosis. Patients with end-stage renal failure show markedly increased mortality due to atherosclerosis. In the above-mentioned study lipopolysaccharide-induced iNOS expression in mononuclear leukocytes was studied using real-time PCR [28]. The iNOS expression was blocked by addition of plasma from patients with end-stage renal failure, whereas plasma from healthy controls had no effect. Hemofiltrate obtained from patients with end-stage renal failure was fractionated by chromatographic methods. The chromatographic procedures revealed a homogenous fraction that inhibits iNOS expression. Using gas chromatography/mass spectrometry, this inhibitor was identified as phenylacetic acid. Authentic phenylacetic acid inhibited iNOS expression in a dose-dependent manner. In healthy control subjects, plasma concentrations were below the detection level, whereas patients with end-stage renal failure had a substantial increase in phenylacetic acid concentration. It was concluded that accumulation of phenylacetic acid in patients with end-stage renal failure inhibited iNOS expression and that the mechanism may contribute to increased atherosclerosis and cardiovascular morbidity in patients with end-stage renal failure.

Dinucleoside Polyphosphates

Dinucleoside polyphosphates are newly detected protein-bound uremic solutes which are structurally constituted by two nucleosides at both ends of the structure, which are linked together by a variable number of phosphate moieties [29]. Diadenosine polyphosphates (Ap_xA) had previously been demonstrated to be accumulated in renal failure, especially in the platelets, and to induce smooth muscle cell proliferation [29].

In this group of compounds, recently a new moiety was detected: uridine adenosine tetraphosphate (Up_4A) [30]. It was shown that this compound was present in endothelial cells and was released from those cells when they were stimulated. It appeared that this compound was a much stronger vasoconstrictor than endothelin. Its effect could be neutralized by specific purine receptor antagonists. Uremic concentrations of these compounds, either intra- or extracellular, are not yet known. If we extrapolate the data known for Ap_xA, it might however be possible that we are confronted here with a new strong uremic toxin.

Unidentified Peptides

Reviewing the until-now identified uremic-retention compounds reveals that to date, only close to 100 uremic-retention solutes have been described [1]. It is very likely that much more compounds are retained in uremia. To examine unknown uremic substances thoroughly, the identification of as many compounds as possible in the ultrafiltrate and/or plasma of patients would lead to a less biased definition of the uremic-retention process compared with what is proposed today. Proteomic analysis is a novel tool for the identification of a large number of molecules present in biological fluids. In a study on ultrafiltrate from uremic and normal plasma obtained with high- or low-flux dialysis membranes, Weissinger et al. [31] used separation by capillary electrophoresis coupled to online mass spectrometry, yielding identification of as yet unknown polypeptides based on their molecular weight [31]. Between 500 and >1,000 polypeptides with a molecular weight ranging from 800 to 10,000 Da could be detected in individual samples, and were identified via their mass and their particular migration time in capillary electrophoresis. In ultrafiltrate from uremic plasma, 1,394 polypeptides were detected in the high-flux versus 1,046 in the low-flux samples, while 544 polypeptides versus 490 were found in ultrafiltrate from normal plasma obtained from membranes with comparable cutoff. In addition, polypeptides >5 kDa were virtually only detected in the uremic ultrafiltrate from the high-flux membrane (n = 28 vs. n = 5 with the low-flux membrane). To demonstrate the feasibility of further characterizing the detected molecules, polypeptides present exclusively in uremic ultrafiltrate were chosen for sequencing analyses. A 950.6-Da polypeptide was identified as a fragment of the salivary proline-rich protein and a 1,291.8-Da fragment was a derivative from α-fibrinogen. These data strongly suggest that the application of proteomic approaches such as capillary electrophoresis and mass spectrometry will result in the identification of many more uremic solutes than those known at present.

It is conceivable that similar strategies might be useful as well for the analysis of smaller, nonpeptidic compounds.

Conclusions

Uremic retention is characterized by the accumulation in the body of a host of compounds. Some of these are relatively innocent, such as urea, whereas others exert a myriad of biological and biochemical activities, resulting at the clinical level in the uremic syndrome. Unfortunately, our knowledge about the true dimensions of this condition is, however, incomplete. On the one hand, many retention compounds with a biological effect remain unknown. On the other hand, we still detect new activities for compounds that have been identified years before. There is an urgent need for a complete mapping and classification according to importance of uremic-retention compounds, before the development of therapeutic strategies will be in the possibility to exceed the stage of empiricism.

It might be interesting to transfer these ideas to the clinical problem which is as of today one of the greatest concerns for the nephrological community: the enhanced cardiovascular risk in uremia. Patients with renal failure are prone to develop cardiovascular problems and these start long before the dialysis stage has been reached. Moreover, the weight of kidney failure in its relation to cardio-vascular risk remains maintained, even if corrections are applied for traditional as well as less-traditional risk factors. Subsequently, it is acceptable that renal failure per se is also a cause of vascular lesions, and if this is the case, it is conceivable that vascular damage is provoked by the factors retained in renal failure.

Some of these factors, as well as their potential role in cardiovascular damage, are known and might be related to functional changes in white blood cells, platelets, smooth muscle cells and endothelium [18]. While reviewing these factors in 2001 [18], it became clear already that most factors involved were either larger 'middle' molecules or protein-bound molecules. The data reviewed at present add to this perception. The only solutes which do not conform with this definition, are the guanidino compounds, but according to kinetic studies, these substances also show a kinetic behavior that is divergent from the standard small water-soluble compound, which is urea. Hence, most or at least a substantial part of molecules at play in uremic toxicity and vascular damage seem to be difficult to remove, so that standard dialytic approaches might be less relevant for their removal. To this should be preferred alternative approaches such as convective strategies, treatment with extremely open membranes which are permeable for proteins, adsorption, alternative timeframes (long slow dialysis, daily dialysis) as well as medicamentous approaches aiming at blockage of receptors and activation pathways induced by toxins, as well as at modification of the metabolic generation or breakdown of those toxins.

In parallel to these efforts, a more complete mapping of responsible toxins and a classification of their importance should be pursued as well.

References

1 Vanholder R, De Smet R, Glorieux G, Argiles A, Baurmeister U, Brunet P, Clark W, Cohen G, De Deyn PP, Deppisch R, Descamps-Latscha B, Henle T, Jorres A, Lemke HD, Massy ZA, Passlick-Deetjen J, Rodriguez M, Stegmayr B, Stenvinkel P, Tetta C, Wanner C, Zidek W: Review on uremic toxins: Classification, concentration, and interindividual variability. Kidney Int 2003;63:1934–1943.

2 De Deyn PP, D'Hooge R, Van Bogaert PP, Marescau B: Endogenous guanidino compounds as uremic neurotoxins. Kidney Int Suppl 2001;78:S77–S83.

3 De Deyn PP, Marescau B, Cuykens JJ, Van Gorp L, Lowenthal A, De Potter WP: Guanidino compounds in serum and cerebrospinal fluid of non-dialyzed patients with renal insufficiency. Clin Chim Acta 1987;167:81–88.

4 Zoccali C, Benedetto FA, Maas R, Mallamaci F, Tripepi G, Malatino LS, Boger R: Asymmetric dimethylarginine, C-reactive protein, and carotid intima-media thickness in end-stage renal disease. J Am Soc Nephrol 2002;13:490–496.

5 Vallance P, Leone A, Calver A, Collier J, Moncada S: Accumulation of an endogenous inhibitor of nitric oxide synthesis in chronic renal failure. Lancet 1992;339:572–575.

6 Marescau B, Nagels G, Possemiers I, De Broe ME, Becaus I, Billiouw JM, Lornoy W, De Deyn PP: Guanidino compounds in serum and urine of nondialyzed patients with chronic renal insufficiency. Metabolism 1997;46:1024–1031.

7   Glorieux GL, Dhondt AW, Jacobs P, Van Langeraert J, Lameire NH, De Deyn PP, Vanholder RC: In vitro study of the potential role of guanidines in leukocyte functions related to atherogenesis and infection. Kidney Int 2004;65:2184–2192.

8   Ross R: Atherosclerosis – An inflammatory disease. N Engl J Med 1999;340:115–126.

9   Stenvinkel P, Heimburger O, Paultre F, Diczfalusy U, Wang T, Berglund L, Jogestrand T: Strong association between malnutrition, inflammation, and atherosclerosis in chronic renal failure. Kidney Int 1999;55:1899–1911.

10 Hirayama A, Noronha-Dutra AA, Gordge MP, Neild GH, Hothersall JS: Inhibition of neutrophil superoxide production by uremic concentrations of guanidino compounds. J Am Soc Nephrol 2000;11:684–689.

11 Perna AF, Ingrosso D, Satta E, Lombardi C, Galletti P, D'Aniello A, De Santo NG: Plasma protein aspartyl damage is increased in hemodialysis patients: Studies on causes and consequences. J Am Soc Nephrol 2004;15:2747–2754.

12 Eloot S, De Smet R, Torremans A, De Wachter D, Marescau B, De Deyn P, Verdonck P, Vanholder R: Urea kinetics are not representative for the behavior of the guanidino compounds. Kidney Int, in press.

13 Vanholder RC, De Smet RV, Ringoir SM: Assessment of urea and other uremic markers for quantification of dialysis efficacy. Clin Chem 1992;38:1429–1436.

14 Miyata T, Wada Y, Cai Z, Iida Y, Horie K, Yasuda Y, Maeda K, Kurokawa K, van Ypersele de Strihou C: Implication of an increased oxidative stress in the formation of advanced glycation end products in patients with end-stage renal failure. Kidney Int 1997;51:1170–1181.

15 Miyata T, Iida Y, Ueda Y, Shinzato T, Seo H, Monnier VM, Maeda K, Wada Y: Monocyte/macrophage response to beta 2-microglobulin modified with advanced glycation end products. Kidney Int 1996;49:538–550.

16 Witko-Sarsat V, Friedlander M, Nguyen KT, Capeillere-Blandin C, Nguyen AT, Canteloup S, Dayer JM, Jungers P, Drueke T, Descamps-Latscha B: Advanced oxidation protein products as novel mediators of inflammation and monocyte activation in chronic renal failure. J Immunol 1998;161:2524–2532.

17 Glorieux G, Helling R, Henle T, Brunet P, Deppisch R, Lameire N, Vanholder R: In vitro evidence for immune activating effect of specific AGE structures retained in uremia. Kidney Int 2004;66: 1873–1880.

18 Vanholder R, Argiles A, Baurmeister U, Brunet P, Clark W, Cohen G, De Deyn PP, Deppisch R, Descamps-Latscha B, Henle T, Jorres A, Massy ZA, Rodriguez M, Stegmayr B, Stenvinkel P, Wratten ML: Uremic toxicity: Present state of the art. Int J Artif Organs 2001;24:695–725.

19 Lesaffer G, De Smet R, Lameire N, Dhondt A, Duym P, Vanholder R: Intradialytic removal of protein-bound uraemic toxins: Role of solute characteristics and of dialyser membrane. Nephrol Dial Transplant 2000;15:50–57.

20 Dou L, Bertrand E, Cerini C, Faure V, Sampol J, Vanholder R, Berland Y, Brunet P: The uremic solutes p-cresol and indoxyl sulfate inhibit endothelial proliferation and wound repair. Kidney Int 2004;65:442–451.

21 Cerini C, Dou L, Anfosso F, Sabatier F, Moal V, Glorieux G, de Smet R, Vanholder R, Dignat-George F, Sampol J, Berland Y, Brunet P: P-cresol, a uremic retention solute, alters the endothelial barrier function in vitro. Thromb Haemost 2004;92:140–150.

22 D'Hooge R, Van de Vijver G, Van Bogaert PP, Marescau B, Vanholder R, De Deyn PP: Involvement of voltage- and ligand-gated Ca^{2+} channels in the neuroexcitatory and synergistic effects of putative uremic neurotoxins. Kidney Int 2003;63:1764–1775.

23 Canalejo A, Almaden Y, De Smet R, Glorieux G, Garfia B, Luque F, Vanholder R, Rodriguez M: Effects of uremic ultrafiltrate on the regulation of the parathyroid cell cycle by calcitriol. Kidney Int 2003;63:732–737.

24 Bammens B, Evenepoel P, Verbeke K, Vanrenterghem Y: Removal of middle molecules and protein-bound solutes by peritoneal dialysis and relation with uremic symptoms. Kidney Int 2003;64: 2238–2243.

25 De Smet R, Van Kaer J, Van Vlem B, De Cubber A, Brunet P, Lameire N, Vanholder R: Toxicity of free p-cresol: A prospective and cross-sectional analysis. Clin Chem 2003;49:470–478.

New Insights in Uremic Toxicity

26   Ingrosso D, Cimmino A, Perna AF, Masella L, De Santo NG, De Bonis ML, Vacca M, D'Esposito M, D'Urso M, Galletti P, Zappia V: Folate treatment and unbalanced methylation and changes of allelic expression induced by hyperhomocysteinaemia in patients with uraemia. Lancet 2003;361: 1693–1699.

27   Jankowski J, Luftmann H, Tepel M, Leibfritz D, Zidek W, Schluter H: Characterization of dimethylguanosine, phenylethylamine, and phenylacetic acid as inhibitors of Ca2+ ATPase in end-stage renal failure. J Am Soc Nephrol 1998;9:1249–1257.

28   Jankowski J, van der GM, Jankowski V, Schmidt S, Hemeier M, Mahn B, Giebing G, Tolle M, Luftmann H, Schluter H, Zidek W, Tepel M: Increased plasma phenylacetic acid in patients with end-stage renal failure inhibits iNOS expression. J Clin Invest 2003;112:256–264.

29   Jankowski J, Hagemann J, Yoon MS, van der GM, Stephan N, Zidek W, Schluter H, Tepel M: Increased vascular growth in hemodialysis patients induced by platelet-derived diadenosine polyphosphates. Kidney Int 2001;59:1134–1141.

30   Jankowski V, Tolle M, Vanholder R, Schonfelder G, van der Giet M, Schluter H, Paul M, Zidek W, Jankowski J: Identification of uridine adenosine tetraphosphate (UP4A) as an endothelial-derived vasoconstrictive factor. Nat Med 2005;11:223–227.

31   Weissinger EM, Kaiser T, Meert N, de Smet R, Walden M, Mischak H, Vanholder RC: Proteomics: A novel tool to unravel the patho-physiology of uraemia. Nephrol Dial Transplant 2004;19: 3068–3077.

R. Vanholder
Nephrology Section, Department of Internal Medicine
University Hospital, De Pintelaan 185
BE–9000 Ghent (Belgium)
Tel. +32 92404525, Fax +32 92404599, E-Mail raymond.vanholder@ugent.be

Ronco C, Brendolan A, Levin NW (eds): Cardiovascular Disorders in Hemodialysis.
Contrib Nephrol. Basel, Karger, 2005, vol 149, pp 325–333

....

Continuous Renal Replacement Therapy for End-Stage Renal Disease

The Wearable Artificial Kidney (WAK)

Victor Gura[a], *Masoud Beizai*[b], *Carlos Ezon*[b], *Hans-Dietrich Polaschegg*[b]

[a]Cedars Sinai Medical Center, UCLA, and Geffen School of Medicine, Beverly Hills, Calif., USA. [b] National Quality Care Inc., Beverly Hills, Calif., USA

Abstract

Daily dialysis offers many benefits but is difficult to implement. CRRT allows dialysis 24/7 but is not suitable for ESRD patients. Thus, the need for a miniaturized ambulatory CRRT device those patients can wear permanently. We report the feasibility, safety and efficiency in uremic pigs, of such a wearable artificial kidney (WAK) that can be worn as a belt, operated with batteries, and weights less than 5 lbs. We used a hollow fiber dialyzer with a surface area of 0.2 sqm. Dialysate was continuously regenerated by a series of cartridges containing several sorbents allowing the use of approximately 375 ml of dialysate. The device includes reservoirs with heparin and electrolytes. Average fluid removal was 100 ml/hr. The Creatinine was 25 ml/min. In 8 hrs the total Creatinine removed was 1 gr, Urea 12 gr, P0.8 gr and K 72 mEq. Weekly st kt/v was extrapolated to approximately 7. There were no side effects. The WAK can be operated safely and continuously 168 hr/week. This would allow for all the advantages of daily dialysis and reduce morbidity and mortality in the ESRD population. It will also reduce cost and manpower utilization.

Introduction

A growing body of literature points out that increased dialysis frequency and prolonged dialysis (preferably daily) are conducive to numerous improvements in quality of life and potentially increased longevity of end-stage renal disease (ESRD) patients. These advantages of daily dialysis are summarized in Table 1.

On the other hand, implementation of daily dialysis encounters several obstacles that make its implementation in a large scale practically impossible. Between these are the inability or unwillingness of most patients to dialyze at

Table 1. The many advantages of daily dialysis

Improved volume control	Improved appetite and nutrition
No hyperphosphatemia	Eliminate phosphate binders
Less hypertension	Reduce hypertension medication
No hypokalemia	No metabolic acidosis
No sodium retention	Improved serum albumin
Reduce bone disease	Improve anemia
Reduce cardiovascular disease	Decrease incidence of stroke
Reduce morbidity	Reduce mortality

home, the lack of manpower both in nurses and technicians to provide more treatments in the dialysis units, and the reluctance of governmental payers to shoulder the expense of additional procedures [1–7].

Even if payers would agree to pay for the additional costs of daily dialysis, it would not only take time, but there would also be major capital investment requirements to build additional capacity in or near the existing dialysis units.

Thus, the need for a technological solution that will allow for increased dialysis time without incurring additional costs or necessitating additional manpower.

Continuous renal replacement therapy (CRRT) in its different versions definitely gives us the possibility of rendering significantly higher doses of dialysis but these modalities are very labor intensive. CRRT machines allow for the delivery of dialysis therapies 24 h a day, 7 days a week. These machines are not suitable to treat ESRD patients because they are heavy, attached to a wall electrical outlet and require many gallons of water, rendering such patients unable to ambulate and perform their activities of daily life. A miniaturized and wearable CRRT machine or wearable artificial kidney (WAK) might solve these problems. In order to build a WAK we had to overcome the following issues:

(1) The device needs to have an ergonomically suitable energy source that though small and light will provide enough energy to power all the necessary systems for a significant period of time and make it independent of a fixed electrical outlet, and all these, however, without creating an excessive weight burden on the patient.

(2) The device has to have enough dialysate with all the necessary additives, and the capability to purify and recirculate such dialysate to avoid requiring many gallons of fluid. However, the volume has to be small enough so that it will not create an intolerable weight burden.

(3) The device must be light and ergonomically adapted to the body contour so that it can be worn continuously without impinging on the patient's ability to ambulate and performs the activities of daily life.

As all these problems were solved and a WAK original model that can be battery operated, worn as a belt and weighing less than 2.27 kg was built and tested on initial bench trials. The purpose of this study was to determine the feasibility, safety and efficiency of the WAK in uremic animals.

Materials and Methods

Twelve pigs underwent ligation of the ureters. Next day, they were anesthetized and dialyzed for 8 h using a double lumen catheter, with a WAK (fig. 1) that can be battery operated, worn as a belt and weighs less than 2.27 kg.

A dialyzer weighing less than 100 g made of hollow capillary fibers with a surface area of 0.2 sqm was used. The dialysate utilized was sterile normal saline. It was continuously regenerated by recirculation through a series of cartridges containing activated carbon, urease, zirconium hydroxide and zirconium phosphate [8]. This allowed the use of approximately 375 ml of dialysate. The device includes several reservoirs containing heparin and supplemental electrolytes such as calcium, magnesium, sodium bicarbonate, etc. Volumetric micropumps were used to deliver 0.5–10 cc per hour of these solutions from the reservoirs to the blood or the dialysate circuit. The removal of fluid was controlled using a volumetric micropump at flow rates ranging between 0 and 700 ml per hour with an average rate of fluid removal of 100 ml/h. The animals were divided into 2 groups: Group I: 6 pigs (weight 74.9 ± 1.2 kg) were dialyzed with a blood flow of 44 ml/min and Group II: 6 pigs (weight 47.9 ± 1.7 kg) were dialyzed with a blood flow of 75 ml/min. Creatinine, urea and electrolytes were measured with an i-STAT Portable Clinical Analyzer from Abbott Laboratories. Effective urea and creatinine clearances and standard weekly urea Kt/V were calculated as follows:

Effective clearance = Blood flow × [Δ solute]/[solute in];

Standard weekly Kt/v = Effective clearance × Time/total body water.

Where Δ solute is the difference between urea or creatinine concentrations in and out of the dialyzer. Time was 480 min in the actual tests and 10,080 min for a weekly extrapolation. Total body water was estimated as 60% of body weight.

Results

No adverse events were observed in the animals included in these studies. The average dialysate flow was 73 ml/min in Group I and 85 ml/min in Group II. The results are summarized in table 2.

The measured values of effective creatinine clearance, total creatinine removed, effective urea clearance, total urea removed, standard weekly Kt/V, phosphorus and potassium removed per 8 h are summarized in table 2. The removal of potassium during the treatment was 71.9 ± 13.3 mmol in Group I and 89.1 ± 25.7 mmol in Group II, per 8 h. The removal of phosphorus was 0.8 ± 0.2 g in Group I and 0.84 ± 0.4 g in Group II, per 8 h. The dialysis doses achieved with the WAK were effective creatinine clearance of 25.5 ± 1.4 ml/min in Group I and 24.7 ± 3.2 ml/min in Group II. The average effective urea

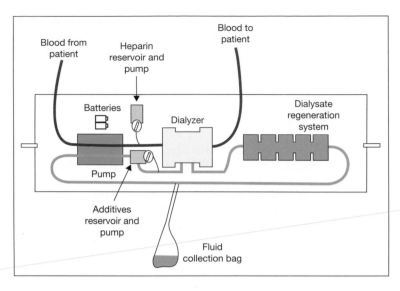

Fig. 1. Schematic draft of the wearable artificial kidney belt.

Table 2. The results of our animal studies (mean ± SD)

Results	Group I	Group II
Effective urea clearance, ml/min	24.3 ± 1.4	23.9 ± 3.5
Effective creatinine clearance, ml/min	25.5 ± 1.4	24.7 ± 3.2
Total urea removal, g	12.7 ± 2.8	12.0 ± 2.9
Total creatinine removal, g	0.9 ± 0.2	1.0 ± 0.1
Total phosphorus removal, g	0.8 ± 0.2	0.84 ± 0.4
Total potassium removal, mmol	71.9 ± 13.3	89.1 ± 25.7
Extrapolated standard Kt/V urea	5.4 ± 2.4	8.4 ± 1.5

clearance was 24.3 ± 1.4 ml/min in Group I and 23.9 ± 3.5 ml/min in Group II. The standard weekly urea Kt/V was 5.4 ± 2 in Group I and 8.4 ± 1.5 in Group II.

The amounts of ultrafiltrate removed are shown in table 2. The volume of fluid removed was changed arbitrarily during the experiment from 0 to 700 ml/h. The limiting factor for the removal of larger amounts of fluid per hour was a progressive decrease in blood flow in the dialyzer with increasing fluid removal rate. The blood flow was reinstated immediately as we diminished the ultrafiltrate removal rate. There were no difficulties, however, in maintaining an average fluid removal of 100 ml/h.

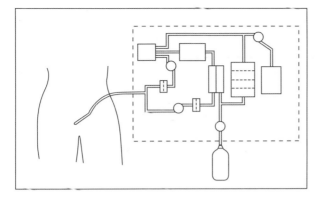

Fig. 2. Wearable peritoneum-based system for continuous renal function replacement [15].

Discussion

The need for increased and more frequent dialysis is well documented. It is abundantly clear by now that this may be the way to improve the currently dismal outcomes in the treatment of ESRD. However, implementing daily dialysis is plagued by several apparently insurmountable limitations, not the least of these being the need for increased funding and nursing manpower. Thus, the nephrology community has to come up with a different technical solution that will answer the need to improve the outcomes in ESRD, without higher costs. The WAK might be the answer.

Several attempts to build a WAK (fig. 2, 3) have been previously described [9–19] but none was ever brought to market.

The lack of suitable light energy sources, the low efficiency of the available membranes, and the inability to miniaturize the components to the extent that the device would perform appropriately with a low-energy budget were main limitations in the past. Similarly, the use of REDY sorbents was incorporated in some of these initial attempts. However, the REDY sorbent cartridge weighs about 2.27 kg and is ergonomically unsuitable for a device that can be worn on a patient's body. The system was never modified or configured to make it light and ergonomically suitable for a WAK. As we addressed those issues one by one, and came up with practical solutions, building a working WAK prototype became feasible.

The use of a sterile dialysate, free of bacterial toxins, would avoid the presumed complications attributed to the current use of water that is not pyrogen free. The relatively small amount of dialysate required by this device, would make the provision of such a dialysate, practical and cost effective.

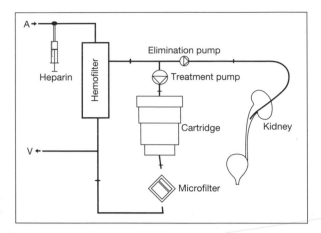

Fig. 3. Continuous arteriovenous hemofiltration in a wearable device to treat end-stage renal disease [17].

The results shown above indicate that the WAK can remove solutes and excess fluids from uremic animals. The amounts removed indicate that if similar results are accomplished in ESRD patients, a major increase in dialysis dose could be provided. The standard weekly Kt/V, about 7 on average, is significantly higher than that commonly administered in the current schedules (fig. 4). The device would provide 168 h per week of dialysis, about 16 times more dialysis hours than that currently used schedules, without interference in the patient's activities of daily life, or increasing the cost of treatment. It seems that the one factor the nephrology community never identified and modified in 6 decades of treatment of ESRD patients, is dialysis time. Yet, the emerging data on daily dialysis seem to indicate that incremental dialysis time is the key for improving the dismal outcomes we obtain today in our ESRD population. The concept that filtering blood with a typical schedule of 12 h a week can accomplish the same task that native kidneys do by filtering the blood 24 h a day, 7 days a week, appears to be erroneous. Molecules of different size, submitted to the same kinetic forces, would travel through membranes at different speeds, according to their molecular weight. Therefore, heavier molecules may not transit in adequate amounts from the blood stream to the dialysate unless they are given enough time to do so. It seems that 168 h a week of blood filtration may be far more physiological than 12 h a week. This may explain, at least partially, the unacceptably high morbidity and mortality of the ESRD population.

The amounts of salt and water removed would make patients free from limitations in the oral intake of water and salt. In addition, this would result in a much-improved control of hypertension and fluid overload and a reduction in

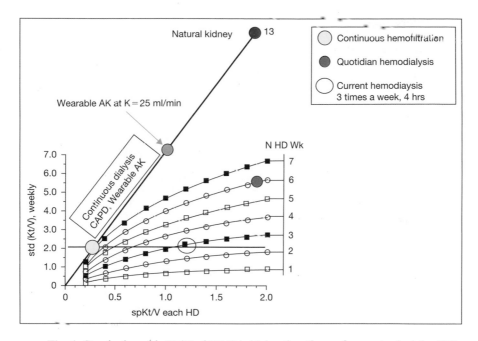

Fig. 4. Standard weekly Kt/V of WAK is higher than those of current schedules [25]. ○ = [26]; ● = [27].

the need for antihypertensive drugs. This notion is further supported by the results obtained with daily [20, 21] or prolonged [22–24] dialysis. Similarly, the amounts of potassium removed in these animals, indicate that if similar results were obtained in ESRD patients, there would be no dietary restriction of potassium, and the risk of hyperkalemia would be greatly reduced.

The amounts of phosphorus we removed with the WAK would eliminate the need for phosphate binders in ESRD. Again, a similar result has been shown in ESRD patients treated with daily dialysis. The beneficial effects of the elimination of hyperphosphatemia on bone disease and arterial calcifications shall be the subject of further studies. In addition, the elimination or reduction of phosphate binders would be an additional, welcome cost reduction in the treatment of ESRD.

Furthermore, the WAK would decrease significantly the amount of capital investments and nursing manpower needed today to provide dialysis to ESRD patients.

Human clinical studies are now needed to corroborate these results in ESRD patients. If successful, the WAK may be the way to provide daily dialysis in an efficient and cost-effective manner.

References

1   Lockridge RS Jr: The direction of end-stage renal disease reimbursement in the United States. Semin Dial 2004;17:125–130.

2   Lockridge RS Jr, McKinney JK: Is HCFA's reimbursement policy controlling quality of care for end-stage renal disease patients? ASAIO J 2001;47:466–468.

3   Manns BJ, Johnson JA, Taub K, Mortis G, Ghali WA, Donaldson C: Dialysis adequacy and health related quality of life in hemodialysis patients. ASAIO J 2002;48:565–569.

4   Mapes DL, Lopes AA, Satayathum S, McCullough KP, Goodkin DA, Locatelli F, Fukuhara S, Young EW, Kurokawa K, Saito A, Bommer J, Wolfe RA, Held PJ, Port FK: Health-related quality of life as a predictor of mortality and hospitalization: The Dialysis Outcomes and Practice Patterns Study (DOPPS). Kidney Int 2003;64:339–349.

5   McFarlane PA, Bayoumi AM, Pierratos A, Redelmeier DA: The quality of life and cost utility of home nocturnal and conventional in-center hemodialysis. Kidney Int 2003;64:1004–1011.

6   Mohr PE, Neumann PJ, Franco SJ, Marainen J, Lockridge R, Ting G: The case for daily dialysis: Its impact on costs and quality of life. Am J Kidney Dis 2001;37:777–789.

7   Patel SS, Shah VS, Peterson RA, Kimmel PL: Psychosocial variables, quality of life, and religious beliefs in ESRD patients treated with hemodialysis. Am J Kidney Dis 2002;40:1013–1022.

8   Marantz LB, Giorgianni MG, inventors: Urease in insoluble form for converting urea present in liquid. US patent 3989622, 11/2/1976.

9   Beltz AD, inventor: Wearable, portable, light-weight artificial kidney. US patent 5284470, 2/8/1994.

10 Bonomini V, Roggeri G, inventors: Hemodialysis and/or ultrafiltration apparatus. US patent 4269708, 5/26/1981.

11 Henne W, inventor: Artificial kidney. US patent 4212738, 7/15/1980.

12 Horiuchi T: [Wearable artificial kidney, portable artificial kidney and implantable artificial kidney]. Nippon Rinsho 2004;62(suppl 5):182–188.

13 Kolff WJ, Jacobsen S, Stephen RL, Rose D: Towards a wearable artificial kidney. Kidney Int 1976;(suppl 7):S300–S304.

14 Mineshima M: [Artificial kidney therapy in next generation]. Nippon Rinsho 2004; 62(suppl 6): 606–609.

15 Roberts M, Lee DB-N, inventors: Wearable peritoneum-based system for continuous renal function replacement and other biomedical applications. US patent 5944684, 8/31/1999.

16 Scott RD, inventor: Wearable, continuously internally operable and externally regenerable dialysis device. US patent 4765907, 8/23/1988.

17 Murisasco A, Reynier JP, Ragon A, Boobes Y, Baz M, Durand C, Bertocchio P, Agenet C, el Mehdi M: Continuous arterio-venous hemofiltration in a wearable device to treat end-stage renal disease. Trans Am Soc Artif Intern Organs 1986;32:567–571.

18 Senoo S, Otsubo O, Watanabe T, Yamauchi J, Yamada Y, Inou T, Takai N, Takahashi H, Fukui H, Kawata Y: [The wearable artificial kidney: Development of a small blood pump]. Jinkou Zouki 1982;11:48.

19 Tada Y, Horiuchi T, Ohta Y, Dohi T: A new approach for the filtrate regeneration system in the wearable artificial kidney. Artif Organs 1990;14:405–409.

20 Lockridge RS Jr: Daily dialysis and long-term outcomes – The Lynchburg Nephrology NHHD experience. Nephrol News Issues 1999;13:16, 19, 23–26.

21 Lockridge RS Jr, Spencer M, Craft V, Pipkin M, Campbell D, McPhatter L, Albert J, Anderson H, Jennings F, Barger T: Nocturnal home hemodialysis in North America. Adv Ren Replace Ther 2001;8:250–256.

22 Charra B, Chazot C, Jean G, Hurot JM, Vanel T, Terrat JC, Vovan C: Long 3 × 8 hr dialysis: A three-decade summary. J Nephrol 2003;16(suppl 7):S64–S69.

23 Charra B, Chazot C, Jean G, Laurent G: Long, slow dialysis. Miner Electrolyte Metab 1999;25: 391–396.

24 Charra B, Terrat JC, Vanel T, Chazot C, Jean G, Hurot JM, Lorriaux C: Long thrice weekly hemodialysis: The Tassin experience. Int J Artif Organs 2004;27:265–283.

25 Gotch FA: The current place of area kinetic modelling with respect to different dialysis modalities. Nephrol Dial Transplant 1998;13:10–14.

26   Murisasco A, Baz M, Boobes Y, Bertocchio P, el Mehdi M, Durand C, Reynier JP, Ragon A: A continuous hemofiltration system using sorbents for hemofiltrate regeneration. Clin Nephrol 1986;26(suppl 1):S53–S57.
27   Suri R, Depner TA, Blake PG, Heidenheim AP, Lindsay RM: Adequacy of quotidian hemodialysis. Am J Kidney Dis 2003;42(supp 1):S42–S48.

Victor Gura, MD
Cedars Sinai Medical Center, UCLA
Geffen School of Medicine, 9033 Wilshire Blvd. #500
Beverly Hills, CA 90211 (USA)
Tel. +1 310 550 6240, Fax +1 310 276 4276, E-Mail Vgura@cs.com

Ronco C, Brendolan A, Levin NW (eds): Cardiovascular Disorders in Hemodialysis.
Contrib Nephrol. Basel, Karger, 2005, vol 149, pp 334–342

....

Slow Continuous Intravenous Plasmapheresis (SCIP™): Clinical Applications and Hemostability of Extracorporeal Ultrafiltration

Harold H. Handley[a], *Rey Gorsuch*[a], *Harold Peters*[a], *Thomas G. Cooper*[b],
Richard H. Bien[c], *Nathan W. Levin*[d], *Claudio Ronco*[e]

[a]Transvivo, Inc., Napa, Calif., [b]Cooper Consulting Services, Friendswood, Tex.,
[c]Queen of the Valley Hospital, Napa, Calif., and [d]Renal Research Institute,
New York, N.Y., USA; [e]St. Bortolo Hospital, Department of Nephrology,
Vicenza, Italy

Abstract

An intravenous plasmapheresis catheter which excludes >99.4% of platelets from external ultrafiltration circuits is currently undergoing safety and efficacy trials for fluid removal from NYHA class II-IV congestive heart failure patients resistant to diuretic drug therapy. In animals, the SCIP™ catheter allowed a four fold increase in ultrafiltration efficiency without hemolysis, hemoinstability or external cartridge changes in 72 hours of treatment. Further, systemic anticoagulation was not required. These techniques might be envisioned for treatment of fluid overload in heart failure, surgery or trauma and may have applications in therapeutic apheresis, venous thrombosis, liver disease or autologous tissue engineering.

Introduction

The practice of intermittent blood filtration, particularly dialysis in the USA, has experienced only incremental conceptual changes over the past 20 years despite the numerous improvements to filtration cartridges and membranes [1]. While the hemodynamic benefits of slower, more frequent or continuous renal replacement therapies have promoted experimentation with daily, nightly and continuous filtration therapies [2, 3], extracorporeal filters require regular replacement or anticoagulation regimens to offset a rapid decline in filtration efficiency over time. We have proposed that the exclusion of cells from extracorporeal circuits during purification may provide increased longevity of the filtration circuit, improve hemostability, decrease requirements for heparin, diminish red cell lysis and minimize attendant labor [4–7]. In this paper, we show the platelet exclusion efficiency of an in situ plasmapheresis catheter and the hemodynamic impact on pigs during 72 h of uninterrupted, slow continuous ultrafiltration (SCUF) of plasma removed by the slow continuous intravenous plasmapheresis (SCIP™) system. Clinical plans will be outlined and additional theoretical applications will be discussed for SCIP™ in nephrology, cardiology, and neurology by ultrafiltration, dialysis, therapeutic apheresis, edema, autoimmune disease, trauma or tissue engineering.

SCIP

The SCIP™ system can provide blood plasma directly from the vena cava to an extracorporeal filtration or adsorption unit by plasmapheresis of the blood in situ. The system comprises a filtration catheter and a fluid control module which regulates both plasma removal rates and extracorporeal ultrafiltration rates [7]. The plasmapheresis catheter also receives an intermittent backflush cycle containing heparin [6]. This combination maintains the porosity and efficiency of the catheter and extracorporeal filter over longer periods of use than might be expected from use of intermittent or daily ultrafiltration systems processing blood at higher flow and pressures than in the SCIP system.

During 72 h of continuous experimental use, SCIP™ catheters continuously excluded greater than 99.4% of blood platelets from extracorporeal access plasma samples (table 1). Further, the localized addition of intermittent heparin through backflush and the absence of platelets in the extracorporeal circuit diminished clogging of the extracorporeal filter and allowed its uninterrupted use beyond 72 h [7]. Despite anticoagulant in the backflush cycle, the peripheral blood coagulation times returned to normal immediately following removal of IV heparin for catheter insertion [7]. Backflush is therefore conceived to provide both an intermittent mechanical disruption and a localized inhibition of coagulation around the SCIP™ catheter pores as well as within the extracorporeal filter without raising the coagulation time of the peripheral blood of normal animals.

Assessments of porcine blood for chemical or cellular changes supported hemostability during the course of plasma SCUF and are shown in figures 1 and 2. All data points represent the mean value for two animals. Two samples

Table 1. Efficiency of plasmapheresis during 72 h of continuous use of a SCIP™ catheter

Sample time (h)	Catheter #Y0254			Catheter #Y0255		
	Blood platelets $(\times 10^{-3}/\mu l)$	Plasma platelets $(\times 10^{-3}/\mu l)$	Efficiency (% platelet exclusion)	Blood platelets $(\times 10^{-3}/\mu l)$	Plasma platelets $(\times 10^{-3}/\mu l)$	Efficiency (% platelet exclusion)
0	313	–	–	321	–	–
6	397	0.35	99.91	309	0.30	99.90
24	233	0.08	99.97	205	0.27	99.87
36	198	1.11	99.44	287	0.04	99.99
48	169	0.05	99.97	278	0.94	99.66
60	260	0.07	99.97	225	0.47	99.79
72	405	2.09	99.48	198	n.d.	n.d.

n.d. = Not determined.

were taken from each individual per time point. Although platelet counts did diminish slightly, dropping to low normal values during the course of treatment, anesthesia and sedation of the animals was likely to have contributed to this mild thrombocytopenia, making its interpretation difficult. However, hematocrit, hemoglobin and plasma free-hemoglobin levels remained stable throughout the treatment and confirms the absence of hemolysis (fig. 1).

Peripheral blood concentrations of sodium, calcium, potassium, chloride and phosphorus remained stable during treatment. Additionally, cholesterol and creatinine remained stable while blood urea nitrogen dropped but remained within normal levels for pigs. Low mean glucose levels at 72 h occurred in one animal which returned to low normal levels upon recovery and feeding. Baseline bicarbonate concentrations were elevated (33 mEq/l) in both animals upon initiation and remained constant throughout the study (normal ≤27 mEq/l). All data suggest that hemostability was maintained in the animals receiving therapy throughout the 72 h treatment.

Thus, we do not expect that plasma SCUF with the SCIP™ system will induce serious instability in blood cell counts or chemistry in clinical subjects.

Clinical Feasibility Study – Plasma SCUF in CHF Patients

The SCIP™ system will undergo clinical feasibility studies this year to demonstrate its safety and efficiency. These trials will investigate the use of plasma SCUF to alleviate fluid overload (overhydration) in class II–IV congestive heart

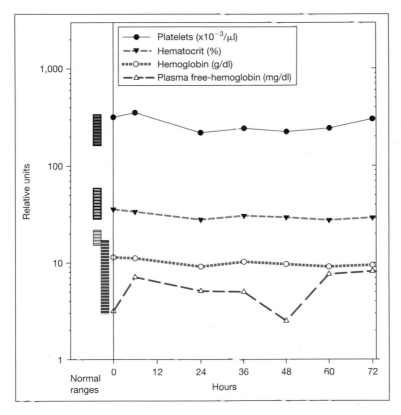

Fig. 1. Hematocrit, plasma free-hemoglobin and blood hemoglobin concentrations remained within normal ranges for Yorkshire pigs during SCIP™-plasma SCUF. These values indicate the absence of hemolysis during SCIP™ treatment. Low normal platelet counts and transient mild thrombocytopenia were observed during therapy. Low normal platelet counts can be induced by either anesthesia during insertion or the mild sedation provided during therapy.

failure (CHF) patients who fail to urinate more than 500 ml over 12 h following diuretic therapy.

Study subjects must have mild-to-moderate fluid excess, defined as ≥ 1.8 kg over expected body weight, a serum creatinine ≤ 3.5 mg/dl and a stable systolic blood pressure 90 mm Hg with or without inotropic drug support. Individuals with CHF secondary to congenital defects, myocardial infarction or arrhythmia within the past 6 months, chronic sepsis, polycythemia, a hematocrit $>40\%$, thrombocytopenia (platelet count $<75,000/\mu l$), thrombocytosis (platelet count $\geq 450,000/\mu l$), hemoglobin <9.0 g/dl, a history of a bleeding disorder or a prothrombin time ≥ 15 s or INR ≥ 2.0 will be excluded from this study.

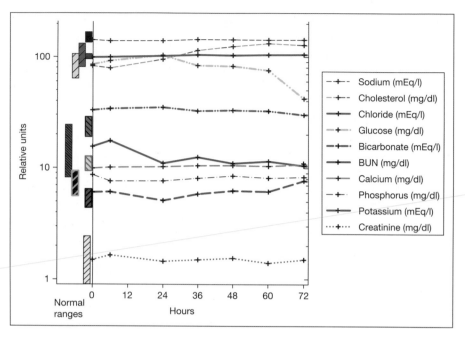

Fig. 2. Hemostability of serum electrolytes and protein during SCIP™-plasma SCUF treatment. Mean values for sodium, chloride, calcium, phosphorous and potassium remained within normal ranges over 72 h of treatment. Bicarbonate was unchanged from a starting high baseline value. Cholesterol and creatinine remained within normal levels, blood urea nitrogen dropped to low normal ranges after 24 h of therapy. Mean blood glucose levels fell below normal at 72 h.

Additionally, pregnancy, deep venous thrombosis or a history of an allergy to contrast agents are exclusion criteria for the study. Study subjects must be at least five feet tall and weigh 55 kg or more to accommodate the size of the catheter selected for this study.

In the safety phase of this study, each of 10 patients will receive no more than 24 h of plasma SCUF therapy. While the primary intent of the study will be to safely introduce the SCIP™ system to patients, normal ICU monitoring, including blood pressure, blood cell counts and blood chemistry measurements comparable to those shown in figures 1 and 2, will be conducted every 8 h and at 1 and 7 days following treatment. Additionally, B-type natriuretic peptide and electrocardiogram analysis will be performed before and after treatment as indicators of heart function. Glomerular filtration rate and blood urea nitrogen will be assessed for kidney function. Rales assessments will be reported for changes in pulmonary edema. The fluid control system will be asked to clear up to 2 l of plasma water per 24 h. A general assessment of the results following the initial 5 patients will be performed and all results will be reviewed for safety by an independent medical data safety review board.

Provided that the device and treatment prove safe and capable of clinically relevant fluid removal, additional subjects are planned for enrollment in a 72-hour course of treatment. More significant improvements in kidney and heart function might be expected in this longer treatment study group.

CHF patients with fluid overload were chosen for this study because of the large numbers of patients and the relative uniformity of this patient population in the USA. Currently, over 4.9 million persons in the USA and nearly 6.5 million persons in Europe suffer from CHF. Nearly 20% are hospitalized annually, with a majority suffering from chronic or acute fluid overload. In most of these individuals, diuretics will fail at some point in their therapy. Further, in patients with prerenal failure or kidney distress, diuretics may be hazardous or contraindicated. Indeed, the 1-year mortality rate from diagnosis of CHF patient is 20% [8]. SCUF has already been shown to improve kidney and heart function in CHF patients [9] and may be superior to diuretic drug therapy [10]. The SCIP™ system promises to deliver an alternative method for removal of plasma water with all the hemodynamic benefits of slow, continuous renal replacement therapy without the attendant labor expenses or the trauma caused by diuretic drugs in a weakened kidney.

Plasma SCUF for CHF, Trauma or Segmental Edema

The proposed investigation of in situ plasmapheresis prior to ultrafiltration with the SCIP™ system in fluid overload CHF patients may provide a foundation for additional applications of in situ blood purification. Acute, chronic or segmental fluid overload following surgery, trauma or deep venous thrombosis might also be treated by SCIP™-based ultrafiltration. Further, in situ plasma removal should allow alternative extracorporeal purification devices to be employed with the benefits of longer continuous uninterrupted use and hemodynamic stability of the patient while minimizing coagulation and clogging of the external circuits.

Therapeutic Apheresis

In situ plasmapheresis systems (without ultrafiltration) have clear applications for therapeutic apheresis as a slow, continuous alternative therapy. Its conceptual impact might be compared to that of continuous renal replacement therapy in kidney failure. Indeed, like intermittent dialysis, most apheresis therapies require repetitive, intermittent treatments of plasmapheresis followed by plasma exchange or adsorption. A large number of autoimmune diseases and diseases of abnormal protein have been treated by plasma exchange or specific protein adsorption [11]. Table 2 was adapted from Sueoka [11] as a partial list of some disease indications for which plasmapheresis, plasma exchange and protein or immunosorption are performed. Diseases with improved clinical outcome following only a few intermittent exchanges, such as thrombotic thrombocytopenic purpura or sudden hearing loss, are likely to show similar or greater benefits from SCIP™ where both plasma *and interstitial fluid* might be cleared of the offending molecular constituents [12].

Discussion

SCUF of CHF-fluid overload patients has been shown by others to have a hemodynamically stable and beneficial impact on the resolution of symptoms and perhaps even organ recovery [10]. Current technologies should be improved to minimize anticoagulation, maintain vascular access and reduce attendant labor required to maintain therapeutic efficiency. We propose that removal of the cells from the extracorporeal circuit alone may eliminate many of these complications.

Until now, pragmatic alternatives to centrifugal plasmapheresis were unavailable. Characteristic milestones required for the clinical efficiency of in situ plasmapheresis are the long-term maintenance of filter porosity (prevention of clogging), flow rate and prevention of sepsis. SCIP™ has overcome or abated some of these obstacles through its intermittent backflush cycle. This backflush cycle is currently used to provide a localized heparinization of the filter surfaces, but might also be expanded in use to include fiber-bound or soluble antiseptic agents. Tunneling of the SCIP™ catheter may also diminish its potential for septic complications. We had previously shown that backflush substantially extended catheter longevity in situ in animals [6].

Initiation of human clinical trials are expected to confirm the hemodynamic and vascular access safety profile of in situ plasmapheresis and extend our clinical efficacy observations. The current clinical SCIP™ system devices may allow continuous therapeutic blood purification techniques to alleviate symptoms of fluid overload, trauma, autoimmune disease and elevated toxic protein more efficiently and with longer time between treatments.

In the future, the potential for smaller fiber pore size in the catheter may theoretically allow in situ ultrafiltration and even perhaps in situ dialysis. It has even been proposed that a 'life-vest', based on this technology, be

Table 2. Disease indications for use of therapeutic apheresis with SCIP™ [adapted from 11]

Organ system	Abnormal protein-related disease	Autoimmune antibody-related disease	Immune complex-related disease
Collagen and rheumatological disease	Raynaud's syndrome Rheumatoid arthritis (RA) Cryoglobulinemia	Systemic lupus erythematosis (SLE) RA Scleroderma Sjogren's syndrome	SLE RA Scleroderma
Neurological diseases	Polyneuropathy	Myasthenia gravis Multiple sclerosis Guillain-Barre syndrome (GBS) Polyradiculoneuropathy	GBS
Liver	Fulminant hepatitis Hepatic failure Primary biliary cirrhosis (PBC) Overdose poisoning	Antimitochondrial Ab PBC Chronic hepatitis	
Hematological diseases	Paraproteinemia Macroglobulinemia Cryoglobulinemia Hemolytic uremic syndrome	Idiopathic thrombocytopenia purpura (ITP) Thrombotic thrombocytopenic purpura (TTP) Autoimmune hemophilia ABO incompatibility Rh incompatibility Autoimmune hemolytic anemia Pernicious anemia AIDS	TTP Immune complex glomerulonephritis Schönlein-Henoch purpura (IgA complex) Gp120 complexes
Renal diseases		Antiglomerular basement membrane glomerulonephritis (GBM)	
Others	Hyperlipidemia: LDL, VLDL, familial hypercholesterolemia Sudden hearing loss Macular degeneration Poisons Toxins	Anti-insulin Ab Asthma (IgE) Urticaria (IgE) Ulcerative colitis (a-LPS) Grave's disease (IgG) Autoimmune thyroiditis Hashimoto's disease Addison's disease Autoimmune atrophic gastritis Pemphigus	

envisioned as the foundation for the development of a wearable, continuous artificial kidney.

References

1 Clark WR, Gao D, Ronco C: Membranes for dialysis. Composition, structure and function. Contrib Nephrol 2002;137:70–77.
2 Mehta RL: Indications for dialysis in the ICU: Renal replacement vs. renal support. Blood Purif 2001;19:227–232.
3 Paganini EP, Kanagasundaram NS, Larive B, Greene T: Prescription of adequate renal replacement in critically ill patients. Blood Purif 2001;19:238–244.
4 Ronco C: Extracorporeal therapies: Can we use plasma instead of blood? Int J Artif Organs 1999;22:342–346.
5 Ronco C, Ricci Z, Bellomo R, Bedogni F, Handley H, Gorsuch H, Levin N: A novel approach to the treatment of chronic fluid overload with a new plasma separation device. Cardiology 2001;96:202–208.
6 Handley HH, Ronco F, Gorsuch R, Peters H, Cooper TG, Levin N: Artificial in vivo biofiltration: Slow continuous intravenous plasmafiltration (SCIP™) and artificial organ support. Int J Artif Organs 2004;27:186–194.
7 Handley HH, Gorsuch R, Levin NW, Ronco C: I.V. catheter for intracorporeal plasma filtration. Blood Purif 2002;20:61–69.
8 Heart Disease and Stroke Statistics. 2005 update: www.americanheart.org.
9 Agostoni P, Marenzi G, Lauri G, Perego G, Schianni M, Sganzerla P: Sustained improvement in functional capacity after removal of body fluid with isolated ultrafiltration in chronic cardiac insufficiency: Failure of furosemide to provide the same result. Am J Med 1994;96:191–199.
10 Ronco C, Bellomo R, Ricci Z: Hemodynamic response to fluid withdrawal in overhydrated patients treated with intermittent ultrafiltration and slow continuous ultrafiltration: Role of blood volume monitoring. Cardiology 2001;96:196–201.
11 Sueoka A: Present status of apheresis technologies: Part 2. Membrane plasma fractionator. Ther Apher 1997;1:135–146.
12 Bosch T: Recent advances in therapeutic apheresis. J Artif Organs 2003;6:1–8.

Harold H. Handley, Jr., PhD
1100 Lincoln Ave., Suite 108
Napa, CA 94558 (USA)
Tel. +1 707 254 9597, Fax +1 707 254 9599, E-Mail hhandley@transvivo.com

Ronco C, Brendolan A, Levin NW (eds): Cardiovascular Disorders in Hemodialysis.
Contrib Nephrol. Basel, Karger, 2005, vol 149, pp 343–353

....

Membraneless Dialysis – Is It Possible?

Edward F. Leonard[a], *Stanley Cortell*[b], *Nicholas G. Vitale*[c]

[a]Departments of Chemical and Biomedical Engineering, Columbia University,
[b]Division of Nephrology, St. Luke's – Roosevelt Hospital Center, New York,
[c]Infoscitex Inc., Albany, N.Y., USA

Abstract

Direct contact between uremic blood and a fluid capable of receiving uremic toxins is possible. Such contact by itself is, however, not beneficial because the selection of molecules that are removed is dependent on diffusion coefficients in blood. This selection is inadequate and would result in the exhaustion of a patient's albumin pool before useful reduction in the urea pool was achieved. Direct contact that is accomplished by sandwiching blood between two layers of a 'sheathing' fluid, followed by diafiltration of the sheathing fluid through conventional membranes and recirculation of the sheathing fluid, is possible. This adaptation of membraneless transport of molecules from blood eliminates almost all contact of blood with solid artificial surfaces and the subsequent diafiltration and recirculation of the sheathing fluid allows precise control of what is removed from the system. Slightly hyperosmotic protein is carried back by the recirculating sheathing fluid. Only solutes and water that pass the diafilter, which operates on a cell-free fluid, are able to leave the system. The system depends strongly on the ability to keep cells out of the sheathing fluid. Preliminary results and earlier reports indicate that this separation is possible and more precise measurements are underway. A quantitative design of a wearable dialyzer based on a circulating sheathing fluid is presented.

All biological transport between phases uses membranes. This is true at the submicroscale where each organelle of a cell possesses a membrane, as does the cell itself. It is true at all higher scales where we encounter the vascular endothelium, the alveolar membrane, the intestinal wall, and many epithelial surfaces. Faced with the prevalence of membranes throughout biological systems, one must ask the very serious question, why would one even attempt mass transfer from one fluid to another without a membrane? The answer to this question lies in the imperfections of man-made membranes.

(1) A typical dialysis membrane is at least 1,000 times thicker than its natural counterpart.

(2) Its interface foments a largely inappropriate set of chemical reactions with blood components.

(3) In long-term use it becomes fouled – clogged with molecular aggregates that impede transport.

We have been concerned with moving toward a dialysis system that would go back to old ideas about a wearable dialyzer through which blood would flow constantly [1–7]. Such a device needs to be small, which means that its exchange rates between blood and dialysate need to be fast. Its blood-wetted surfaces should be highly biocompatible and should not require heparinization of the blood flowing continuously past them. Since the blood-wetted surface is essentially permanent, it should be much more resistant to fouling than any artificial membrane now known. These qualities could be obtained if it were possible to have direct, liquid-liquid contact of blood with dialysate.

This concept is, however, naïve. It is wise to remind ourselves what special things membranes do.

(1) They select – so that some molecules pass through them and others do not. In biological systems, transport without molecular selectivity at some point in the transport path is essentially useless.

(2) They offer a mechanical barrier that prevents gross mixing of two otherwise miscible fluids, and they permit the use of a pressure difference to extract water.

(3) They define the boundaries of different compartments.

In the balance of this paper, we describe a system which offers the advantages of both membrane-moderated and membraneless transport, and show how this system might serve as a wearable hemodialysis system.

Microfluidic-Membraneless Transport

It has been known for many years that one fluid could be made to flow beside another without convective mixing [8, 9]. However, the concept became useful only with the advent of reliable means of microfabricating thin fluid channels, usually through the use of photolithographic and micromachining techniques that were developed first for manufacturing large-scale, integrated electronic devices on silicon chips. Several proposals have hinted at the use of direct contact for therapeutic hemodialysis [10–12]. In the system considered here, we examine the flow of three liquid layers, each – nominally – 100 μm thick. To a very good approximation the outer, sheathing layers will have a mean speed that is half that of the center (blood) layer (fig. 1), so that, overall,

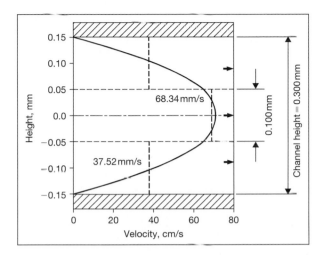

Fig. 1. Cross-section of blood flowing at a mean velocity of 68.34 mm/s, sandwiched between two layers of sheath fluid, each flowing at a mean velocity of 37.52 mm/sec. Each layer is flowing from left to right. Each layer is 100-μm (0.1 mm) thick. The blood layer is isolated from wall contact by the layers of moving sheath fluid and experiences a very low rate of shear, evidenced by its very flat velocity profile.

the two liquids will have the same flow rate. Systems with different flow rates are possible but are generally less desirable. It is important to notice that the flows in direct contact must be in the same direction, as shown in figure 1. Some recent proposals [12] have failed to recognize this requirement. Nonetheless, any attempt at counter- or cross-flow will inevitably lead to gross mixing of the streams.

How quickly will two streams, contacted in this manner, come to equilibrium? Fortunately, a good answer to this question was obtained, for other purposes, by Boltzmann, almost 150 years ago. Boltzmann's result is a complex formula but can be represented graphically as in figure 2. For small molecules, with diffusion coefficients, D, equal to about 10^{-5} cm^2/s, such as ions, sugars, and urea in blood, 90% equilibration will occur in about 2.1 s, a remarkably short time that reflects the thinness of the layers and the absence of a membrane. For larger molecules, e.g. albumin (D equal to ~5 × 10^{-7} cm^2/s), 95% equilibration will occur in 42 s. This difference might seem large enough to afford discrimination between small molecules and proteins, but others and we have shown, beyond any doubt, that it is not. (For the same exposure time, which is the condition to be met in any one device and flow, the best selectivity between small and large molecules depends on the square root of their diffusion coefficients, 1/4.5 for the case considered here. In most practical situations, the

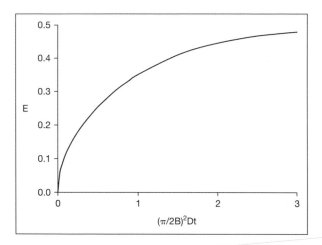

Fig. 2. Boltzmann's relationship expressed as extraction fraction, E, between two stagnant solute layers, each of thickness, B, as a function of time, t [18, 19]. The solute has a diffusion coefficient of value D. In the present paper we apply this result to a different but related situation: A central blood layer of thickness 2B diffuses into two sheathing streams, also of thickness, 2B. Each half of the blood layer feeds one of the two sheathing streams. The approximate calculations reported here assume that each half of the blood layer, whose thickness is B equilibrates with the sheathing layer, whose thickness is 2B, but whose velocity is half that of the blood, and – notwithstanding the operational differences – behaves according to the Boltzmann result. Thus a 100-μm blood layer is treated as two 50-μm layers, each communicating with a 50-μm layer of sheathing fluid that is traveling with the same velocity as the blood. Urea extraction is 90% when E equals 0.45. This corresponds to a value Dt/B^2 of 0.848 and requires, for molecules whose diffusion coefficients are in the neighborhood of 10^{-5} cm^2/s a blood residence time in the contact area of 2.1 s. The approximation has been validated with more precise finite-element calculations and is accurate within 5%.

selectivity is, in fact, worse. In a simple, direct contact blood treatment, the plasma albumin pool would be completely removed long before there was reasonable depletion of urea in body water [13]). A membraneless interface, by itself, is indiscriminate.

The Concept of a Recirculating Sheath

If the blood-contacting fluid is not a dialyzing fluid but rather a continuously circulating fluid that is dialyzed in a small, conventional dialyzer, one has the system shown in figure 3. The governing principle of this system is that both the blood and the sheath fluid must transport what the conventional dialyzer permits them to transport – no more, no less. Any solute not removed by the

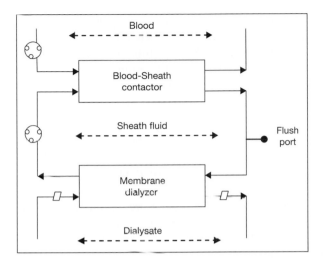

Fig. 3. Blood and sheath fluid circulate together through the blood-sheath contactor, driven by a 2-headed peristaltic pump. Blood returns to patient. Sheath fluid enters membrane dialyzer and is then recirculated to the blood-sheath contactor. Dialysate, when connected, flows countercurrent to sheath fluid in membrane dialyzer. Blood and sheath fluid flow at all times. Dialysis is effective when dialysate is connected. With a device to maintain transmembrane pressure in the dialyzer, ultrafiltration can occur in the absence of dialysate. Flush port allows for periodic injection of sterile saline into sheath stream in order to force cells back into bloodstream.

conventional dialyzer accumulates in the sheath fluid and returns to the blood contactor. The same principle applies if the device that extracts material from the sheath fluid is a hemodiafilter, absorber, or chemical reactor.

The extraction device also controls volume transport, although the resultant situation in the blood contactor is a little harder to interpret. After a short start-up period, the protein concentration in the sheath fluid exiting the blood-sheath contactor will come close to the protein concentration in blood. If water is ultrafiltered from the conventional dialyzer, the sheath fluid leaving the conventional dialyzer and returning to the blood-sheath fluid contactor will have a higher protein concentration that that in blood, and will thus be hyperosmotic. Water then passes from blood into the sheath fluid osmotically, without a hydrostatic pressure difference, at a rate set precisely by the ultrafiltration rate of the conventional dialyzer. How water is removed from the device that extracts materials from the sheath fluid is immaterial; only the amount removed is consequential.

Thus, in this system, control of transport in the conventional dialyzer defines precisely both volume and solute transport out of the blood layer. One may, however, well ask, if there is a conventional dialyzer in the system, what overall advantage has been gained? There are three advantages:

(1) Whole blood does not contact an artificial membrane; the blood interface is a pair of moving liquids. The interface should be highly biocompatible and cannot be fouled; anything that might deposit at the interface would be swept away. Because volume transport is osmotic, there is no tendency to draw cells to the interface.

(2) Transport is very rapid. While one might ask whether transport in the conventional dialyzer is not now the limiting factor, the absence of cells in this device allows for higher rates of shear and ultrafiltration than would be possible if whole blood were present.

(3) The shear stresses imparted to the blood layer are extremely low. The highest shear stress in the system is low and occurs in the sheath fluid where it contacts the wall.

Preventing Cells from Entering the Sheath Fluid

The concepts presented here are only effective if blood cells, particularly erythrocytes whose conservation is a recognized criterion of good dialysis, do not migrate into the sheath fluid. It is important to understand how stringent this requirement must be, although some ways of relaxing it will be discussed below. If 100 l of patient blood is dialyzed per week, and cell loss is kept at 1:10,000, the volumetric blood loss per week would be 10 ml, probably less than losses encountered in current therapies, and probably acceptable. Cells entering the sheath fluid must either remain there or return to the bloodstream, which they are unlikely to do, unaided. The proposed system, as described so far, provides no place for the small number of cells that may migrate into the sheath fluid to escape. Figure 3 shows a flush port which allows for the intermittent injection of a small volume of sterile saline into the sheath fluid. Such an injection would force return of entrapped cells to the circulation. Since sheath fluid volume is less than 5 ml, displacing the entire sheath fluid volume into the bloodstream is equivalent to the volume removed in about 2 min of ultrafiltration.

The general tendency for cells to migrate to the center of a flowing stream is well documented [14–16]. In addition to this general phenomenon, it has been shown that when shear rates are low enough to allow rouleaux formation, the migration is even more pronounced [17]. Early measurements of cell migration in a prototype, membraneless device confirm that cells migrate well to the center of a smooth, steadily operated flow channel [13], but we do not yet have quantitative data.

How Would a Real Ambulatory Dialyzer Perform?

An ambulatory dialyzer, through which blood is flowing at all times, need not be dialyzing at all times. We have suggested that such a device might have a continuous blood flow of 40 ml/min and be attached to a source of dialysate 50% of the time, 84 h/week. Thus, just over 200 l of patient blood would be dialyzed per week. Because the flows in this device are concurrent and, in present designs, approximately equal, the maximum urea clearance will be relatively low, approximately 18 ml/min, leading to an estimated Kt/V of about 1.8. We have assumed that volume transport might reach 15 l/week, which could be accomplished during dialysis (3 ml/min) or continuously (1.5 ml/min), since dialysate is not required to produce ultrafiltrate through the conventional dialyzer unit.

The rapid equilibration times for small molecules cited above lead to en-face contact areas between blood and sheath fluid of only about 50 cm^2 [13]. (The actual contact area, because the sheath fluid contacts both sides of the bloodstream, is about 100 cm^2.) The change in pressure from entrance to exit of the blood-sheath contactor is about 5 mm Hg.

It is important to recall that blood flow is continuous in this system and that the starting and stopping of actual dialysis requires only the not-necessarily sterile attachment and detachment, respectively, of dialysate leads. We imagine that a patient is given a dialysis prescription in the form of the required number of hours per week of dialysis, and is left largely free to decide when to be attached to, or detached from, a source of dialysate. Most patients would accomplish the bulk of their dialysis overnight.

We anticipate that the conventional dialyzer will foul, although more slowly because it operates in a cell-free, high-shear environment. We anticipate that it will need to be replaced every other day. We are testing prototypes of this dialyzer, with areas of 500 and 1,000 cm^2, for diffusive and hydraulic permeability and for the rate at which performance decrements. The shell for these devices looks, at present, like a mini-conventional dialyzer (fig. 4).

What Would a Real Ambulatory Dialyzer Look Like?

In current concept, the ambulatory dialyzer would be worn on the lower arm. The unit is comprised of five elements: (1) The blood-sheath contactor, a 2-layer plate whose dimensions are about $5.5 \times 8 \times 0.6$ cm. (2) The conventional dialyzer of figure 4 reconfigured to have an elliptical or rectangular cross-section. (3) A small, battery-operated, 2-tube peristaltic pump that maintains both blood and sheath fluid flows. (4) An exchangeable, rechargeable

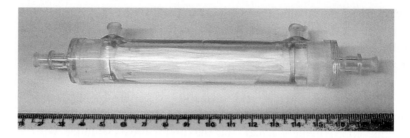

Fig. 4. Miniaturized dialyzer: Fiber length is about 9 cm, with total surface area in the unit shown of 500 cm² of polysulfone hollow fiber (Saxonia BioTec, Radeberg, Germany).

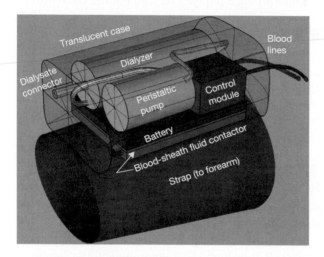

Fig. 5. Schematic drawing of wrist-size wearable dialysis system. The blood-sheath fluid contactor, shaped as a plate about 5.5 × 8 cm is sized to lie on the forearm. A 2-headed peristaltic pump, dialyzer, and control module are mounted above the plate. Power for the pump can be provided by a battery shaped to match the dimensions of the contactor. The assembly would be placed under a smooth cover in actual patient use.

battery with dimensions similar to that of the blood sheath contactor. (5) A small control module. These elements are shown conceptually, as they might be worn on a patient's forearm in figure 5. We have not attempted, at this early stage of research and development, to design the blood access system. One approach is the use of two catheters similar to those that are used in long-term total parenteral nutrition.

An important concomitant of the proposed system is the ability of the patient to secure maintenance of his device by going to a service center. Some patients may be able to change the dialyzer module; others may wish assistance. Assistance could be provided at a walk-in service unit. With or without assistance, the proposed system is designed to empower patients in the management of their disease.

Other Realizations of the Membraneless Concept

Plasmapheresis

The flowpath of the blood-sheath contactor appears to afford an excellent geometry for achieving plasmapheresis. Under these circumstances, an initial charge of sheath fluid would rapidly become equivalent to plasma and, during apheresis, a fraction of the circulating sheath fluid, equivalent to plasma, would be continuously withdrawn. Present plasmapheresis devices require either membrane contact or the application of centrifugal force. In this application of the membraneless flowpath, neither would be required.

Studies of Molecular Transport in Blood

Membraneless transport offers a potentially important, non-clinical opportunity. The movement of many molecules through blood and, sometimes, through dialysate is poorly understood. This is true of molecules distributed across extra- and intracellular space if their performance is not trivialized because they are either instantly equilibrated or completely unaffected as transport occurs. It is also true for molecules that are bound to slower-moving molecules or cell surfaces and dissociate from these havens as transport occurs. Finally it is true of molecules that change their molecular shape, or charge, or degree of aggregation with concentration. We know little of the transport of these molecules in blood, in large part because the study of this transport often occurs in the presence of a membrane whose dominant resistance obscures anomolous intraphase transport. For example, the factors limiting transport of phosphate, bilirubin and fatty acids in blood are poorly understood. Membraneless transport emphasizes exchange within phases, not across their boundary, and can even be conducted between contiguous layers of blood, only one of which, initially, contains the solute of interest.

Summary

Throughout this paper it has been argued that any value for membraneless transport arises from the shortcomings of artificial membranes and that transport without molecular discrimination is valueless. Membraneless dialysis is, in fact, not possible, but membraneless transport coupled with sheath dialysis is possible and probably practical. It may represent an advance over current membrane systems, especially in the much desired, but difficult to achieve, modality of long, slow, safe ambulatory renal replacement therapy.

Acknowledgment

The authors thank Foster Chen for figure 5 and Prof. Alan West for many provocative discussions and a careful reading of the manuscript.

References

1   Blackshear, PL: Two new concepts that might lead to a wearable artificial kidney. Kidney Int Suppl 1978;S133–S137.
2   Blaney TL, Lindan O, Sparks RE, Adsorption: A step toward a wearable artificial kidney. Trans Am Soc Artif Intern Organs 1966;12:7–12.
3   Henne W, Scheuren J, Saleh A, Bandel W: Wearable artificial kidney. Artif Organs 1977;1:126–126.
4   Kolff WJ, Jacobsen S, Stephen RL, Rose D: Towards a wearable artificial kidney. Kidney Int 1976;10:S300–S304.
5   Saito A, Takagi T, Sugiura K, Ono M, Minakuchi K, Teraoka S, Ota K: Maintaining low concentrations of plasma β_2-microglobulin through continuous slow hemofiltration. Nephrol Dial Transplant 1995;10(suppl 3):52–56.
6   Seo S, Otsubo O, Horiuchi T, Kuzuhara K, Sugimoto H, Takai N, Takahashi H, Inou T: Improvement of the wearable artificial kidney. Artif Organs 1981;5:321–321.
7   Vanholder R, Ringoir S: Pitfalls of wearable artificial kidney. Int J Artif Organs 1990;13:715–719.
8   Giddings JC: Continuous separation in split-flow thin (SPLITT) cells – potential applications to biological materials. Sep Sci Technol 1988;23:931–943.
9   Levin S, Tawil G: Analytical SPLITT fractionation in the diffusion mode, operating as a dialysis-like system devoid of membrane. Application to drug-carrying liposomes. Anal Chem 1993;65: 2254–2261.
10 Fazio F: Artificial kidney and methods of using same, in European Patent Register. EU, Renal Plant Corp, 2002.
11 Ronco C: Microfluidic, membrane-free dialysis. Annual Meeting, of the American Society of Nephrology 2002.
12 Wellman PS, Substantially Inertia Free Hemodialysis. US, 2004.
13 Leonard EF, West AC, Shapley NC, Larsen MU: Dialysis without membranes: How and why? Blood Purif 2004;22:92–100.
14 Moger J, Matcher SJ, Winlove CP, Shore A: Measuring red blood cell flow dynamics in a glass capillary using Doppler optical coherence tomography and Doppler amplitude optical coherence tomography. J Biomed Opt 2004;9:982–994.
15 Singh M, Ramesh ATV: Hematocrit dependence of cellular axial migration and tubular pinch effects in blood-flow through glass capillaries. Current Science 1990;59:223–226.
16 Uijttewaal WSJ, Nijhof EJ, Heethaar RM: Lateral migration of blood cells and microspheres in 2-dimensional poiseuille flow: A laser-Doppler Study. J Biomech 1994;27:35–42.
17 Goldsmith HL, Spain S: Margination of leukocytes in blood flow through small tubes. Microvasc Res 1984;27:204–222.

18   Loschmidt J: Experimental-Untersuchungen uber die Diffusion von Gasen ohne porose Scheidewande. Sitzungsber Kais Akad Wiss Wien Math Naturwiss Kl II 1870;61:367.

19   Wakeham WA, Kestin J: The measurement of diffusion coefficients; in Ho CY (ed): Transport Properties of Fluids. New York, Hemisphere, 1988, pp 225–228.

Edward F. Leonard
Departments of Chemical and Biomedical Engineering
Columbia University, Mail Code 4721
500 West 120 St, New York, NY 10027 (USA)
Tel. +1 212 854 4448, E-Mail leonard@columbia.edu

Author Index

354

Subject Index